MARINE SLIDES
AND OTHER
MASS MOVEMENTS

NATO CONFERENCE SERIES

I   Ecology
II Systems Science
III Human Factors
IV Marine Sciences
V   Air – Sea Interactions
VI Materials Science

IV MARINE SCIENCES

MARINE SLIDES AND OTHER MASS MOVEMENTS

Edited by
Svend Saxov
Aarhus University
Aarhus, Denmark

and

J. K. Nieuwenhuis
Rijkswaterstaat, Deltadienst
The Hague, The Netherlands

Published in cooperation with NATO Scientific Affairs Division

PLENUM PRESS · NEW YORK AND LONDON

Library of Congress Cataloging in Publication Data

NATO Workshop on Marine Slides and Other Mass Movements (1980 : Algarve, Portugal), Marine slides and other mass movements.

(NATO conference series. IV, Marine sciences ; v. 6)
Bibliography: p.
Includes index.
1. Marine sediments—Congresses. 2. Land-slides—Congresses. 3. Mass wasting—Congresses. I. Saxov, Svend Erik, 1913-   . II. Nieuwenhuis, J. K. III. North Atlantic Treaty Organization. Division of Scientific Affairs. IV. Title. V. Series.
GC377.N37 1980   551.46'08   81-19874
ISBN-13: 978-1-4613-3364-7   e-ISBN-13: 978-1-4613-3362-3 AACR2
DOI: 10.1007/978-1-4613-3362-3

Proceedings of a NATO Workshop on Marine Slides and Other Mass Movements, held in Algarve, Portugal, December 15-21, 1980

© 1982 Plenum Press, New York
Softcover reprint of the hardcover 1st edition 1982
A Division of Plenum Publishing Corporation
233 Spring Street, New York, N.Y. 10013

PREFACE

Svend Saxov* and J.K. Nieuwenhuis**

*Laboratory of Geophysics, Aarhus University
 Finlandsgade 6, DK-8200 Aarhus N, Denmark

**Ministry of Transport and Public Works
  Rijkswaterstaat, Deltadienst
  Van Alkemadelaan 400, 2597 AT the Hague
  The Netherlands

The NATO Science Committee Special Programme Panel on Marine Sciences decided in September 1979 that a working group should prepare a document dealing with the problems of marine slides.

The working group submitted to the Panel a proposal for the arrangement of an interdisciplinary workshop on Marine Slides and other Mass Movements, and the Panel decided to fund such a workshop.

Mass movement of soils and rocks on subaerial slopes is a subject which has been intensely studied over the past fifty years, and its literature is becoming extensive. The submarine phenomenon, although similar in its physical basis, is much less studied, but has become an important subject (particularly during the past decade) because of the increase of man's use of the coastal environment and the intensified exploration for oil and minerals on the continental margins and ocean basins. There are numerous examples which underscore the importance of this phenomenon and suggest an urgency for increased studies of submarine slope failure. The success of drilling on the steeper slopes of the outer continental shelf and upper continental slope where the potential for mass movements increases could have a large economic impact. This has been demonstrated in the Gulf

of Mexico, in the Mississippi Delta, where several oil
production platforms have been destroyed by the movement of the upper seabed stratum. Such occurrences and
tragedies such as the one at Nice, France, where nine
persons died in a local tsunami apparently caused by the
sudden failure of a man-made submarine waste dump, could
be averted in the future with a better understanding of
submarine mass movements.

It is also apparent that submarine slides and associated phenomena are widespread both on modern and ancient slopes, and a better understanding of the process
is important to diverse disciplines in the earth sciences.
The literature on submarine slope failure shows that a
variety of investigators from disciplines such as marine
sedimentology, marine geophysics, soils engineering,
paleontology, marine biology, coastal engineering, coastal geomorphology, and physical-chemical oceanography
have either published papers on or have a strong interest in the subject.

The working group decided to include the following
subjects/sessions:

1. Geologic and Geodynamic Settings

2. Coastal Geomorphology

3. Mechanics of Slides and Other Mass Movements

4. Geophysics

5. Earthquake Seismology

6. Oceanography

7. Marine Sedimentology

8. Marine Biology

9. Environmental Impact and Coastal Engineering

We wish to thank Professor, Dr. Adrian Scheidegger,
Institut für Geophysik, Technische Universität Wien,
Austria, and Dr. Robert W. Embley, National Ocean Survey,
National Oceanic and Atmospheric Administration, Rockville, Md., USA, who as members of the working group
have contributed substantially to the workshop programme. They have also been very instrumental at the pre-
paratory stage of the workshop. Professor, Dr. Carlos
Almaça, Lisbon, took very kindly upon himself to make
the arrangement concerning meeting place. Our thanks go
to him and also to the Marine Science Panel for the
financial support.

CONTENTS

MARINE SLIDES - SOME INTRODUCTORY REMARKS

Svend Saxov

Laboratory of Geophysics
Aarhus University
Finlandsgade 6
DK-8200 Aarhus N - Denmark

DEFINITION

It is not possible to find a decent definition of marine slides in the Glossary of Geology and Related Sciences by the American Geological Institute; however, a sudden movement of earth and rocks down a steep slope is defined as a landslide. Further, concerning slumping the following is said, "When the soil and earthy material on a steep slope become charged with water, their weight is greatly increased. At the same time the water makes them more mobile. Under these circumstances the material sometimes slides down the slopes. Such movements are known as slumping or sliding. If the movement is on a large scale, it is landslide."

In a recent paper, Herzer (1979) defines the term slide as any slope failure under stress and a slump refers to a slide in which the mass of material moves as a unit or as several subsidiary units along one or several curved slip surfaces, usually with backward rotation of the mass. A bedding-plane slide is a slide in which the mass moves more or less intact, along a planar slip-surface parallel to the bedding, without backward rotation of the mass. By referring to the literature during the last 25 years (among others: Embley and Jacobi, 1977; Gill and Kuenen, 1957; Heezen and Drake, 1964; Jacobi and Mrozowski, 1979; Lewis 1971; Moore et al., 1970; Normark, 1974; Stanley and Silverberg, 1969; Uchupi, 1967; Walker and Massingill, 1970) it is stated that submarine

1

sliding is a common feature of continental slopes, particularly where the rate of sedimentation is high and
earthquakes are frequent.

CAUSES

    Suboceanic changes are defined by John Milne in his
papers in the Geographical Journal (1897) as caused by
bradyseismical actions represented by secular folding,
thrust, or crush, e.g. seismic and volcanic activity or
by sedimentation and erosion, and the consequent creation
of unstable contours which may be destroyed by seismic
action, facial sliding, basal crush by overloading, or
the action of submarine springs, ocean currents effects
or changes in shallow water. He also lists that during
the period 1616 to 1886 some 333 disturbances due to
earthquakes and volcanic shocks have occurred in the
oceans, and some of them have caused damage to submarine
cables.

    Milne furthermore states that one form of submarine
landslips which may be considered is that which may be
found on the submerged faces of a delta. However, he
concludes that because earthquake origins are more numerous beneath the sea than upon the land, it is fair to
assume that the bradyseismical operations resulting in
the folding, bending, crushing, faulting, and thrusting
of rock masses are more active in the recesses of the
ocean, than they are upon our continents. Suboceanic
volcanic activity as, for example, that which is met
within the mid-Atlantic, probably indicates the existence
of bradyseismic movement and a relief of strain (mid-
Atlantic Ridge!). That submarine landslides of great
magnitude have had a real existence is proved for certain localities by the fact that after an interval of a
few years very great differences in depth of water have
been found at the same place, whilst sudden changes in
depth have taken place at the time of and near to the
origin of earthquakes. More recently Herzer (1974) has
discussed an area of uneven bathymetry off New Zealand
(Vergan Bank). He concludes that groups of magnetic,
acoustically opaques knolls occur in association with
submarine slides, and that it is possible that their volcanic emplacement was responsible for touching off the
slides. While Milne had the opinion that submarine earthquakes were the cause for marine slides, Benest (1899)
instead favoured submarine gullies, channels or river
outlets.

A similar approach is taken by Henkel (1970) who points out that the pressure differences on the sea floor associated with waves may be important in the development of underwater landslides and geomorphology. Breaks in a submarine telephone cable in 1966 and 1968 have been studied by Krause et al. (1970), and they suggest that the damage was caused by earthquake-triggered turbidity currents. As early as in 1952 (Heezen and Ewing) a new explanation was given to the 1929 Grand Bank earthquake and to the breaks of all submarine telegraph cables. The cables were broken by a turbidity current originating as a slump on the continental slope. This current travelled across the continental slope, continental rise, and ocean basin floor and continued far out on the abyssal plain well over 450 miles from the continental shelf.

Striem and Miloch (1976) have studied the extreme changes in the sea level along the coast of the Levant finding that more often a recession of the sea than a flooding of the shore has occurred. Such events may have been caused by slumping on the continental slope.

It was mentioned earlier that submarine sliding is a common feature of continental slopes. Winterer (1980) writes that in virtually all Atlantic-margin drill holes, a very large share of the terrigenous sediments actually cored showed evidence of transport by turbidity currents, debris flows, or other essentially down-slope mechanisms. Disturbed beds - so-called slump structures - are especially common at drill sites located on the upper part of the continental rise. In some places the disturbance has proceeded to disruption of the bedding, and finally to the pebbly, mudstone fabric. Modern debris flows have been identified (Embley, 1976) on seismic profile records and in piston cores on the continental rise in the same area off North Africa. Drilling on the continental rise off Morocco showed a sequence of repetitions of middle Cenomanian beds some 200 m thick, suggestive of gravitational slide sheets originating higher on the continental rise. Reflection profile records in the same area show the slide mass as a discrete body spreading out northward and seaward from the drill site, where it was cored.

Breaks of submarine telegraph cables have been mentioned as one of the consequences of marine slides in the ocean. Marine slides, however, also occur in fjords (Bjerrum, 1971), usually in the postglacial deltas and estuaries which have accumulated where the rivers from the nearby mountains debouch the fjord. These slides, which occur intermittently in subaqueous slopes in sand

and silt unsurpassed in dimensions by slides above the
ocean level, have been surrounded by considerable mystery
as they frequently occur in slopes which, according to a
conventional stability analysis, should be stable beyond
any doubt, and in addition the extent of the failures
observed above water along the shore line is generally
very small compared with the volume of the slides in submarine slopes. Marine slides occurring near to coastlines
may cause damage of the coastal area (de Moor, 1979),
e.g. on the dikes.

With the increasing number of pipelines for oil and
natural gas, and even for water, the engineering aspect
of marine slides has grown considerably, and much con-
sideration is given to the problem. In a recent paper
Crans et al. (1980) have developed a theory which leads
to an idealized model - the Multi-Unit Delta Model -
which may be useful for the interpretation of subsurface
deltaic data. They conclude that growth faulting in such
regions can be explained solely as the result of gravity
sliding of overpressured clay and silt.

PORTUGAL

Let me close this exposé by very shortly mentioning
some of the characteristics of the Portugall off-shore
conditions. The area off the coast of Portugal is very
interesting. The Azores are situated on the mid-Atlantic
ridge and have all the characteristics like earthquakes,
volcanoes and hot springs. The East Azores Fracture Zone
stretches from south of the Azores in the west to the
Madeira-Tore Rise in the east (Laughton and Whitmarsh,
1974), the Madeira-Tore Rise is characterized by sea-
mounts, e.g. the Tore seamount. Closer to the coast of
Portugal we have the Interior Basin with Galicia Bank,
Vigo seamount and Oporto seamount. A study of the magnetic results is interesting, because it reveals a tripple-
point (Sibuet et al., 1980).

In November 1980 a seminar on actual oceanographic
problems in Portugal was arranged by Professor Almaça
who is also a contributor to this volume. On that occasion we learned that there is a changing current from
north to south and from south to north along the western
coast of Portugal. Even if a number of rivers are running
to the coast, no marine slides are to be expected there.
That may be different when we consider the Galicia Bank
(Stride et al.).

When on the other hand we look at the southern
coastline - along Algarve - the picture is different,
and we may expect marine slides along this coast.
Actually, we have at the Bay of Alvor where the Alvor
and Odiaxere rivers reach the sea evidence of marine
slides (Ambar, 1980).

REFERENCES

Ambar, I.L. Soares de Albergaria, 1980, Mediterranean
    influence off Portugal, Paper presented at "Semia-
    rio Problemas actuais da Oceanografia em Portugal",
    November 19-20, 1980, Lisbon.
Benest, Henry, 1899, Submarine gullies, river outlets,
    and fresh-water escapes beneath the sea-level, The
    Geographical J., XIV: 394-413.
Bjerrum, L., 1971, Subaqueous slope failures in Norwe-
    gian fjords, Norw. Geotechn. Inst. Publ., 88: 1-8.
Crans, W., G. Mandl and J. Harembourue, 1980, On the
    theory of growth faulting: a geomechanical delta
    model based on gravity sliding, J. Petrol. Geol.,
    2: 265-307.
de Moor, Guy, 1979, Recent beach evolution along the
    Belgian North Sea coast, Bull. Soc. belge de Géol-
    ogie, 88: 143-157.
Embley, Robert W., 1976, New evidence for occurrence of
    debris flow deposits in the deep sea, Geology, 4:
    371-374.
Embley, Robert W. and Robert D. Jacobi, 1977, Distribu-
    tion and morphology of large submarine sediment
    slides and slumps on Atlantic continental margins,
    Marine Geotechnology, 2: 205-228.
Gill, W.D. and P.H. Kuenen, 1957, Sand volcanoes on
    slumps in the carboniferous of county Clare, Ireland,
    Quat. J. Geolog. Soc., London, 113: 441-460.
Heezen, Bruce C. and Charles L. Drake, 1964, Grand Banks
    slump, Bull. Amer. Ass. Petrol. Geologists, 48:
    221-233.
Heezen, Bruce C. and Maurice Ewing, 1952, Turbidity cur-
    rents and submarine slumps, and the 1929 Grand
    Banks earthquake, Amer. J. Sci., 250: 849-873.
Henkel, D.J., 1970, The role of waves in causing submar-
    ine landslides, Géotechnique, 20: 75-80.
Herzer, R.H., 1974, Uneven submarine topography south of
    Mernoo Gap - the result of volcanism and submarine
    sliding, New Zeal. J. Geol. and Geoph., 18: 183-188.
Herzer, R.H., 1979, Submarine slides and submarine cany-
    ons on the continental slope off Canterbury, New
    Zealand, New Zeal. J. Geol. and Geoph., 22: 391-406.

Jacobi, R.D. and C.L. Mrozowski, 1979, Sediment slides
    and sediment waves in the Bonin Trough, Western
    Pacific, Marine Geology, 29: M.1-M.9.
Krause, D.C., W.C. White, D.J.W. Piper and B.C. Heezen,
    1970, Turbidity currents and cable breaks in the
    Western New Britain Trench, Geol. Soc. Amer. Bull.,
    81: 2153-2160.
Laughton, A.S. and R.B. Whitmarsh, 1974, The Azores-
    Gibraltar Plate Boundary, in: Geodynamics of Ice-
    land and the North Atlantic Area, Proceedings of
    the NATO Advanced Study Institute, Reykjavik, Ice-
    land: 63-81.
Lewis, K.B., 1971, Slumping on a continental slope in-
    clined at 1°-4°, Sedimentology, 16: 97-110.
Milne, J., 1897, Sub-oceanic changes, The Geographical
    Journal, X: 129-146 and 259-289.
Moore, T.C., T.H. van Andel, W.H. Blow, and G.R. Heath,
    1970, Large submarine slide off northeastern con-
    tinental margin of Brazil, Amer. Ass. Petrol. Geol.
    Bull., 54: 125-128.
Normark, W.R., 1974, Ranger submarine slide, Northern
    Sebastian Vizcaino Bay, Baja California, Mexico,
    Geol. Soc. Amer. Bull., 85: 781-784.
Sibuet, J.C., W.B.F. Ryan, M. Arthur, R. Barnes, G.
    Blechsmidt, O. de Charpal, P.C. de Graciansky,
    D. Habib, S. Iaccarino, D. Johnson, B.G. Lopatin,
    A. Maldonado, L. Montadert, D.G. Moore, G.E. Morgan,
    G. Mountain, J.P. Rehault, J. Sigal, and C.A. Wil-
    liams, Deep drilling results of Leg 47b (Galicia
    Bank area) in the framework of the early evolution
    of the North Atlantic Ocean, Phil. Trans. R. Soc.
    London, A294: 51-62.
Stanley, D.J. and N. Silverberg, 1969, Recent slumping
    on the continental slope off Sable Island Bank,
    southeast Canada, Earth and Planet. Sci. Let., 6:
    123-133.
Stride, A.H., J.R. Curray, D.G. Moore, and R.H. Belder-
    son, 1969, Marine geology of the Atlantic continen-
    tal margin of Europe, Phil. Trans. R. Soc. London,
    264: 31-75.
Striem, H.L. and T. Miloch, 1976, Tsunamis induced by
    submarine slumpings off the coast of Israel, in:
    Proceedings on earthquake risk for nuclear power
    plants, A.R. Ritsema, ed., Eur. Seismol. Comm.,
    de Bilt: 133-137.
Uchupi, E., 1967, Slumping on the continental margin
    southeast of Long Island, New York, Deep-Sea Re-
    search, 14: 635-639.

Walker, J.R. and J.V. Massingill, 1970, Slump features
     on the Mississippi Fan, Northeastern Gulf of Mexico,
     Geol. Soc. Amer. Bull., 81: 3101-3108.
Winterer, E.L., 1980, Sedimentary facies on the rises and
     slopes of passive continental margins, Phil. Trans.
     R. Soc. London, A294: 169-176.

GEOLOGY AND GEODYNAMICAL SETTINGS

Landslides represent the rapid collapse of a slope. "Rapid" means in this connection that the slide occurs in a short time, e.g. a quarter of an hour.

Landslides can be studied in nature and in the laboratory as well, and it is possible to formulate mathematical expressions and physical conditions for the slides.

Marine Slides may represent the rapid collapse of a slope, however, it may also represent the very slow collapse of a slope. That means that subaqueous slides may have similar dimensions as landslides, but they may also be several times larger.

There is one distinct difference between Landslides and Marine Slides, and that is that the submarine soil is water-saturated, and that makes laboratory measurements more difficult.

A.E. Scheidegger has attempted to investigate the possibility of geological influences (in addition to the well-established control by deposition, toe-erosion, and pore pressure conditions) on the occurrence of submarine slides.

He concludes that the correspondence principle between subaerial and submarine phenomena leads one to expect that the mass movements in either case are significantly influenced by geotectonic factors, and that the morphology of the submarine slopes may also be influenced by geological factors.

In their paper, Prior and Coleman are presenting series of data from geological and geophysical surveys of the Mississippi River Delta covering a very long range of area. It is shown that the rate of sedimentation varies from a few millimetres per year on the shelf to more than 30 metres per year at the front of the delta slope.

It is concluded that sedimentation rates are extremely rapid in delta regions, which indicate that submarine mass movement processes are extremely signifant; however, many aspects of the processes are not well understood and need more research.

ON THE TECTONIC SETTING OF SUBMARINE SLIDES

Adrian E. Scheidegger

Institut für Geophysik
Gußhausstraße 27 - 29
A-1040 Wien - Austria

ABSTRACT

It is suggested that in the occurrence of submarine slides
there is a geological control present in addition to the well-
established control by "exogenic" agents, such as deposition, toe-
erosion and pore pressure conditions. This is inferred from the
existence of a principle of correspondence between subaerial and
subaqueous phenomena based on the notion of effective pressure.
Inasmuch as it can be shown that subaerial slides are co-deter-
mined by geotectonic conditions, the same must be held to be true for
submarine slides.

INTRODUCTION

Submarine slides are quite generally mass movements that occur
below the shore line of a body of water. The analysis of the incep-
tion of such phenomena has generally been confined to the investigation of the mechanisms of instability build-up and triggering.
The analysis of the further progress of submarine slides is based
on an investigation of gravity sliding generally.

In this paper, an attempt is to be made to investigate the
possibility of geological influences on the occurrence of submarine
slides. Whilst it is clear that the availability of sediment and
triggering effects constitutes the main factors in the generation
of submarine slides, it may be expected that the general geological
(including the geotectonic) conditions are also contributing factors.

MORPHOLOGICAL ANALYSIS

As noted in the Introduction, the standard analysis of submarine slides is mainly based on morphology. Embley and Jacobi (1978)
have given a review of the distribution and morphology of large
submarine slides on Atlantic coastal margins (see Fig. 1), in which
they show that there is no apparent consistent relationship between
the presence of the slides and the sedimentary environment in which
they occur. Naturally, slides may be expected to arise more frequently in areas of rapid sediment-deposition, such as in the vicinity of river mouths(Rio Magdalena, Mississippi), than elsewhere;
however, they occur nevertheless also in pelagic sediments on isolated oceanic rises (e. g. Madeira Rise) where sedimentation rates
are less than 2 cm/1000 years: Eventually, unstable conditions are
created even in such cases.

As failure mechanisms, tectonics, overloading, erosion of a
slope at the toe, and in-situ changes in the internal pore pressures have been invoked. The Grand Banks Slide of 1929 is known
to have been triggered by an earthquake (Heezen and Ewing, 1952).

The slide-mechanisms have been classified into two types: rotational and translational. The former are often referred to as
"slumps", the latter as "slides" sensu proprio. The "slumps" are
similar to rotational subaerial landslides (model of Terzaghi,
1943), the "slides" sensu proprio are similar to subaerial debris
flows in which material is translated along a glide plane over a
long distance. In underwater slides, such events evolve into turbidity currents which may have a very long reach indeed (Grand
Banks: 800 km).

The morphology of submarine slides, thus, does not appear to
give any immediate clues as to their geological setting. An attempt
at defining the geological setting of submarine slides has been reported by Moore (1978) who discussed the correlations between the
occurrence of submarine slides and plate tectonics in general terms.
However, again no definite statements could be made: slides occur
on active as well as on passive continental margins. Subduction
zones may cause a decrease of the availability of sediments (by
subducting them) or an increase (by scraping them off and piling
them up). Generally speaking, the most likely places for submarine
slides are the continental slope regions where the continental
crust passes into the oceanic crust: Here are the areas of maximum
deposition rates of sediments on Earth.

THE PRINCIPLE OF ANTAGONISM

If one wishes to separate those features of a phenomenon which
are caused by geological agents from those which are caused by morphological agents, it is first of all necessary to have an unequi-

vocal characterization of the two types of features. This can be achieved by taking cognizance of the "principle of antagonism" (Scheidegger, 1979): It is a rather well-known fact that the present-day morphology of the Earth is caused as the instantaneous outcome of the action of two types of agents; these may be referred to in turn as endogenic and exogenic. The characteristic difference in those two types of agents is that the endogenic agents are inherently non-random, the exogenic agents, on the other hand, are inherently random.

If, therefore, non-random features can be detected in the morphology of mass-movements, these can be assumed to be of endogenic, i. e. geological, origin. One has here an unequivocal means of separating the geological from the morphological controls in a phenomenon.

THE PRINCIPLE OF CORRESPONDENCE

The literature on submarine slides is not very large. Most studies concern rather morphological descriptions of single events. A recent summary of such descriptions is contained in the review of Moore (1978) which has already been mentioned earlier. From these single descriptions, it is rather difficult to arrive at general mechanical relationships.

The literature on subaerial slides is much larger than that on submarine slides (see e. g. Scheidegger, 1975). Hence, if results from the study of subaerial slides could be transferred to the mechanics of submarine slides, much could be gleaned from the former that could be of significance with regard to the latter.

In this instance, it may be noted that a general principle of correspondence exists between mass movements in an aerial and in an aqueous environment. The difference in the comportment of the two environments lies in the fact that the water in the subaqueous environment can sustain pressures between the mass-particles which are much greater than those supported by the air in the subaerial environment. The principle of correspondence was enounced by Terzaghi (1943) who introduced the concept of "effective pressure" P_e in a liquid-saturated mass given by the following relation

$$P_e = P_{actual} - P_f$$

where P_{actual} is the really acting pressure and P_f the pressure in the fluid. The principle of correspondence then states that the deformation of a mass in an aqueous environment is the same as that in a subaerial environment if the total pressure acting on the mass in a subaerial environment is replaced by the effective pressure in the aqueous medium. The reason for the operation of this principle is the fact that the forces between the mass-particles are diminished by the forces supported by the permeating liquid.

The principle of correspondence refers mainly to the mechanics of sliding. It states that inferences can be made from the study of the mechanics of subaerial slides on that of submarine slides. Thus, it has been noted that subaerial slides are mostly "flow", not "slide"-phenomena (Hsü, 1978), caused by the bouncing-interaction of the individual slide-"particles". The same must be held to be true for submarine slides inasmuch as the ambient water tends to enhance flow phenomena over slide-phenomena.

Furthermore, the stability considerations for subaerial slope banks can be simply extended to submarine slope banks by referring everything to the "effective", rather than the total pressure. This approach had already been taken by Terzaghi (1943).

GEOLOGICAL CONTROL OF SLIDES

The main aim of the present paper is to uncover possible geo-logical, i. e. endogenic, controls in submarine slides. As noted, pertinent general investigations of submarine slide phenomena are relatively few in number, and thus we shall present here some in-vestigations on subaerial slides in the hope that the principle of correspondence allows one to transfer the results to submarine conditions.

The occurrence of subaerial slides, like that of submarine slides, is generally ascribed to a local instability that may be induced by the oversteepening of a slope bank by an undercutting river, by the accumulation of debris on the slope from above and such like. The actual triggering may be effected by an increase in pore water pressure or even an earthquake. In any case, the primary cause (not the trigger) is ascribed to exogenic effects.

However, evidence has been accumulating recently that the predesign of mass-movements is not entirely of exogenic origin, but may have a co-origin in endogenic, mostly neotectonic, pheno-mena and may thus be subject to a geological control.

Let us mention here tree cases supporting the above thesis, which were analysed by the present author and his co-workers. In these studies, the directions of mass movements observed in the field have been compared with the principal stress directions of the present-day neotectonic stress field. The latter have been de-duced from local joint-orientation measurements according to the method of Kohlbeck and Scheidegger (1977). This method consists of a statistical procedure, in which two probability distributions of the type $\exp(k \cos^2 \delta)$ about a mean direction are fitted to the data; the two best-fitting mean directions are determined by computer using a function-minimization procedure. From the position of the two best-fitting orientations the principal stress directions are calculated as the respective bisectrices. This procedure corresponds to the interpretation of recent joints as shear surfaces in the neotectonic stress field.

The first of the cases referred to above concerns a slide area in the Felber Valley (Salzburg Province, Austria) (Carniel et al., 1975). For its analysis, tachymetric, seismic, geomorphological and geological studies were carried out. The slide was an actual rapid slide that occurred in May, 1965 after a lengthy rainy period. The slope above and below is constantly being subject to slow mass movements from the ledge atop the vally (presenting the aspect of "mountain fracture") to the valley bottom. It is significant to note that the motions of the slide area as well as those of the creeping slope fit with the joint-orientations into a single geophysical stress pattern with a maximum horizontal compression in the N 15°W and a minimum compression in the N 65°E direction. The mountain-fracture "scars" are roughly normal to the minimum compression direction (Fig. 2). The displacements thus, are not random but are aligned to the local geotectonic stress field.

The second of our cases refers to a slow mass displacement in the Valley of Gastein, Austria (Hauswirth and Scheidegger, 1980). Many houses in this resort area show traces of movements (cracks in walls and foundations). The displacement rates could be calculated from data made available by the Austrian Geodetic Survey (Fig. 3). A careful analysis of the statistical distribution of the displacement vectors and the joint orientations shows that the displacement is parallel to one of the preferred joint orientations and orthogonal to the other. In the present case, therefore, one has the evidence for geotectonic shearing motions occurring at the present time which give rise to the mass displacements.

The third case refers to an unstable valley slope in the Lesach Valley in Eastern Tyrol, Austria (Hauswirth et al., 1979). A careful study of the motion of this slope was made by means of repeated surveys. The result showing the displacements found to have taken place over a period of 3 years (1973 to 1976) is given in Fig. 4. The displacement vectors may be compared with measurements of joint orientations on the area. It turns out that the average displacement direction corresponds closely to one of the principal stress directions calculated from the orientations of the joints, which, incidentally, is the regional smallest principal "intraplate" stress direction of Europe.

The three cases presented above show clearly that subaerial
mass movements are partially pre-designed by geological-tectonic
factors. Naturally, the triggering of the rapid slide in the first
of the instances considered, was due to hydrological conditions.
Nevertheless, the general "lay-out" of the displacement vectors
appears to have been pre-designed by the geotectonic stress field.
In this connection, it should not be forgotten that the orientation
of the valleys on whose sides the slides occur may also be designed by the geotectonic stress field (Scheidegger, 1980).

CONSEQUENCES FOR SUBMARINE SLIDES

It remains to draw consequences for submarine slides from the
exposition given in the previous sections.

As noted earlier, little is known about possible geological
controls of submarine slides. It is well known that the actual
triggering of such slides is caused by the instantaneous exogeni-
cally controlled conditions, such as availability of sediment, pore
water pressure, stability of slope bank etc. Nevertheless, the subaerial mass movements have been found to be significantly influenced by geotectonic factors, and the correspondence principle between subaerial and submarine phenomena leads one to expect that
the same holds true in the latter case as well. In this connection,
it should be recalled that the very morphology of the submarine
slopes on which the slides occur, may also be influenced by geological factors. Inasmuch as it has been found that the orientation
structure of subaerial valleys is largely "predesigned" by tectonic
forces, the same may be held to be true of submarine canyons. Unfortunately, it is difficult to make joint-orientation measurements on the continental slope, not to speak of in-situ stress
measurements. However, it may be a fitting task for a workshop on
submarine slides to propose suitable procedures for solving this
problem.

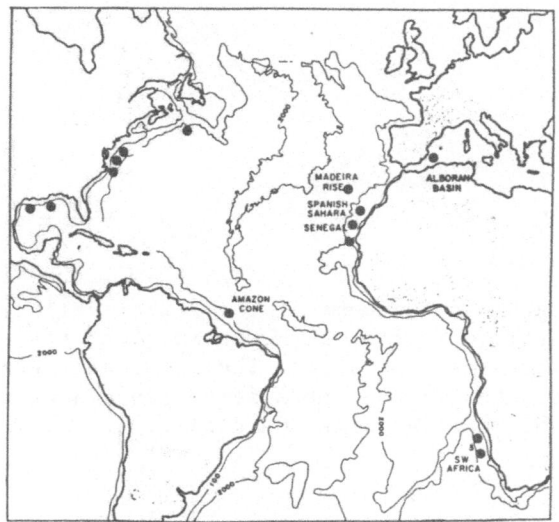

Fig. 1: Locations of large submarine slides in the
        Atlantic (after Embley and Jacobi, 1978)

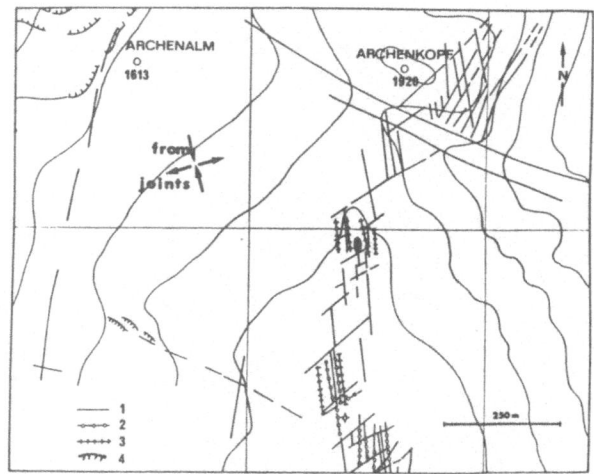

Fig. 2: Geomechanical scheme of the Felber Valley slide.
1: Surface faults; 2: Mountain fractures; 3: Ledges,
4: Tear scars of slide. The principal stress directions
calculated from joint-orientation measurements are also
shown. Modified after Carniel et al. (1975)

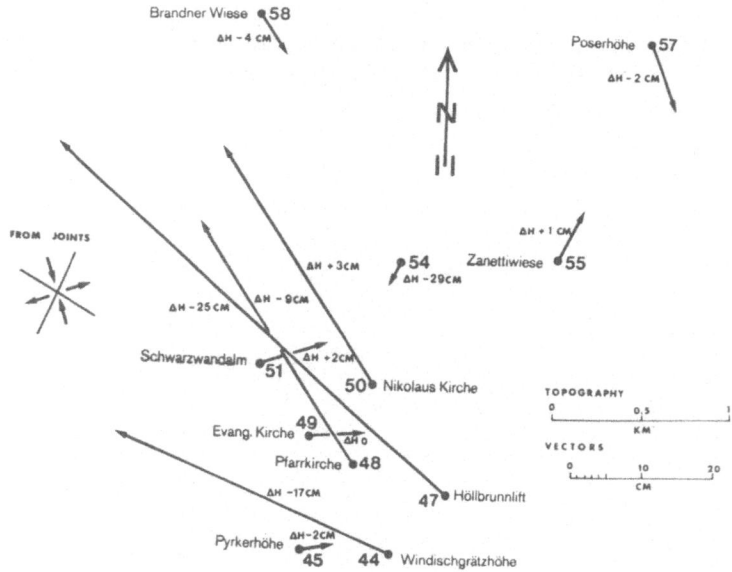

Fig. 3: Motions in Bad Gastein during a period of 40 years
(1933-1973), compared with the preferred joint strikes and
principal stress directions calculated therefrom. Modified
after Hauswirth and Scheidegger (1980)

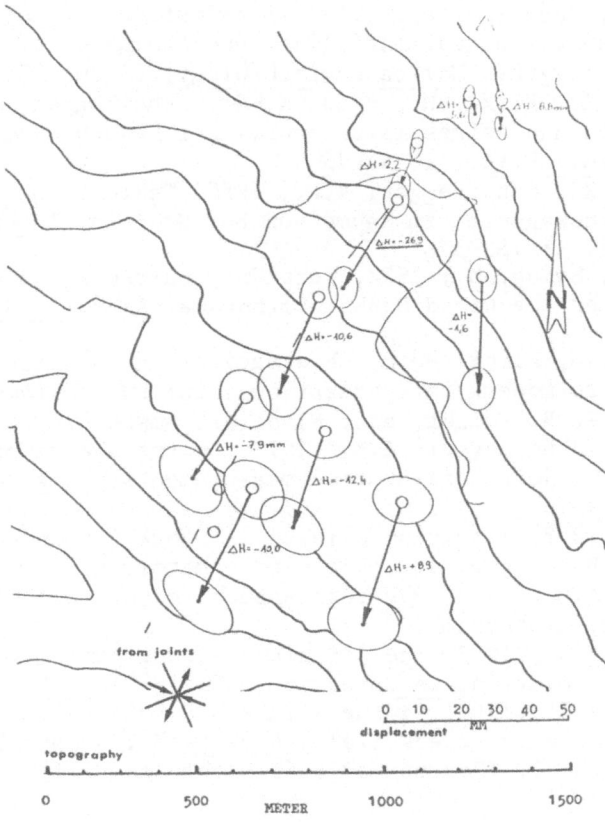

Fig. 4: Topography and slide motion in the Lesach Valley compared
with preferred joint strikes and principal stress directions calculated therefrom (modified after Hauswirth et al.
1979)

REFERENCES

Carniel, P., Hauswirth, E.K., Roch, K.-H., Scheidegger, A.E.,
    1975, Geomechanische Untersuchungen in einem Rutschungs-
    gebiet im Felbertal in Österreich, Verh. Geol. B.-A.,
    Wien 1975, 4: 305-330
Embley, R.W., Jacobi, R.D., 1975, Distribution and morphology of
    large submarine sediment slides and slumps on Atlantic
    coastal margins, Marine Geotechnology, 2: 205-228
Hauswirth, E.K., Pirkl, H., Roch, K.-H., Scheidegger, A.E., 1979,
    Untersuchungen eines Talzuschubes bei Lesach (Kals, Osttirol),
    Verh. Geol. B.-A., 2: 51-76
Hauswirth, E.K., Scheidegger, A.E., 1980, Tektonische Vorzeichnung
    von Hangbewegungen im Raume von Bad Gastein, Interpraevent
    1980, Bd. 1: 159-178
Heezen, B.C., Ewing, M., 1952, Turbidity currents and submarine
    slumps and the Grand Banks earthquake, Amer. J. Sci., 250:
    849-873
Hsü, K.J., 1978, Albert Heim: Observations on landslides and re-
    levance to modern interpretations, in: Rock Slides and Ava-
    lanches I, B. Voight, ed., Elsevier, Amsterdam
Kohlbeck, F., Scheidegger, A.E., 1977, On the theory of the eva-
    luation of joint orientation measurements, Rock Mech.,
    9: 9-25
Moore, D.G., 1978, Submarine slides, in: Rock Slides and Avalan-
    ches I, B. Voight, ed., Elsevier, Amsterdam
Scheidegger, A.E., 1975, Physical Aspects of Natural Catastrophes,
    Elsevier, Amsterdam
Scheidegger, A.E., 1979, The principle of antagonism in the
    Earth's evolution, Tectonophys., 55: T7-T10
Scheidegger, A.E., 1980, Alpine joints and valleys in the light of
    the neotectonic stress field, Rock Mech., Suppl. 9: 109-124
Terzaghi, K., 1943, Theoretical Soil Mechanics, Wiley, New York

ACTIVE SLIDES AND FLOWS IN UNDERCONSOLIDATED MARINE SEDIMENTS

ON THE SLOPES OF THE MISSISSIPPI DELTA

David B. Prior and James M. Coleman

Coastal Studies Institute
Louisiana State University
Baton Rouge, LA   70803

ABSTRACT

On the continental shelves off large deltas, rapid progradation and deposition result in highly underconsolidated marine sediments. These deposits, which are often also rich in interstitial methane gas, can be subject to widespread and active mass movement downslope. For example, the submarine slopes of the Mississippi River delta are affected by a variety of sediment instability processes. Geologic and geophysical surveys using side-scan sonar, subbottom profilers, and precision depth recorders have been completed for the entire subaqueous delta. Survey lines were spaced at 240-m intervals, and water depths ranged from 5 m to 20 m. Bottom morphology, including sediment deformations indicative of instability, has been mapped at a scale of 1:12,000, and large-area, scale-corrected sonar mosaics have been constructed. The features identified include collapse depressions, bottleneck slides, shallow rotational slides, mudflow gullies, overlapping mudflow lobes, and a wide variety of faults. The slides and mudflows are extremely active, and movement rates of several hundred metres per year have been recorded. Damage to offshore oil and gas pipelines and platforms has occurred. Also, the concept of slow, continuous deltaic progradation must be modified to include the effects of these processes. For example, on the shelf, normal settling of suspended clays averages only a few millimetres per year, whereas at the front of the delta slope more than 30 m of sediment has been deposited by mudflows and slides since 1875.

These deltaic processes are the result of complex temporal and spatial combinations of different factors, including the generation of excess pore pressures by rapid sedimentation, methane gas within the sediments, wave-induced stresses, and localized slope oversteepening. These conditions are not unique to the Mississippi Delta, and indeed similar processes, which affect geologic deposition models and provide design constraints for offshore engineering, appear to be common in many deltaic environments.

INTRODUCTION

Recent detailed investigations on continental shelves and shelf slopes have revealed that subaqueous gravity-induced mass movements of sediment, whether active or relict, are extremely common phenomena and should be considered integral components of normal shelf and shelf slope transport processes. In some shelf environments, especially those seaward or downdrift of large river discharges, sediment transport and deposition by subaqueous mass movement accounts for a large proportion of the shelf deposits. Continental shelves such as those bordering the north-central Gulf of Mexico, northern coasts of South America (Magdalena River, Esmeraldas River, Orinoco River, Surinam, the Guianas, and the Amazon River), Alaskan shelves (Yukon, McKenzie, and Copper Rivers), and other deltaic areas such as the Niger, Congo, Orange, Ganges-Brahmaputra, Indus, Nile, Yangtse, Red, and Hwang Ho Rivers all display sediment and sea-floor morphology characteristic of downslope mass movement and slope failure. The following generalizations can be made about instabilities in these regions: a) instability can occur on very low angle slopes (generally less than 1°); (b) large quantities of sediment are transported from shallow water depths into deeper water offshore along well-defined landslide channels; (c) individual failures, although variable in size, generally possess three morphological components: a source area with subsidence and rotational slumping, a central transport zone, often defined by a channel or chute, and a composite depositional area composed of overlapping lobes of remolded debris; (d) although movement areas are not generally known, it appears that displacements of sediment can accompany the initiation of new features on a previously stable part of the shelf or reactivation of previously existing unstable areas.

The Mississippi River delta and adjacent shelf region have been investigated for many decades, but recently there have been many improvements in the systematic utilization of various techniques for underwater exploration and sea-floor mapping. The application of side-scan sonar and high-resolution seismic techniques has led to substantial progress in the documentation and mapping of the subaqueous region of the delta. These techniques, aided by the history of problems encountered in foundation design for offshore oil and gas structures and pipelines, have permitted the identification of a variety of active slope and sediment movements.

THE OFFSHORE MISSISSIPPI DELTA

Sediment Distribution and Properties

        The annual sediment discharge from the Mississippi River
into the Gulf of Mexico is estimated at 6.2 x 10^{11} kg. The bedload
consists of 90% fine sand, and the suspended load is characterized
by 65% clays and 35% silt. The coarser material is deposited at or
near the distributary mouths because of rapid effluent decelera-
tion and saltwater entrainment as the plume leaves the distributary. The fine-grained sediment is kept in suspension and spreads
laterally far beyond the immediate ends of the distributaries.
Deposition of the widely disseminated fine-grained sediments
builds a platform fronting the delta that consists of clays that
were rapidly deposited, have an extremely high water content, and,
because of abundant fine-grained organics, rapidly degraded by
bacteria, include large accumulations of sedimentary gases (primarily methane and carbon dioxide). Seaward progradation rates of
the distributary mouths (Fig. 1) vary from more than 100 m/year to
less than 50 m/year, depending on specific distributaries. Sedi-

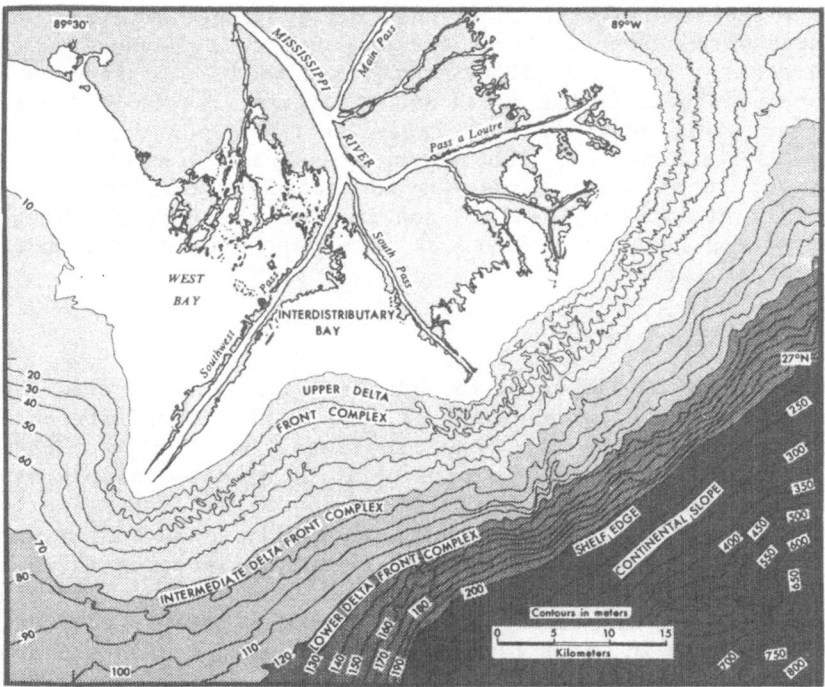

Fig. 1. Hydrographic map of the Mississippi River delta.

mentation rates seaward of the river mouth are extremely high,
averaging 1 m/year, but during periods of high flood, accumulations of 3-5 m over a 4-month period have been measured.   In
adjacent interdistributary bays accumulation rates rarely exceed a
few centimetres per year, and in some cases bay bottom erosion is
occurring. In offshore waters in front of the delta, accumulation
rates vary considerably, from a few centimetres per year in 50-m
water depth to fractions of a centimetre in water depths near the
shelf-edge break. As will be described later, this spatial pattern of sedimentation by settling from the water column is substantially modified by mass-movement processes.

    The geotechnical properties of the sediments from in situ
borehole measurements and analysis of cores show that the delta
front is composed largely of heavily underconsolidated clays, with
water contents at or near the liquid limit in the upper 15 m of
sediment. Undrained shear strength profiles determined by vane
tests commonly show strengths less than 0.05 kg/cm^2 at the mudline; they may remain less than 0.1 kg/cm^2 to depths of more than
30 m. Alternatively, a "crust" of somewhat higher strength is
sometimes present in the upper 14-15 m of sediment in which
strengths may exceed 0.25 kg/cm^2, but overlies a zone of strength
cutback at approximately 20-25-m depth, where values can again be
as low as 0.05 kg/cm^2 (Hooper, 1980). These low strength values in
what is sometimes referred to as the "failure zone" are apparently
associated with extremely large excess pore-water pressures within
the sediments. Terzaghi (1956) postulated that low strengths and
high pore pressures in the delta are the direct result of the rate
of deltaic sedimentation, especially on the upper delta slopes
near the distributaries. Subsequently, direct piezometric measurement of in situ pressures has indicated pore-water pressure
ratios (r_u) of greater than 0.9 and in some cases approaching or
even exceeding geostatic values at some depths below the mudline
(Dunlap et al., 1979).

Offshore Slopes

    The submarine slopes of the entire delta front have extreme-
ly low angles, rarely exceeding 1.5°, and in the interdistributary
bays the bottom slopes are generally less than 0.5° and rarely
greater than 0.2°. In water depths of 10-80 m bottom slopes range
from 0.7° to 1.5°, and in depths of 80-200 m the slopes are less
than 1°. At the shelf break, which generally occurs in a water
depth of 200 m, the slopes increase slightly, averaging 1.7-2.2°.
In general, hydrographic maps indicate extremely irregular topography, the bottom displaying a large number of radially trending
submarine gullies in water 10-80 m deep and broad terraces seaward
to 200-m water depth. At the shelf edge and on the upper continental slope, abrupt scarps are found on the sea floor, some as
high as 20 m and with slopes approaching 2.5-3.0°.

Storms and Waves

The offshore delta region is affected by winter storms and hurricanes, which generate surface water waves. Seaward of the delta, in 200-m water depths, 10- and 11-second waves with heights in excess of 20 m have been recorded. For example, Hurricane Camille, one of the most destructive hurricanes ever experienced in the region, generated waves of 20-24 m. More recently, in 1979 Hurricane Frederic, which passed more than 100 km east of the delta, was responsible for generating waves in excess of 10 m seaward of the delta. As such waves approach the delta, the sea-floor sediments begin to influence their propagation and they can be strongly attenuated (Forristall et al., 1980). For example, with Hurricane Frederic waves the low-frequency components were attenuated by factors between one and two orders of magnitude. The significant wave height of approximately 9.0 m at the shelf edge was reduced to less than 2.0 m over 28 km into shallower water. Similarly, during Hurricane Camille maximum wave heights varied from 20 m in 100-m water depths to 3-5 m in 20-m water depths (Bea and Audibert, 1980).

The effects of surface waves on sediment properties have been addressed using instrument packages including accelerometers and piezometers installed on the sea floor (SEASWAB experiment). Suhayda et al. (1976) and Dunlap et al. (1979) have reported some of the results, and it was demonstrated that substantial bottom sediment motions and pore-water pressure increases accompany large surface water waves.

ENGINEERING PROBLEMS RESULTING FROM SOIL MOVEMENTS

In the 1950s offshore exploration commenced in the shallow-water regions of the delta, and data from soil foundation corings and hydrographic surveys were utilized to develop the first regional assessments of the geology of the delta (Fisk et al., 1954; Fisk and McClelland, 1959; Shepard, 1955). In particular, Shepard (1955) drew attention to the delta-front valleys or gullies which occur in profusion, radiating out toward deeper water from the mouths of the passes. Shepard (1955) suggested that they were due to submarine sediment sliding and mass movement on the low-angled (0.5°) slopes. Terzaghi (1956), considering this hypothesis, showed how high pore-water pressures could be satisfactorily explained in relation to sedimentation and agreed with Shepard that the delta-front gullies represented slope instability mechanisms.

By the early 1960s the offshore oil and gas industry had begun to experience directly the effects of sediment mass movement down the delta-front slopes. Since that time, with the construction of more than 500 platforms and thousands of kilometres

of pipelines in the area, various engineering problems have
occurred.

(a) Movement of permanent platforms. Pile-supported structures have been damaged by substantial sea-floor movements. Bea
and Audibert (1980) document a number of examples. One of the
earliest structural losses was a well jacket in 17 m of water
during Hurricane Betsy (1965), apparently due to a retrogressing
slide that had earlier caused pipeline damage nearby. The best
documented examples of platform failure are those associated with
the passage of Hurricane Camille in 1969 (Fig. 2). Sediment instability extended to a depth of 21 m below the surface in 100-m water
depth, and two platforms were destroyed. One moved a distance of
30 m downslope and toppled on its side. Another platform was
damaged by small lateral displacements of 1-2 m. This damage
apparently accompanied soil movements extending 1,200 m downslope.

(b) Movement of jack-up rigs. Jack-up rigs experience two
problems associated with bottom sediment instability. Firstly,
difficulties have been encountered in establishing stable foundation during initial setup. As the platform legs are emplaced on
the sea floor they encounter insufficient support because of both
the low bearing capacity of the sediments and actual downslope

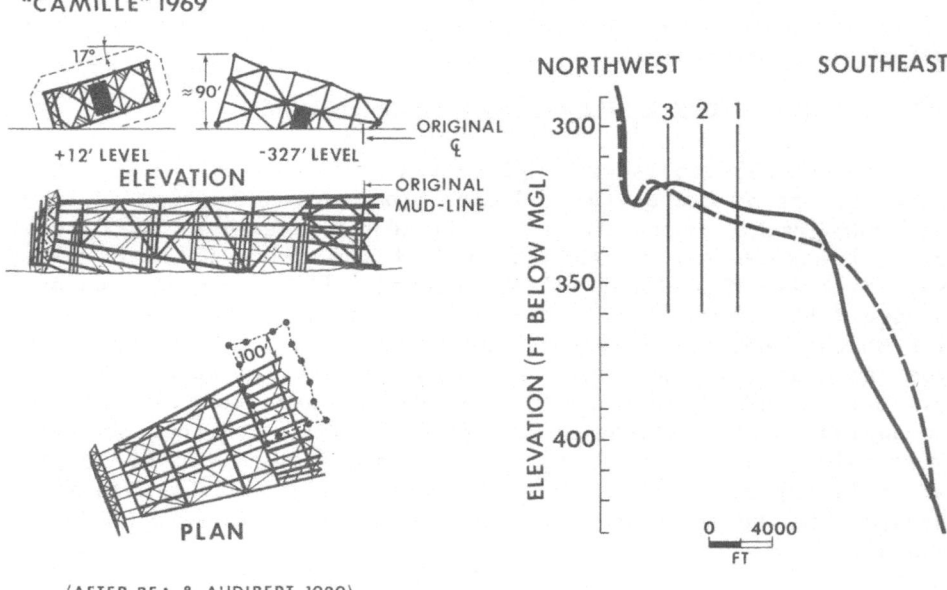

(AFTER BEA & AUDIBERT, 1980)

Fig. 2. Rig damage and profile changes associated with bottom
          sediment movements during Hurricane Camille, 1969.

movement. Secondly, after emplacement rigs have suffered loss of
support and differential settlement in which individual legs
apparently foundered vertically within the sediment to depths of
several meters.

(c) Pipeline damage. Considerable numbers of pipeline
breaks have occurred, especially on the upper delta slopes in
water depths of less than 30 m. These breaks often cannot be
attributed to routine hardware maintenance problems. Survey and
divers' reports indicate that pipelines are both locally displaced
in a downslope direction and appear to have sunk within the sediment. Repeated pipeline breaks frequently occur in particular
areas, especially where there is recognizable irregular bottom
topography.

While structural damage and pipeline breaks often accompany
severe weather conditions, when surface wave energies are large
this is not exclusively the case. Indeed, there are numerous
instances of breaks that do not accompany storms. Similarly, at
one location a platform is subject to irregular, spasmodic movements not related to sea state.

TYPES OF SEDIMENT INSTABILITY PROCESSES

In view of the increasing evidence for sea-floor mass-
movement activity in the delta region, the U. S. Geological Survey
instigated a major project: mapping the entire submarine delta
and identifying and describing the subaqueous instability features
and processes (Coleman et al., 1980). Geophysical surveys utilized high-frequency (~110 kHz) acoustic sources for bathymetry,
3.5- to 12-kHz frequency sources for near-surface (<70 m) subbottom penetration, and lower frequency (50-100-kHz) sparkers or
other acoustic sources for deeper subsurface penetration. Side-
scan sonar data were acquired with both conventional, uncorrected
systems and with the digital E. G. and G. SMS 960. The digital
system involves processing for corrections of ship speed and slant
range and results in almost undistorted acoustic views of seafloor objects (Prior et al., 1978a, b; Prior and Coleman, 1980).
The range setting on the sonar was primarily 150 m, giving swath
widths of 300 m. Track line spacing of 240 m gave overlapping
side-scan coverage, eliminating the need for interpretation between lines and allowing the construction of large-area mosaics of
sea-floor and instability morphology. Some 14,000 km of data were
acquired for an area of approximately 100 lease blocks. Maps
depicting track lines, bathymetry, sea-floor morphology, and near-
surface geology at a scale of 1:48,000 have been published for the
entire delta-front slope as BLM Open File Report 80-01 (Coleman et
al., 1980).

The types of subaqueous instabilities commonly encountered in the offshore Mississippi River delta include a) peripheral rotational slumps; b) collapse depressions and bottleneck slides; c) retrogressive, elongate slides and mudflow gullies; d) depositional lobes of mudflows; and e) a variety of types of slumps and faults that are normally of deep-seated origin, yet affect the present sediment surface. Previous literature discussing submarine failures in this region includes papers by Coleman et al. (1974); Coleman and Garrison (1977); Garrison (1974); Henkel (1970); Prior and Coleman (1978a, b); Prior and Suhayda (1979a, b); Prior et al. (1979a, b). Many of the subaqueous features display morphological characteristics that are similar to instabilities that have been documented in the subaerial environment, and it is possible to draw from these analogies some inferences concerning the mechanisms responsible for the formation of the subaqueous failures.

The main types of slope and sediment instability that have been mapped in the 5-300-m water depths are illustrated schematically in Figure 3, which shows their distribution around a single distributary and part of an interdistributary bay. Similar spatial organization can be identified around the entire periphery of the modern Mississippi River delta. In the vicinity of the passes rotational slumps are the most common feature, and in the shallow bays and near shore (5-10 m water depth) collapse depressions and bottleneck slides are the most common instability. Elongate retrogressive slides and coalescing mudflow gullies are the major instabilities on the upper delta front. Remolded debris discharged from the gullies spreads out as overlapping depositional lobes on the seaward ends of the gullies. In many cases the seaward toes of the depositional lobe form abrupt scarps on the sea floor, and pressure ridges and mud volcanoes form adjacent to the terminal edge of the lobe.

Differential Weighting and Diapirism

At the mouths of the Mississippi, vertical mass movement is caused locally by the rapid deposition of dense distributary-mouth bar sands on top of less dense prodelta and marine shelf clays. The differentially weighted clays are squeezed out and forced upward as diapiric intrusions into the overlying material (Morgan, 1961; Morgan et al., 1963, 1968; Coleman, 1976). Such intrusions can result in vertical displacements as great as 200 m, and, correspondingly, bar sands that have subsided into the muds can accumulate to thicknesses as great as 150 m. Thin diapiric spines of mud that result from this process appear as islands, which are locally referred to as mudlumps, and occur near each of the major distributaries of the Mississippi River. During the period 1876-1973 some 105 individual diapiric spines were mapped in a 13-km^2 area at the mouth of South Pass (Morgan et al., 1963). As these

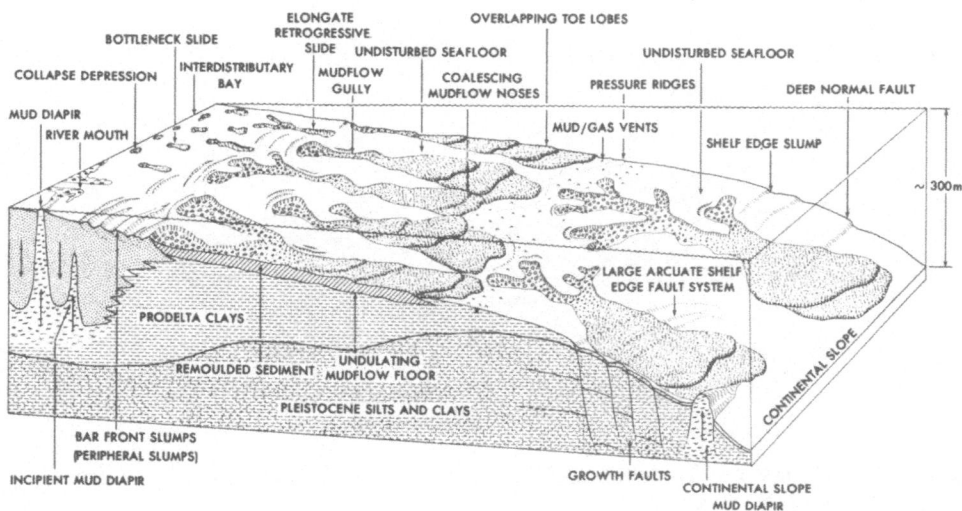

Fig. 3.   Schematic distribution and morphology of subaqueous landslides in the vicinity of a distributary and offshore,
Mississippi River Delta.

diapiric spines of mud are intruded into the overlying bar sands,
wave reworking of the muds results in redeposition of faunal
remains from the muds into the sand bodies. Movements of the
mudlumps coincide primarily with river stage. Following each
major flood of the river, when large amounts of sands have been
added to the bar, mudlump activity increases in intensity; old
mudlumps are rejuvenated, and vertical uplift ensues, as well as
formation of additional new mudlump islands. It is not uncommon
for an individual mudlump to be pushed vertically 6-9 m during a
single flood.

Peripheral Rotational Slumps

Downslope movement of large sediment masses begins high on
the upper delta front, near the distributary mouths of the river.
Bottom slopes at the immediate mouths of the distributaries range
from 0.2° to 0.6°, but often major scarps that display distinctive
curved or curvilinear plan views scar these gentle slopes. The
scarps vary in height from 3 m to 8 m and exhibit slopes of 1-4°;
in many cases they give the bar front a stairstepped appearance in
profile view. Tensional crown cracks are often present upslope
from the major scarp, and frequently mud vents are associated with
these scarps. The surface of the slump block normally displays
extensive hummocky, irregular bottom topography and displaced
clasts of sediment. In most instances the slump blocks have been
downthrown; often they have been rotated in an upslope direction,

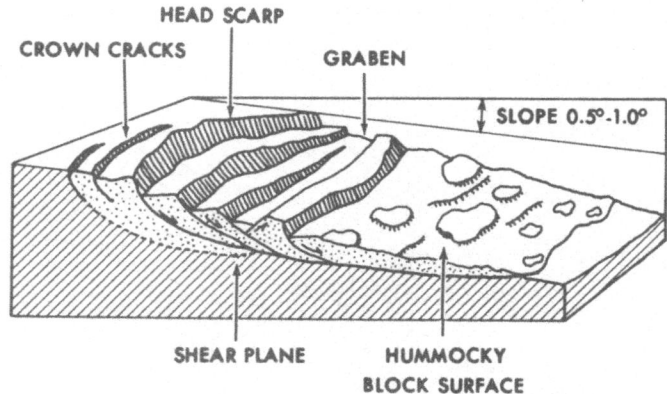

Fig. 4.   Schematic representation of the morphology of a rota-
          tional slump area.

producing a recognizable reverse slope.   Figure 4 shows in schematic form the most common morphological characteristics of these
features.   Figure 5 is a side-scan sonar mosaic across a series of
slump features.   Note in this figure the distinct curved traces of
shear planes, the stairstepped appearance of the bottom, and the
irregular topography on the seaward side of the slump region.

     This type of morphology is indicative of rotational sliding
over slightly curved shear planes that are concave upward, and
subsequent translational movements downslope.   This type of shear
failure occurs when stresses exceed strength because of excess
pore-water pressures, and failure is often enhanced by slight
oversteepening of the bar front as a result of addition of sediment
during floods.   Shear deformation is essentially confined to multiple discrete failure zones which separate large, intact blocks.
These shear planes have a slight concave-upward curvature, and
they tend to coalesce at depth into a single basal shear plane.
High-resolution seismic data confirm the presence of multiple
concave-upward shear planes.   The average depth of movement is
approximately 30-35 m.   Although movement rates have not been
documented in detail, repeated surveys indicate that blocks whose
individual widths ranged from 20 m to 350 m moved well over 1.5 km
downslope over a 1-year period.   Repeated hydrographic surveys in

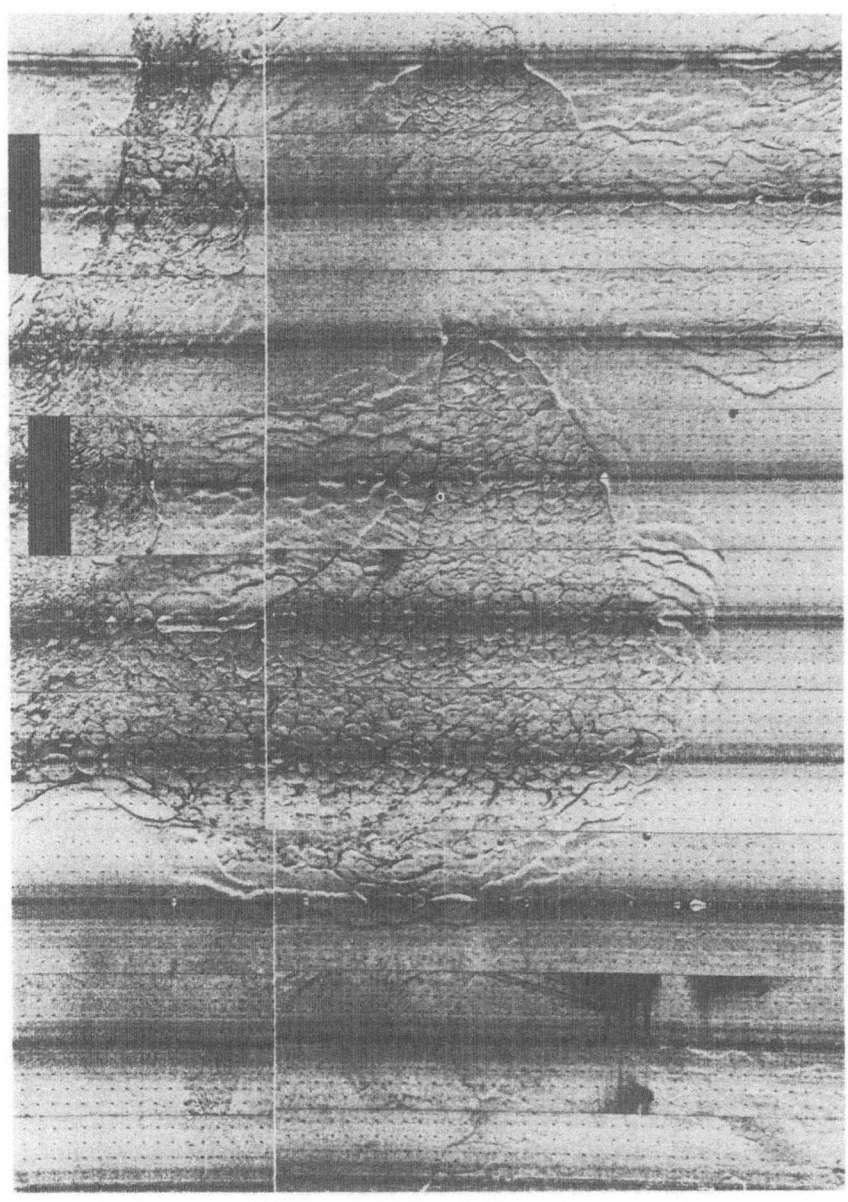

Fig. 5. Scale-true side-scan sonar mosaic across a
slumped zone feeding into a mudslide channel.
Mosaic covers an area of approximately 2 km x 3.5
km.

the region have indicated that this type of deformation is common
and occurs on a nearly continual basis at the mouth of each of the
distributary-mouth channels.

Collapse Depressions and Bottleneck Slides

Collapse depressions and bottleneck slides occur primarily
in the shallow-water areas of interdistributary bays. They are
commonly associated with slopes of 0.1-0.2° and where sedimentation rates are relatively small by comparison with the more
active areas of sediment accumulation in the delta. The collapse
depressions are generally relatively small in relationship to
other mass movement landforms in the delta; however, they are
extremely numerous within any one given area. The features range
in size from 50 m to 150 m and have width to length ratios of 1-
1.5. Typically the depressions are bounded by curved or near-
circular escarpments up to 3 m in height, within which the bottom
is depressed and filled with irregular blocks of sediment. Side-
scan sonar records clearly show that such bowl-shaped areas,
bounded by scarps, have been displaced vertically and represent
distinct depressions on the sea floor. Figure 6 illustrates schematically the morphology of these features. The depressed central
area of the collapsed features displays irregular and hummocky
topography, as seen in the side-scan sonar record in Figure 7. On
the upslope margin, crown cracks often extend into adjacent stable
sediments, and on the downslope side there is a shallow-angle
reverse slope; occasionally a slightly raised rim of sediment is

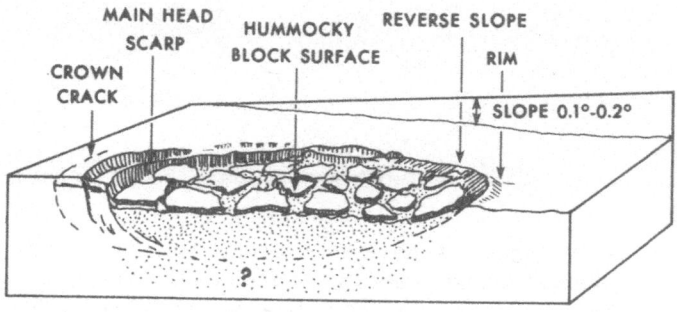

Fig. 6.   Schematic representation of the morphology of a collapse
          depression failure.

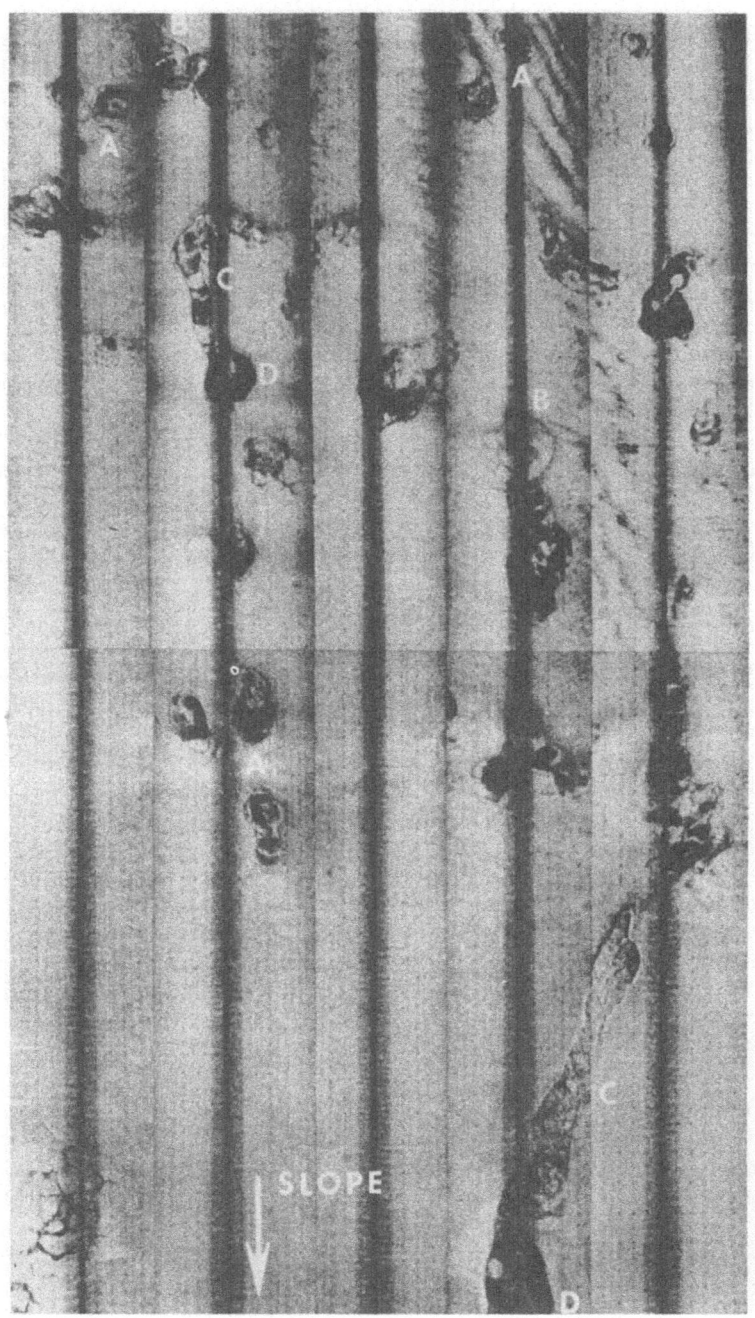

Fig. 7. Sonar mosaic across an area of collapse depressions (A), crown cracks (B), bottle-neck slides (C), and depositional lobes (D). (Area is approximately 1,500 m x 900 m.)

observed, indicating a tendency for downslope translatory movement. The depressed floors of the feature often show no slope and
are horizontal. These features are interpreted to be the result of
subsidence of parts of the sea floor and represent a decrease in
the volume of the sediment-gas-water system. It is likely that
such volumetric changes are accomplished directly by loss of methane gas and pore water from the sediment at the instant of instability. Thus at the instant of failure the deposits would be in a
state of liquefaction and essentially would not support any objects on the sea floor. The major factors responsible for producing these features are undoubtedly sedimentary loading by river
deposition on the adjacent distributaries, cyclic loading by passage of storm waves, and nearly continuous production of methane
gases within the sediment by biochemical degradation of incor-
porated organic debris.

On slightly steeper slopes within the interdistributary
regions and on slopes of 0.1-0.4° are features that are generally
referred to as bottleneck slides. These features are similar
morphologically to collapse depressions, but the boundary scarps
do not form a totally closed perimeter around the instability.
Rather, they may have narrow openings at the downslope margins
through which debris is discharged over surrounding intact slopes.
At the narrow opening of the source area where the depositional toe
begins there are often transverse tensional cracks. The areas of
displaced debris are arranged as distinct undulatory lobate depositional fans, which may have clearly identifiable sharp edges or
may grade out imperceptibly downslope as thin fans. Bottleneck
slides vary in length from 150 m to 600 m and have length to width
ratios of 1.5-3.0.

In this sort of instability there are both subsidence within
the source area as material moves downslope and retrogression or
extension of the source area in an upslope direction. Remolded,
viscous mud moves through the bottleneck before being deposited.

Elongate Retrogressive Slides and Mudflow Gullies

Side-scan sonar records and high-resolution seismic data
make it clear that the delta-front valleys or gullies (Shepard,
1955) emerge from within an extremely disturbed area of slumped
topography high on the delta, and each one has a recognizable area
of rotational instability or shear slumps at its upslope margin.
This feature is the most common type of sediment instability
fronting the Mississippi River delta.

Each possesses a long, narrow chute or channel that links a
depressed, hummocky source area on the upslope end to composite
overlapping depositional lobes or fans on the seaward end. Figure
8 schematically illustrates the major morphology characteristic of

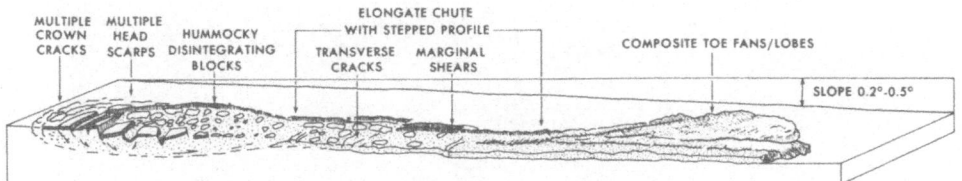

Fig. 8.   Schematic representation of the morphology of an elongate
          retrogressive slide (mudflow gully) and depositional
          lobe.

these features. The unstable area is bounded on its upslope side
by a bowl-shaped depression that serves as the source area. Often
multiple head scarps and crown cracks can be seen on the side-scan
sonar records, indicating their upslope retrogression. Within the
bowl-shaped depression hummocky, disintegrative blocks of various
sizes and attitudes can be discerned. Downslope from the bowl-
shaped source area is an essentially elongate, narrow chute.
These chutes or gullies are bound by very sharp linear escarpments
that are arranged parallel or subparallel to one another. The area
enclosed by the scarps is downthrown and is composed of irregular,
chaotic topography of blocks of debris of varying sizes. Commonly
the blocks within the chute area are smaller toward the central
axis of the gully. The gully floors lie from a few metres to as
much as 20-30 m below the adjacent intact bottom. The slopes along
the sides of the gullies vary considerably and range from less than
1° to as high as 19°. Most of the valleys extend downslope approximately at right angles to depth contours and achieve lengths
in excess of 8-10 km. In some instances, especially in the deeper
water areas, lengths of up to 15-20 km are not uncommon. In plan
view these features are rarely straight and quite commonly display
markedly sinuous plan views, with alternating narrow constrictions
or chutes and wider bulbous sections. Figure 9 is a side-scan
sonar mosaic run across a zone of landslide gullies. The area
covered by the mosaic is 1.0 km by 1.5 km. The mosaic shows three
major elongate gully systems, beginning with a blocky source area
bounded by scarps. Source-area geometry is very irregular, and
there is considerable difference in block size and orientation.
The narrow gullies or channels are relatively deeply incised, and
there is local evidence of side-wall instability, as indicated by
alternations of bulbous source areas and narrow chute regions.
The widths of the individual gullies vary considerably, from 20 m
to 150 m at the narrowest points to 600-800 m where the gullies are
widest. In a few instances, widths of the gully systems can be
extreme, up to 1,200-1,500 m. In many instances, in particular in
shallow water depths, adjacent gullies coalesce to form branching

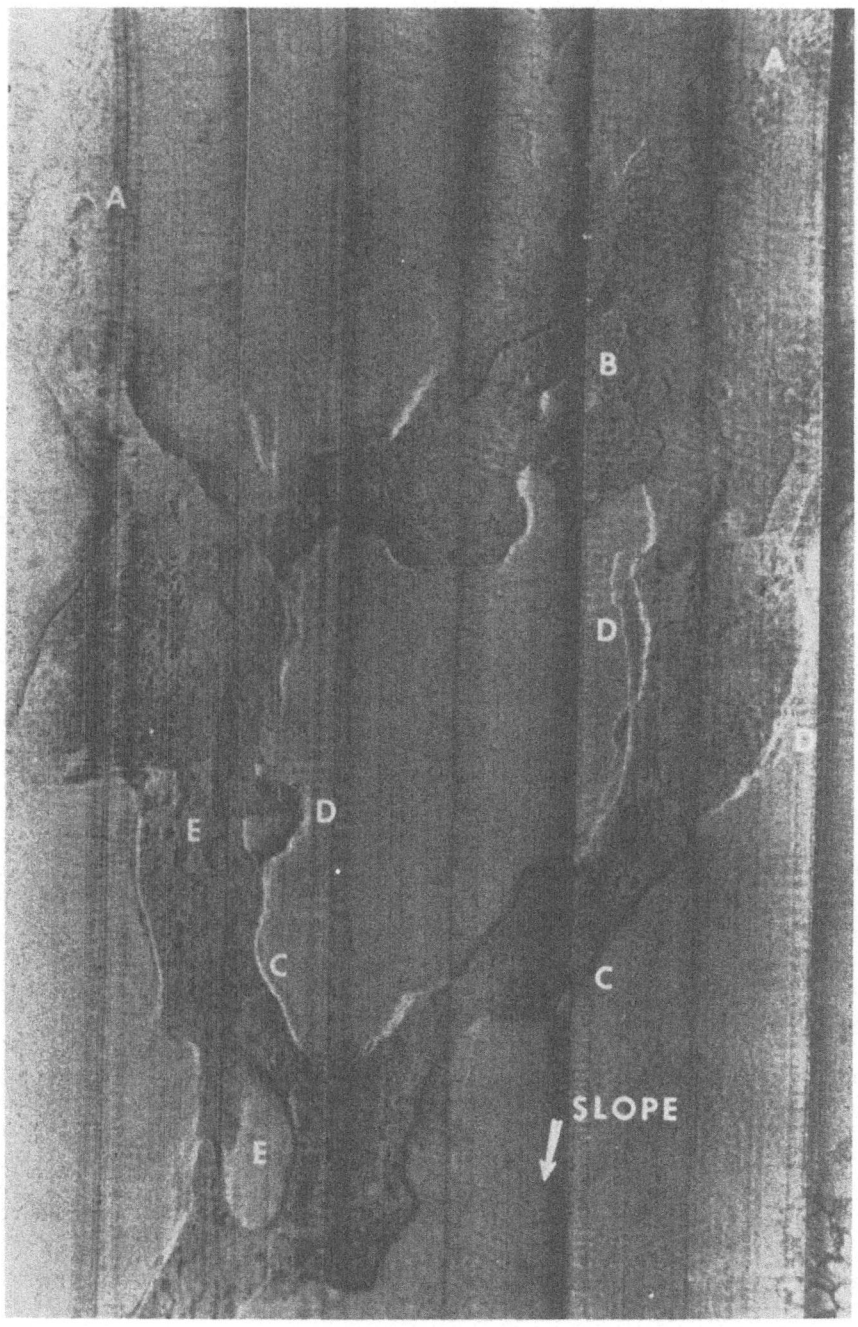

Fig. 9.   Sonar mosaic showing mudslide gullies: (A) source
          areas, (B) retrogressive gully, (C) incision, (D)
          unstable side wall, and (E) large blocks.   Area
          covered is 1,500 m x 1,000 m.

tributary systems, and often their junctions are discordant and are marked by accumulations of lobes of debris discharged from a tributary into the main channel.

The side walls of the landslide gullies are subject to in-stability; this slumping can produce contrasting forms and is probably responsible for localized widening along an individual gully system. Sometimes the sides of the chutes are characterized by elongate crack systems parallel to the main boundary scarps, suggesting linear block faulting toward the gully floor. Alter-natively, bowl-shaped reentrants and arcuate scarps represent shallow rotational side slumping of the neighboring sediments into the gully system. Figure 10 is a high-resolution seismic line run at right axis to a landslide gully. The horizontal scale is 152 m between each fix mark, and the vertical scale is 9.6 m between each timing line. As can be seen from this figure, several rotational slumps are carrying blocks of undisturbed sediment from the chute walls toward the axis.

The chutes or channels generally emanate from the slump zones and constitute transport conduits for disturbed and remolded sediments, together with displaced blocks of various sizes. Remolding develops as a direct consequence of disturbance of the sediment-water-gas system that accompanies slumping and represents fluidization/liquefaction mechanisms. The mechanism of transport is probably adequately characterized as slurry flow, which can be a type of plug flow, in which rigid plugs move over and within a zone of liquefied mud. The presence of partially disintegrated rafted blocks suggests laminar or plug flow rather than turbulent flow. It is likely that these subaqueous features are similar to many subaerial debris flows utilizing different modes of transport at different places within their overall geometry and alternating from one to another at different periods of activity.

Depositional Lobes and Mudflow Noses

At the seaward or downslope ends of mudflow gullies there are extensive areas of irregular bottom topography composed of discharged blocky, disturbed debris. In plan view this discharged debris is arranged into widespread overlapping lobes or fans (Fig. 3).

Each lobe is composed of two major morphological features: an almost flat or gently inclined surface (less than 0.5°) and an abrupt distal scarp representing the downslope "nose" of the dis-placed debris. The scarps display wide variation in height, ranging from only a few metres to in excess of 25 m and slopes as great at 7-10°. In plan view the scarps are generally curved, and adjacent lobes are separated from one another by major reentrants. Because of the large number of chutes or gullies that front the

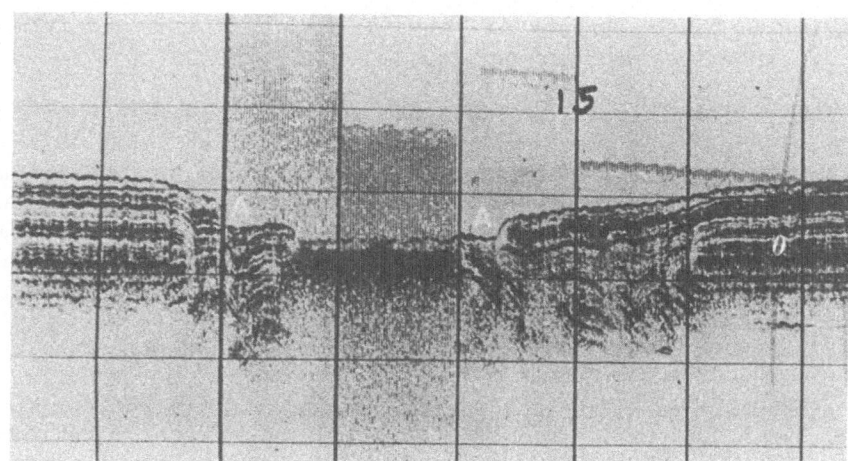

Fig. 10   A 3.5-kHz seismic record run across a landslide
           gully showing lateral rotational slumps.   The
           horizontal scale is 152 m between fix marks, and
           the vertical is 9.6 m between timing lines.

present Mississippi Delta, the displaced debris from adjacent
chutes may coalesce, providing an almost continuous sinuous frontal scarp that may extend peripheral to the delta 20-40 km.   Detailed mapping by side-scan sonar in some of these regions indicates that the depositional areas are composed of several overlapping lobes, each with its own distinctive seaward nose, and are
due to periodic discharge from the gullies farther upslope.   The
more recent the emplacement of a lobe, the more irregular and
blocky the surface topography, whereas in older depositional lobes
the topography is often characterized by small-scale pressure
ridges arranged as sinuous parallel ridges and hollows, and often
it features numerous small mud and gas vents that are similar to
mud volcanoes produced by sedimentary loading.   Figure 11 illustrates a side-scan sonar mosaic across depositional lobes of one
of the mudflows off the mouth of the Mississippi River.   On fathom-
eter records these regions are characterized by extremely irregular topography.   Figure 12 illustrates a high-resolution 3.5-kHz
seismic record run across a mudflow nose.   The line of section runs
essentially perpendicular to the seaward-leading edge of the mudflow.   The horizontal scale is 152 m between fix marks, and the
vertical timing lines are 9.6 m.   In most seismic sections a major
portion of the main body of the surface flow is characterized as an
amorphous unit showing few or no internal reflections.   In some
instances large and highly disorganized reflection events can be
seen, probably representing the large-scale disordered nature of
the bedding associated with these features.   Note in Figure 10 low

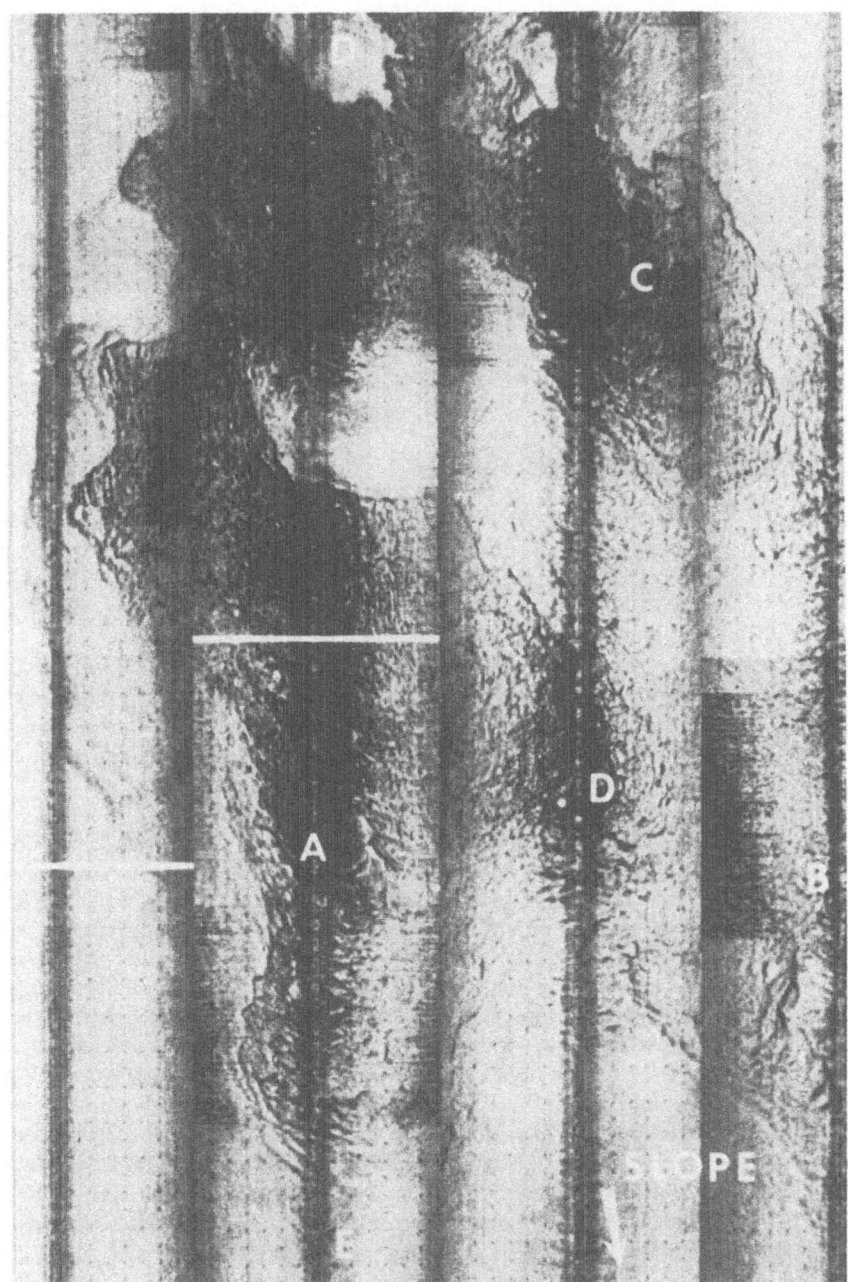

Fig. 11. Scale-corrected side-scan sonar mosaic across
depositional lobes (A, B, C), different block patterns (D), and pressure ridges (E). Mosaic covers
an area of 1,500 x 1,000 m.

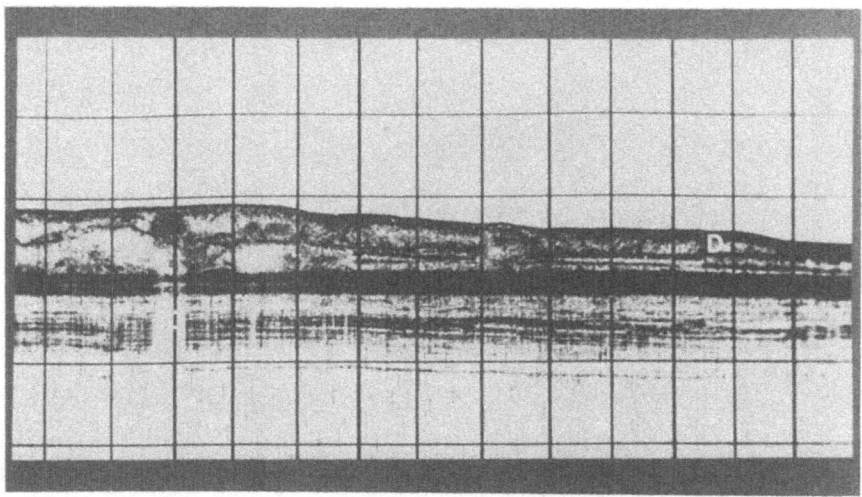

Fig. 12.   Subbottom profiler (3.5 kHz) record across a mud-
           slide deposition system. Vertical lines are 152 m
           apart, horizontal time lines are spaced at 9.6 m.

scarps on the sea floor and, seaward or downslope from the scarps,
thin aprons of mudflow material, which normally also display an
amorphous nature on seismic records. These aprons in front of the
main mudflow nose can extend as far as 3-4 km seaward of the main
scarp.

     The mudflow lobes and noses are capable of advancing down-
slope, and repeated surveys have shown that they may move forward
more than 1,000 m/yr. Forward movement is probably also accompanied by oversteepening of the frontal slope, which produces surficial rotational sliding and could account for some of the large
chaotic blocks associated with the sea-floor scarp. It is highly
possible that movements associated with depositional lobes occur
rapidly but episodically. The presence of smaller blocks within
the overall debris of the mudflow indicates that the material was
in a highly viscous state during movement. Movement undoubtedly
ceases when the momentum is checked by degassing and drainage of
internal water or by the lower slope angles of the stable shelf
across which the mudflows prograde. At the downslope margins and
to a lesser extent at the lateral margins the mudflows encounter
passive pressure from pre-existing shelf sediments, and this
causes some upthrusting. Some of the energy is dissipated by the
forward intrusion of a basal wedge or sole into the downslope
materials. Within the area of deposition, and in some instances
for large distances downslope beyond it, the effects of rapid
sedimentary loading cause localized pore-water/gas pressures to
increase and be released by the development of mud and gas vents.

The thicknesses of the lobes are often difficult to determine precisely, but each distinct lobe is normally on the order of
5-15 m thick. However, because of overlapping, the total thickness of the mudflow can approach 50-60 m. Coleman and Garrison
(1977) estimate that, at the shelf edge off South Pass, in one area
of approximately 877 km^2 the volume of the discharged debris was
11.2×10^6 m^3.

Shelf-Edge Slumps and Depositional Faults

In the deeper waters along the outer continental shelf and
upper continental slope fronting the Mississippi River delta a
variety of slumps and faults are found at the immediate break in
slope at the shelf edge. Large arcuate-shaped families of shelf-
edge slumps and deep-seated contemporaneous or depositional faults
are active presently along the peripheral margins of the delta
front. In most instances these features tend to cut the modern
sediment surface, often forming abrupt scarps on the sea floor.
These surface scarps then provide localized areas for additional
downslope mass movement of material by slumping.

Shelf-edge slumps give a stairstepped appearance to the edge
of the continental shelf and are highly reminiscent of the rotational peripheral slumps higher on the continental shelf, near the
mouths of the modern distributaries. These features, however,
generally display much larger scales and involve larger amounts of
material. Figure 13 is a high-resolution seismic profile run

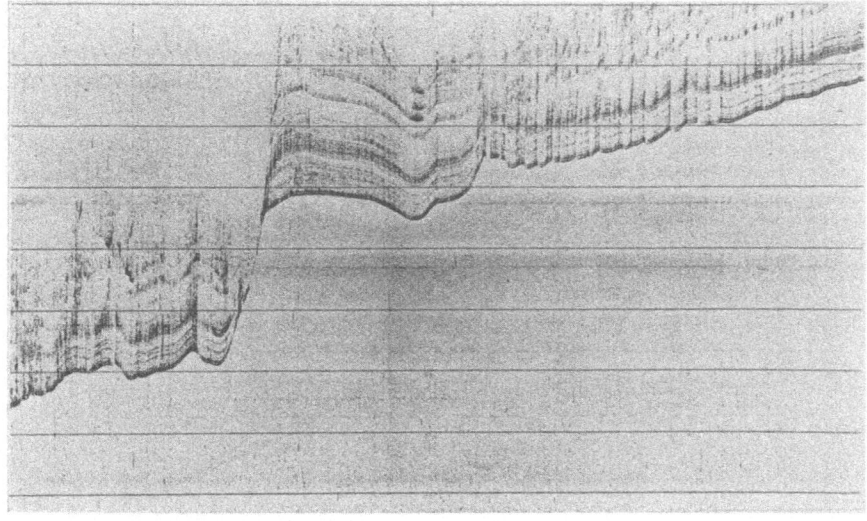

Fig. 13. Seismic profile across shelf-edge slump feature
          showing back scarp, displaced block, and a complex
          sheared zone farther downslope.

across one slump system. In plan view the slumps generally display an arcuate pattern and have lateral continuities that range from a few kilometres to 8-10 km. Material slumped off the shelf edge is moved downslope and lodged as large chaotic blocks on the upper continental slope in water depths of 1,000 m or greater. The rate of movement of these particular features is unknown, primarily because older seismic data run across the shelf edge were sparse and in many instances navigation was poor. Thus comparisons with recently run profiles become unreliable.

A second type of deformational feature commonly found along the shelf edge consists of en echelon contemporaneous depositional faults. A contemporaneous fault is one in which movement is contiguous with deposition, and thus with time and continued movement along the fault plane offsets of individual marker beds increase with depth. Very commonly sedimentation or accumulation rates are higher on the downthrown sides of the fault, and hence the thickness of individual beds between marker units will increase abruptly as one crosses the fault to the downthrown block. This type of feature is illustrated in Figure 14. This particular fault extends from the sediment surface to depths greater than 500 m. Offsets in the uppermost or younger units across the faults generally tend to be on the order of 10-15 m, whereas at a depth of

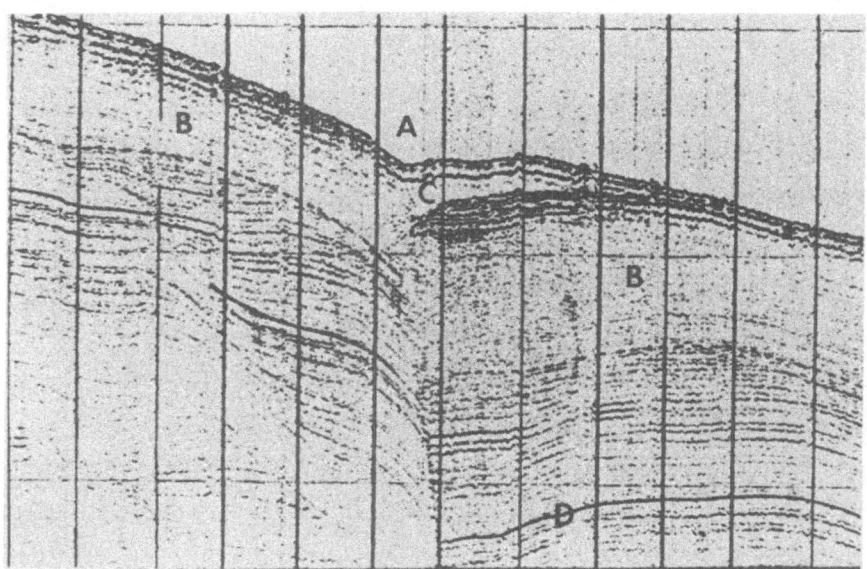

Fig. 14.   Seismic profile across a contemporaneous fault (A) showing offsets, thickening of units (B), and zone of mudflow material (C). Deeper marker horizons (D) show greater displacement with depth.

approximately 500 m marker beds are offset 50-60 m. Note that in
Figure 14 the uppermost unit displays internal amorphism, or no
seismic reflections. Note also its slightly increased thickness
on the downthrown side of the fault. This amorphous zone may
represent a surface mudflow that has progressed seaward beyond the
limits of the fault. As this surface mudflow crosses the fault
zone, it can be seen to thicken slightly, and thus it is highly
possible that the increased thicknesses on the downthrown side of
these faults are in fact the result of the movement of surface
mudflows across the fault zone. Because the fault zone is blanket-
ed by a large mass of rapidly introduced sediment, surface scarps
on the sea floor are eliminated. With continued movement along the
fault, however, a new scarp will tend to form at a later time, and
another mudflow will move across the fault, again adding sediment
to the downthrown side. This type of interaction between surface
mudflows and contemporaneous faults may play a large role in con-
trolling offset and movement along the fault itself.

FACTORS CONTRIBUTING TO SLOPE INSTABILITY

     The analysis of precise mechanisms responsible for the fea-
tures in the Mississippi Delta area is a difficult task because
they are the result of interaction of many variables rather than
the product of any single factor. Even though the area is one of
the most carefully documented of its type in the world, information from bore logs, marine surveys, and monitoring of sea-floor
conditions is still insufficient for construction of a fully predictive model of instability processes. The basic conditions for
failure exist when stresses exerted on the sediment exceed its
strength. This circumstance can be due to stress increases, sediment strength reduction, or a combination of the two. Two basic
analytical approaches have been followed; they look at instability
in terms of either the effects of wave loading (e.g., Henkel, 1970)
or the intrinsic low sediment strength caused by excess pore pressures resulting from rapid sedimentation (Prior and Suhayda,
1979a, b).

Stresses

     Although in general slope angles are small, they do con-
stitute gravitational stresses on the sediment. Henkel (1970)
concluded that these stresses are unlikely to be sufficient to be
the sole cause of failure, but this conclusion is clearly influenced by the assumptions made about strength/pore water pressure
relationships (Prior and Suhayda, 1979a, b).

     Henkel (1970) considered the oil rig damage resulting from
sliding associated with Hurricane Camille and determined theoretically that waves imposed oscillatory motion on the sediments
sufficient to cause downslope movement. The magnitude of dis-

turbing (M_d) and resisting movement (M_r) can be calculated using the equations

$$M_d = \frac{2}{3} \chi^3 \beta\gamma' + \frac{L^2 \Delta P}{2\pi^2} (\sin \alpha - \alpha \cos \alpha)$$

$$M_r = 2 \chi^3 \frac{Cu}{\gamma'z} \gamma' \frac{\sin \theta - \theta \cos \theta}{\sin^3 \theta}$$

where β is the slope angle; Cu is undrained shear strength; γ' is submerged weight of sediments; Z is depth; ΔP is sinusoidal pressure change of wavelenth L; dimensions of slide are given by the 1/2 length χ and the depth d to which sliding extends; 2θ is the angle of the arc sliding; and α is the portion of the sinusoidal loading acting on the chosen length of the slide. When $M_d > M_r$ failure occurs, and Henkel (1970) suggested that the delta-front instability was explained by hurricane-wave-induced downslope stresses. Indeed, Suhayda et al. (1976) have confirmed that substantial bottom pressures are generated during storms; and, further, Dunlap et al. (1979) showed how such waves can also induce pore-water pressure increases.

However, instabilities also occur when wave stresses are low. Coleman et al. (1974) ascribed rotational sliding near the distributary bars to oversteepening as progradation proceeds. This process represents a localized increase in gravitational stresses.

Strength

Sediment strength at a potential failure surface is a function of cohesive and frictional forces, and these are strongly influenced by the weight of sediment over the slip surface. The delta environment provides a number of conditions that progressively alter the sediment strength properties, primarily by increasing the internal pressures that reduce the normal load. Firstly, the highly water saturated sediments exhibit excess pore-water pressures because of underconsolidation. Secondly, cyclic loading of the sediments by surface waves causes pore-water pressure increases. Prior and Suhayda (1979a, b) considered the instability mechanism as an infinite slope problem in effective stress terms. On the assumption that F = 1 (failure condition), the pore-water pressure (u) needed for failure can be calculated by

$$u = \frac{c' - F (\gamma'Z \sin \beta \cos \beta)}{\tan \phi'} + \gamma'Z \cos^2 \beta$$

where c' is cohesion, ϕ' is the angle of friction, γ' is the unit weight of the sediment, Z is the depth of soil above the slip surface, and β is the slope/slip surface angle. For all cases, the pore-water pressure needs to be very large for failure, and must be in excess of hydrostatic, approaching geostatic or a condition of almost zero effective stress. While such conditions are difficult to envisage, they do indeed match the results of in situ piezometer data (Dunlap et al., 1979). Thus, the presence of such large excess pressures means that the delta-front slopes are inherently in a condition of incipient failure, which can be triggered by stresses generated by waves or progradational oversteepening (Prior and Suhayda, 1979a, b).

The sediment/water system is further influenced by the internal generation of large amounts of biogenic methane gas. Whelan et al. (1976) indicated values of as much as 15% volume of methane in the sediments. The exact effects of this process on sediment cohesion and friction are largely unknown, but it is likely that formation of gas bubbles reduces strength as total gas and water pressures increase in the soil voids.

Thus, the initiation of slope instability in the delta is not the result of a single causative mechanism but represents a very complex interaction of processes operating on different time scales, and will produce failures with differing morphologies and magnitudes. The subaqueous failures described result from intricate combinations of factors, summarized schematically in Figure 15. It is emphasized that individual thresholds, when stresess exceed strength and failure occurs, are likely to be achieved by quite different combinations of the same basic factors over time and space. For example, storm waves may be capable of bottom perturbation to failure if strength reduction is well advanced by other factors. Alternatively, rapid generation of in situ methane gas, or its mobility from one zone to another, may result in failure without any external changes in stress conditions. Clearly, much work remains to be done to better document the features and evaluate the initiating and post-failure mechanisms.

CONCLUSION

Investigations in the Mississippi River delta region, as well as in other deltas where sedimentation rates are extremely rapid, indicate that submarine mass-movement processes are extremely significant. Many aspects of these processes are not well understood, and are the focus of continuing research by the U.S. Geological Survey (Mississippi Delta Project) and by various industry groups. Work is in progress on:

1. Monitoring of movement activity to determine rates and volumes of displacements, timing of initial failures, and reacti-

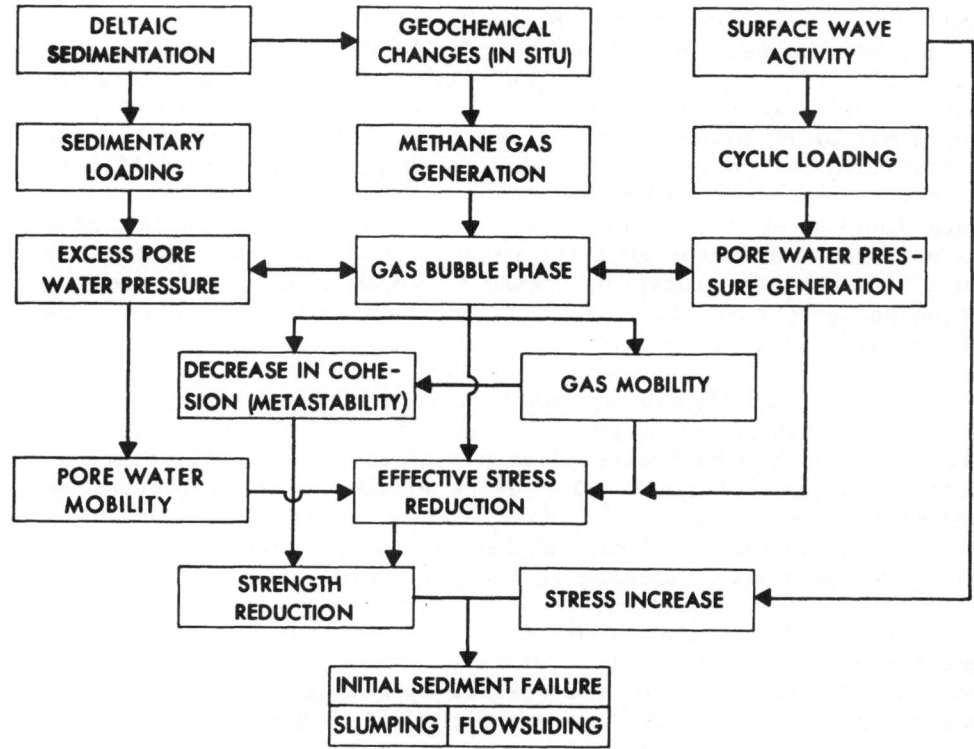

Fig. 15.   Summary of factors and interactions leading to initial
failure of marine sediments on low-angle slopes in the
Mississippi River delta.

vation potential. In particular, periodic resurveys using existing bathymetric and side-scan sonar data as a basis for comparison
(e.g., Coleman and Prior, in press) are in progress and have already shown large-scale short-term mudflow surges and slump retrogression.

     2. Measurement of sediment strength properties using undis-
turbed samples from boreholes. This problem involves isolation of
the effects of interstitial methane gas on sediment strength. A
pressurized core barrel that maintains samples at in situ gas
contents and pressures has been designed by Dunlap and Bryant
(Texas A and M University, personal communication; Denk et al., in
press) and is being used for triaxial and consolidation testing.

     3. Field measurement of environmental parameters that
affect sediment stability. Specifically, instruments have been
designed that allow long-term monitoring of pore-water and gas

pressures with high resolution (Reece et al., 1979). The GISP
(Geotechnically Instrumented Seafloor Probe) will perform its
first full-scale experiment during 1981. Additionally, a Sea Wave
Attenuation Measurement Program (SWAMP) is in progress (Forristall
et al., 1980).

4. Determination of the detailed geological characteristics
of mass-moved sediments. Continuing acquisition of cores from the
delta is allowing the identification of sediment properties for
particular locations. Roberts (in press) uses X-ray radiography
to catalog diagnostic deformational structures.

5. Application of computer modeling to stability analysis,
particularly for factor interaction effects. One approach has
been wave/sea bottom interaction studies using specific soil
parameters and design wave amplitudes and frequencies to predict
downslope sediment motion (Schapery and Dunlap, 1978; Dunlap and
Prior, in press).

As these tasks progress in the Mississippi Delta region, it
is anticipated that methods and concepts will be forthcoming that
will have application to similar submarine mass-movement environments elsewhere.

ACKNOWLEDGMENTS

Funds for support of this research, a part of the Mississippi Delta Project, were provided by the United States Geological
Survey.

REFERENCES

Bea, R. G., and Audibert, J. M. E., 1980, Offshore platforms and
      pipelines in Mississippi River delta, J. Geotech. Engr.
      Div., Proc. Am. Soc. Civil Engr., 106(GT8):853-869 (Paper
      15645).
Coleman, J. M., 1976, Deltas: Processes of Deposition and Models
      for Exploration, Burgess Publishing Co., 7108 Ohms Lane,
      Minneapolis, Minn. 55435.
Coleman, J. M., and Prior, D. B., 1981, Resurveys of active mud-
      slides, Mississippi Delta, Geomarine Letters (in press).
Coleman, J. M., and Garrison, L. E., 1977, Geologic aspects of
      marine slope instability, northwestern Gulf of Mexico, Mar.
      Geotech., 2:9-44.
Coleman, J. M., Suhayda, J. N., Whelan, T., III, and Wright, L. D.,
      1974, Mass movements of Mississippi River delta sediments,
      Proc. 24th Conf. of Gulf Assoc. Geol. Soc., Lafayette, La.,
      pp. 49-68.

Coleman, J. M., Prior, D. B., and Garrison, L. E., 1980, "Sub-aqueous Sediment Instabilities in the Offshore Mississippi River Delta," Bureau of Land Management Open-File Rept. 80-01, New Orleans, La.

Denk, E., Milburger, L. J., Dunlap, W. A., and Bryant, W. R., 1981, A pressurized core barrel for sampling gassy marine sediments, Proc., Offshore Tech. Conf., Houston, Texas (1981).

Dunlap, W. A., and Prior, D. B., 1981, Factors contributing to the initiation of slope instability, Mississippi Delta, Proc., Offshore Tech. Conf., Houston, Texas (in press).

Dunlap, W. A., Bryant, W. R., Williams, G. N., and Suhayda, J. N., 1979, Storm wave effects on deltaic sediments--results of SEASWAB I and II, Proc., Port and Ocean Engr. under Arctic Conditions, 2:899-920.

Fisk, H. N., and McClelland, B., 1959, Geology of continental shelf off Louisiana: its influence on offshore foundation design, Bull. Geol. Soc. Amer., 70:1369-1394.

Fisk, H. N., McFarlan, E., Jr., Kolb, C. R., and Wilbert, L. G., 1954, Sedimentary framework of the modern Mississippi Delta, Jour. Sediment. Petrol., 24:76-99.

Forristall, G. Z., Reece, A. M., Thro, M. E., Ward, E. G., Doyle, E. H., and Hamilton, R. C., 1980, Measurements of sea wave attenuation due to deformable bottoms, Proc., Amer. Soc. Civil Engr., Miami, Florida (in press).

Garrison, L. E., 1974, "The Instability of Surface Sediments on Parts of the Mississippi Delta Front," U.S. Geol. Survey Open-File Rept., Corpus Christi, Texas, 18 pp.

Henkel, D. J., 1970, The role of waves in causing submarine landslides, Geotechnique, 20:75-80.

Hooper, J. R., 1980, Crustal layers in Mississippi Delta mudflows, Proc., Offshore Tech. Conf., Houston, Texas, Paper 3770, pp. 277-287.

Morgan, J. P., 1961, Mudlumps at the mouths of the Mississippi River, in: "Genesis and Paleontology of the Mississippi River Mudlumps," Louisiana Dept. of Conservation Geol. Bull. 35, pp. 1-116.

Morgan, J. P., Coleman, J. M., and Gagliano, S. M., 1963, "Mudlumps at the Mouth of South Pass, Mississippi River: Sedimentology, Paleontology, Structure, Origin, and Relation to Deltaic Processes," Coastal Studies Series No. 10, Coastal Studies Inst., Louisiana State Univ., Baton Rouge, 190 pp.

Morgan, J. P., Coleman, J. M., and Gagliano, S. M., 1968, Mudlumps: diapiric structures in Mississippi Delta sediments, in: "Diapirism and Diapirs," Amer. Assoc. Petrol. Geologists Memoir 8, pp. 145-161.

Prior, D. B., and J. M. Coleman, 1978a, Disintegrating, retrogressive landslides on very low angle subaqueous slopes, Mississippi Delta, Mar. Geotech., 2:37-60.

Prior, D. B., and Coleman, J. M., 1978b, Submarine landslides on the Mississippi delta front slope, in: "Geoscience and Man," XIX:41-53, School of Geosciences, Louisiana State Univ., Baton Rouge.

Prior, D. B., and Coleman, J. M., 1980, Sonograph mosaics of submarine slope instabilities, Mississippi River delta, Mar. Geol., 36:227-239.

Prior, D. B., and Suhayda, J. N., 1979a, Application of infinite slope analysis to subaqueous sediment instability, Mississippi Delta, Engr. Geol., 14:1-10.

Prior, D. B., and Suhayda, J. N., 1979b, Submarine mudslide morphology and development mechanisms, Mississippi Delta, Proc., Offshore Tech. Conf., Houston, Texas, Paper 3482, pp. 1055-1061.

Prior, D. B., Coleman, J. M., and Garrison, L. E., 1979a, Digitally acquired undistorted side scan sonar images of submarine landslides, Mississippi Delta, Geology, 7:423-425.

Prior, D. B., Coleman, J. M., and Caron, R. L., 1979b, Sea floor mapping using micro-computer assisted side-scan sonar, Proc., 13th Internat. Remote Sensing Symposium, Ann Arbor, Mich.

Reece, E. W., Ryerson, D. E., Kestly, J. D., and McNeill, R. L., 1979, The development of in situ marine seismic and geotechnical instrumentation systems, Proc., Conf. on Port and Ocean Engr. under Arctic Conditions, Trondheim, Norway.

Roberts, H. H., 1980, Sediment characteristics of Mississippi River delta-front deposits, Trans., Gulf Coast Assoc. Geol. Soc. (in press).

Schapery, R. A., and Dunlap, W. A., 1978, Prediction of storm-induced sea bottom movement and platform forces, Proc. Offshore Tech. Conf., Houston, Texas, Paper 3259, pp. 1789-1796.

Shepard, F. P., 1955, Delta front valleys bordering the Mississippi distributaries, Bull. Geol. Soc. Amer., 66:1489-1498.

Suhayda, J. N., Whelan, T., III, Coleman, J. M., Booth, J. S., and Garrison, L. E., 1976, Marine sediment instability: interaction of hydrodynamic forces and bottom sediments, Proc., Offshore Tech. Conf., Houston, Texas, Paper 2426, pp. 29-40.

Terzaghi, K., 1956, Varieties of submarine slope failures, Proc., 8th Texas Oil Mech. and Engr. Conf., pp. 1-41.

Whelan, T., III, Coleman, J. M., Roberts, H. H., and Suhayda, J. N., 1976, The occurrence of methane in recent deltaic sediments and its effects on soil stability, Internat. Assoc. Engr. Geol. Bull., 14:55-64.

COASTAL GEOMORPHOLOGY

Historically, geomorphology has dealt largely with the surfaces of the continents; however, it has been extended to include the origin of the relief features of the continental shelves and ocean basins. This latter part is often mentioned as marine geomorphology.

Geomorphology and pedology deal with the surface configuration of the soil (lands) and with the thin but extremely complex soil layers that cover much of the earth's land, respectively. The earth's topographic features or land forms including coast lines and their associated soils develop in the zone of interaction between atmosphere and solid earth. The marine features near shore and on the continental slopes develop in the zone of interaction between ocean and solid earth. Of great interest in the last case is the role of oceanic water and atmosphere, the air-sea interaction.

The physical, chemical, and organic process by which the landscape is influenced is therefore the objective of the geomorphologists by studying the geologic processes and structures of continents and ocean bottoms, e.g. erosion by runoff on slopes may induce and enlarge canyons.

From the engineering standpoint the ability of waves to erode a shoreline is a matter of considerable interest and importance. A rapid cutting back (or retograde activity) will destroy valuable shore properties (houses etc.) and produce larger human unhappiness.

Auffret et al. describe the consequences and the after-effects of the large marine slide triggered by a tsunami which took place in the western Mediterranean in October, 1979. Two telephone cables parallel to the shore line were broken.

No earthquake was registered at the time of the catastrophic event.

RECENT MASS WASTING PROCESSES ON THE PROVENCAL MARGIN (WESTERN MEDITERRANEAN)[°]

Groupe ESCYANICE (G.A. Auffret[+], J.M. Auzende[+], M. Gennesseaux[*], S. Monti[+], L. Pastouret[+], G. Pautot[+], J.R. Vanney[*]

+ Centre Océanologique de Bretagne, B.P.337,29273 Brest
* Université Paris VI, 4 place Jussieu, 75230 Paris cedex 5

On October 19, 1979 off Nice, the upper continental shelf has been affected by a catastrophic event (Gennesseaux et al., 1980).

(a) At 2 p.m. (approximate time), following a lowering of the sea level, a tsunami of several meters amplitude reached the coast line south of Antibes. This was followed during 4 hours by several oscillations of decreasing amplitude, that were felt on a shore length of about 100 km.

(b) About the same time, an embanked area 300 m long and wide, collapsed in a few seconds ; after the event the sea-floor in that area was found at a depth of 50 meters.

(c) At 17h45 TU and 22h00 TU, two telephone cables have been broken 80 and 110 km off Nice respectively (fig. 1).

(d) No earthquake was registered by the Monaco Observatory during this time period.

After this event, we surveyed the area with Sea-Beam and observed the walls and canyons floors with the submersible Cyana.

The Sea-Beam map (Pautot, in press ; fig. 2) shows several morphological units : the Cap Ferrat, Antibes and Nice ridges, the Var and Paillon canyons (fig. 3) and an innerfan related to the Var and Paillon rivers.

[°]Contribution n° 719 du Centre Océanologique de Bretagne.

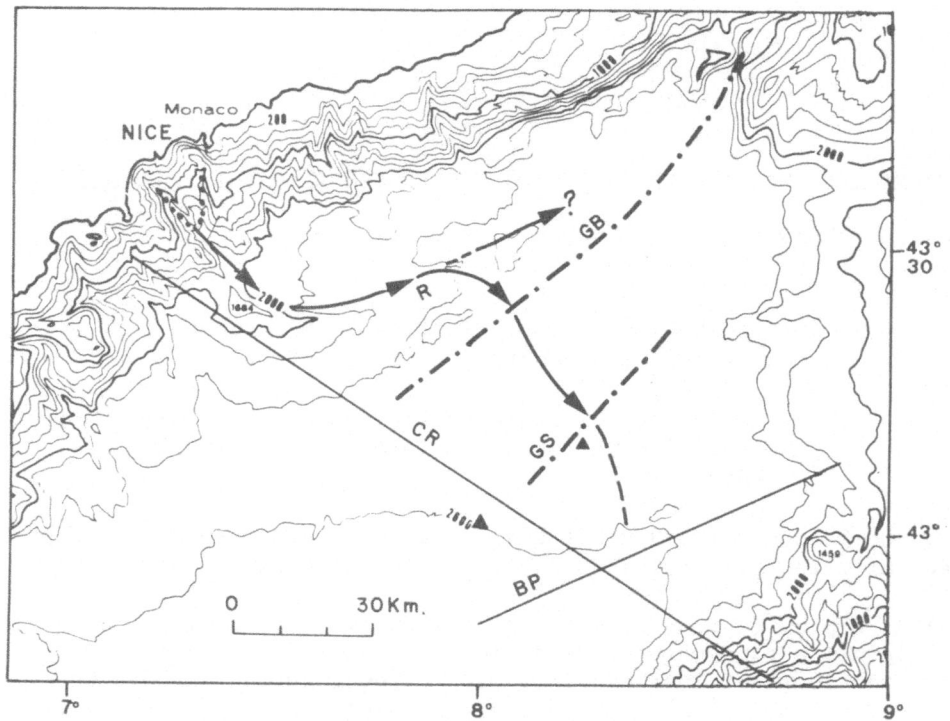

Fig. 1 : Path of inferred turbidity current (black arrow). G B and
        G S : broken cables. BP and CR : not affected cables. R :
        Antibes ridge.

The main characteristics of the area are the narrow shelf and
steep continental slope (average about 5°). The fan and sedimentary ridges sediments consist of a 1 000 m thick section of
plioquaternary conglomerate, mud and marly ooze (Gennesseaux et
Le Calvez, 1960 ; Pautot, 1968), that overlie a continental erosion
surface of upper Miocene (Messinian) age (Olivet et al., 1971).
The present morphology appears as the end product of erosive
processes that have affected a large Pliocene deep sea fan, since
the beginning of the Quaternary.

The canyon walls present morphological features of different
scales (fig. 4). First order features (at the hectometric scale)
are represented by sedimentary aprons (I1) that are affected by
second order erosional features giving an "herring bone" aspect
to the slopes. The same type of morphology has been often observed
in Mediterranean canyons (Vanney et al., 1979). The direction of
major aprons is generally perpendicular to the canyon floor axis.
This landscape recalls us of the so-called "bad-lands" topography.
In some place local mass wasting at a decimetric to a metric scale
affects the slope sediments (I2).

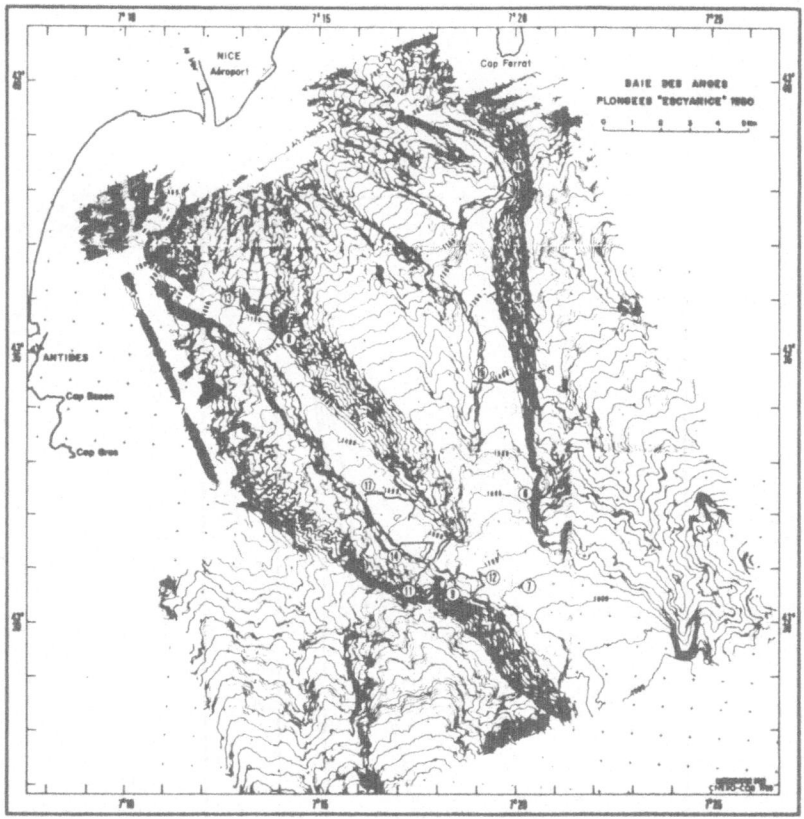

Fig. 2 : Sea-Beam map and dives location.

        Erosional and depositional features have been observed on the
canyon floor : erosional features consist of depressions with step
like walls of metric amplitude(II1a). The vertical outcrops show
interbedded mud and conglomerates of probable Pliocene age. Down
canyon we observed limestones and yellowish conglomerates of
probable Messinian age (Olivet et al., 1971). sometime in over-
hanging wall (II1b).

        Accumulation features consist of pebbles levees, 2 to 3
meters high, aligned parallel to the canyon axis (II2a). They
present most often a dissymetric cross-section. Other accumulations
include angulous blocks(II2b). Down canyon, the levees were predominantly observed on the western side of the inner channel.

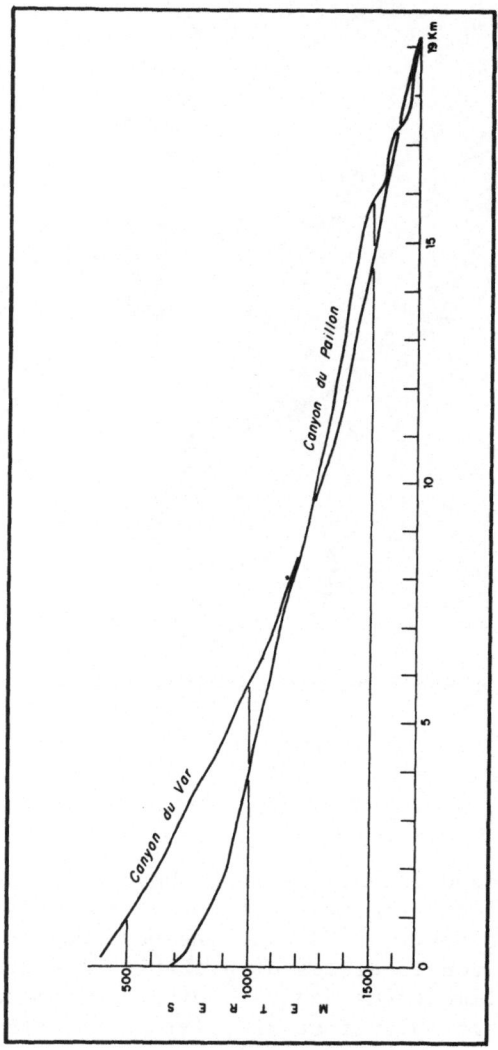

Fig. 3 : Longitudinal profiles of Var and Paillon canyons.

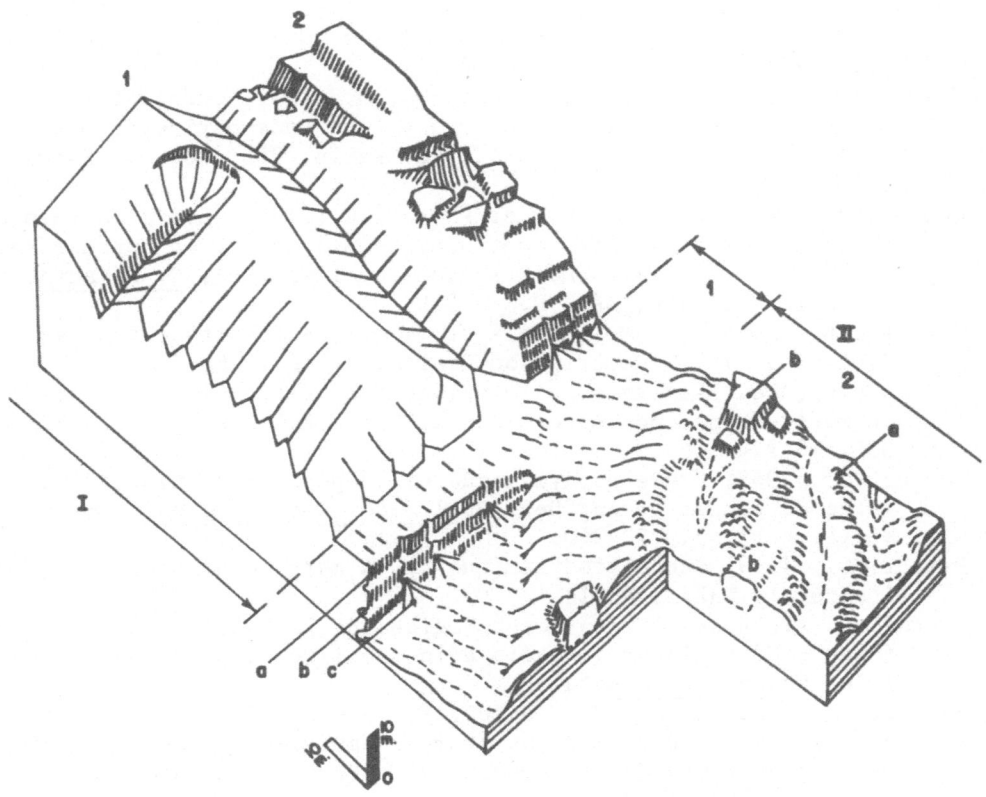

Fig. 4 : Synthetic morphological diagram from in situ observations.
See text for explanations.

The Sea-Beam map and in situ observations during the dives
provide us with valuable information. These and other data will be
used for the reconstitution of the gravity movements that may have
operated in that area.

References

Groupe ESCYANICE (G.A. Auffret, J.M. Auzende, M. Gennesseaux, S.
    Monti, L. Pastouret, G. Pautot, J.R. Vanney), in press,
    Morphologie des versants et des fonds des canyons du Var et
    du Paillon : observations faites avec le submersible Cyana.
    C.R. Acad. Sc. Paris.
Gennesseaux, M., 1966, Prospection photographique des canyons
    sous-marins du Var et du Paillon (Alpes Maritimes) au moyen
    de la troïka. Rev. Géogr. Phys. et Géol. Dyn.,v.8, fasc.I,
    p.3-38.

Gennesseaux, M. et Le Calvez, Y., 1960, Affleurement sous-marin de vases pliocènes dans la baie des Anges (Nice), C. R. Acad. Sc. Paris, t. 251, p. 2064-2066.

Gennesseaux, M., Guibout, P. et Lacombe, H., 1971, Enregistrement de courants de turbidité dans la vallée sous-marine du Var (Alpes Maritimes), C. R. Acad. Sc. Paris, 273, sér. D, 2456-2459.

Gennesseaux, M., Mauffret, A., et Pautot, G., 1980, Les glissements sous-marins de la pente continentale niçoise et la rupture de câbles en mer Ligure (Méditerranée occidentale), C. R. Acad. Sc. Paris, t. 290, sér. D, p. 959.

Olivet J.L., Auzende J.M., Mascle,J., Monti, S., Pastouret L. et Pautot G., 1971, Description géologique de la bordure provençale. Résultats de la campagne de flexo-électroforage en Méditerranée nord-occidentale. C.N.E.X.O., n° 2, p. 375-394, 18 fig.

Pautot, G., 1968, Etude géodynamique de la pente continentale au large de Cannes et Antibes, Bull. Soc. Géol. France, 7ème série, t. X, p. 253-260.

Pautot, G., 1969, Etude géodynamique de la marge continentale de l'Estérel. Thèse, Paris.

Pautot, G., in press, Cadre morphologique de la baie des Anges (Nice - Côte d'Azur). Modèle d'instabilité de pente continentale, Oceanologica Acta.

Vanney, J.R., Bellaiche, G., Coumes, F., et Irr, F., 1979, Types de modelés observés par submersible dans les canyons méditerranéens au cours de la campagne Cyaligure, C. R. Acad. Sc. Paris, t. 288, p. 735-738.

MECHANICS OF MARINE SLIDES AND OTHER MASS MOVEMENTS

In recent years more and more pipelines for extraction of offshore oil and gas deposits have been constructed, and more and more platforms have been built up for exploration, exploitation, and extraction of the energy resources. Damage to pipelines and platforms has occurred.

In order to minimize the risk involved in a marine project and to take effective measures for the protection of lives and properties, the marine engineers need to know the mechanics and the stability of the ocean bottom.

Karlsrud and Edgers are discussing some aspects of submarine slope stability. They summarize existing methods for analysing marine slopes under wave and earthquake loadings, static limiting equilibrium, static deformation, liquefaction, and full dynamic analyses. Examples of coastal slides in Norway in loose sand and soft clays are presented.

The authors conclude that analytic methods for engineering analysis of submarine slopes, although not perfect, are quite well developed. The major uncertainty is the determination of the soil input parameters.

Koning in his paper on marine flow slides in sand points out that for an evaluation of the stability of marine slopes in sand, knowledge of the porosity in situ and of the critical density is essential; other properties like the angle of internal friction, compressibility, and permeability are closely connected to porosity. Examples of flow slides in the Netherlands illustrate the paper.

Submarine slumping and mass movements on the continental slope of Israel is the subject of the paper by Almagor and Wiseman. They discuss by length the continental margin of Israel and describe a geotechnical testing programme. It is concluded that the continental

59

margin of northern Israel and southern Lebanon on one
side and that of northern Sina and Israel on the other
side are different in the way that at the northern margin
the sediments are stronger and form more stable slopes
than at the southern margin. Also the rate of sedimentation is much slower at the northern margin. Earthquakes
of very small magnitude may therefore only affect the
more unstable slopes of the southern margin.

Bennett et al. summarize the results of four in situ
pore pressure experiments conducted in the Mississippi
Delta including a brief description of the piezometer
instrumentation, the geotechnical properties at the probe
sites, and the surface wave effects on pore pressures.

Even if important results are obtained it is stressed that considerably more analyses are necessary in order
to assess the potential time-dependent changes in stress
with depth below the mudline for the short-period wave
activity.

In a clear and distinct essay Richards and Chaney
have described how marine slides or slope instability
may be investigated. The first phase is the construction
of bathymetric and geological maps; both are mainly
based on geophysical data supplemented by bottom observations. Second phase is carried out by geotechnical
expertise and includes soil sampling, testing, modelling,
and analysis.

SOME ASPECTS OF SUBMARINE SLOPE STABILITY

Kjell Karlsrud and Lewis Edgers *

Norwegian Geotechnical Institute
Oslo, Norway

* On leave from Tufts University, Medford, MA

ABSTRACT

This paper critically summarizes existing methods for analysing
marine slopes under wave and earthquake loadings. Static limiting
equilibrium, static deformation, liquefaction, and full dynamic
analyses are summarized. Methods for analyzing the effects of
underconsolidation in rapidly accumulating or gaseous sediments
are described. Recent theoretical developments which account for
the effects of porewater compressibility and seafloor compressibility and permeability on wave induced bottom pressures are reviewed. The behaviour of a submarine slope after an instability
develops remains the area of greatest uncertainty in marine slope
stability problems. In particular, the conditions of grain size,
soil mass density, velocity, slope angle, etc. for transformation
of a limited instability to a flow or turbidity current are very
poorly understood. Examples of coastal slides in Norway, in loose
sands and soft clays are presented. These cases illustrate possible triggering mechanisms, and the importance of progressive and
retrogressive action in the rapid downslope transport of large
masses of material. There is a great need to develop data from
well documented cases of submarine slope instabilities in order
to better evaluate and calibrate the available analyses.

INTRODUCTION

The scope of this paper is to highlight the available methods, and
different aspects, of submarine slope stability analysis, including
the post-failure movements of masses involved in a slide. This will
be illustrated by some examples of coastal slides in Norway in
loose sands and soft clays.

METHODS OF ANALYSIS

In principal there are no great differences between the stability
analysis of subareal and submarine slopes. There are, however, some
differences in what the predominant loading and failure mechanisms
are. For example, overloading of weak, underconsolidated soils due
to rapid sedimentation, or cyclic wave induced pressures may be important factors in the submarine environment. Submarine slope instabilities may rapidly propagate very large distances because of
the weak loose, or sensitive nature of many submarine soils and the
presence of water.

Submarine slopes are commonly fairly gentle, uniform and homogeneous over considerable horizontal distances. Thus in many cases
"infinite slope" limiting equilibrium methods are applicable. A
general expression for the factor of safety, F, of an infinite
homogeneous slope is given by:

$$F = \frac{\bar{C} + \left(1 - \frac{\gamma_t}{\gamma_b}a_v - \frac{\gamma_t}{\gamma_b}a_h \tan i\right)\gamma_b \cdot d \cdot \cos^2 i \cdot \tan\varphi' - \Delta u \tan\varphi'}{\left(1 - \frac{\gamma_t}{\gamma_b}a_v + \frac{\gamma_t}{\gamma_b} \cdot \frac{a_h}{\tan i}\right) \cdot \gamma_b \cdot d \cdot \sin i \cdot \cos i}$$

c, φ' = Shear strength parameters
γ_t = Total unit weight
γ_b = Buoyant unit weight
Δu = Pore pressure in excess of hydrostatic
a_v, a_h = Vertical, horizontal acceleration in fractions of g
d = Depth of sliding surface
i = Slope angle

For the limiting equilibrium case (F = 1,0), the following special
cases might be of interest:

A. $a_h = a_v = \Delta u = 0$

     $\frac{c'}{\gamma_b \times d} = \cos^2 i \, (\tan i - \tan \varphi')$

B. $a_v = \Delta u = 0, \varphi' = 0, c' = s_u$

     $\frac{s_u}{\gamma_b \times d} = \frac{1}{2} \sin 2i + a_h \frac{\gamma_t}{\gamma_b} \cos^2 i$

Numerical solutions to case "B" are presented in Fig. 1. As an example, for a normally consolidated clay with $s_u/\gamma_b \times d$ of 0,25, a
moderate horizontal acceleration of 0,05 g reduces the maximum
stable slope from 15 degrees to only 6 degrees. Also, the undrained
strength under cyclic loading conditions may be significantly lower
than the static strength. See for instance Andersen et al (1976).

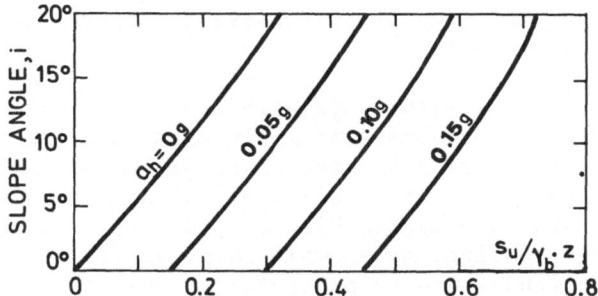

Fig. 1. Limiting equilibrium of infinite slope subjected to earthquake, undrained case (Morgenstern, 1967)

Pore pressures in excess of hydrostatic, Δu, might exist in a submarine sediment for a number of reasons, the most important being:

1. Underconsolidation due to rapid sedimentation
2. Gas generation within sediment
3. "Artesian" gas or water pressures in deeper formations
4. Pore pressures generated by repeated wave- or earthquake
   loading

Underconsolidation due to rapid sedimentation is an important cause
for submarine slope failures, especially near the outlet of major
rivers. The Mississippi Delta is a classical case (see for example
results of the SEASWAB studies reported by Garrison; Bennet; Hirst
et al, 1977). Sangrey (1979) also concluded that rapid sedimentation was a major cause of instability in Alaskan coastal sediments.
Gibson (1958) presented a solution for the consolidation of a soil
layer increasing in thickness with time. With present computer programs, general and rigorous solutions can be found. There is, however, a great need for verification of such excess pore pressure
computations by direct in-situ observations.

In gas-laiden sediments, pore pressure measurements are difficult.
See for instance discussion relating to the SEASWAB project mentioned above. Gas generation and its effect on the shear strength
of sediments was discussed by Esrig and Kirby (1977), but does
certainly require further studies and verification. Sills (1980)
has made some very careful measurements of pore pressure changes
in soft sea beds in the Bristol channel and Holyhead harbour. Both
measurements showed a marked difference from the incompressible
response indicating the presence of gas in the soil. For instance
in one case at 3,5 m depth below mudline, the maximum pore pressure
in the sediment occurred 6 hours after peak tide, at the same time

as low tide. The amplitude of pore pressure changes was only about
50% of the total variation.

That gaseous sediments with compressible pore fluid can have a pro-
nounced difference in response to fully water saturated sediments
has been shown in theoretical studies by Yamamoto (1978) and
Madsen (1978). Describing the soil skeleton as linearly elastic
and the pore fluid as compressible, they derived the response of a
sediment to wave loading. An example is presented in Fig. 2, giving
the maximum amplitude of pore pressure response as function of
depth. (The pressure actually fluctuates periodically with time,
with the same frequency as the wave). They also showed that in some
instances pore pressures generated by a single wave may reduce
effective stresses to zero. It is emphasized that Madsen's and
Yamamoto's solutions do not include any cyclically induced excess
pore water pressures as discussed later.

Infinite slope type stability analysis is of course no longer valid
if external wave loads on the slope are to be included in the analysis. Henkel (1970) studied the undrained stability of a slope
subjected to external wave loading using circular failure surfaces,
as shown in Fig. 3. Finn and Lee (1979) presented results from a
more general effective stress method of slices, which in addition

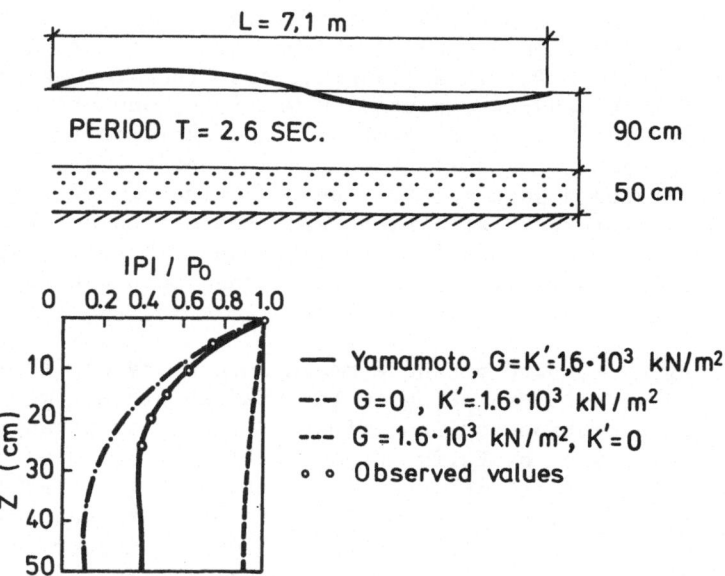

Fig. 2.   Example of wave-induced pore pressure in pervious com-
          pressible bed (Yamamoto, 78)

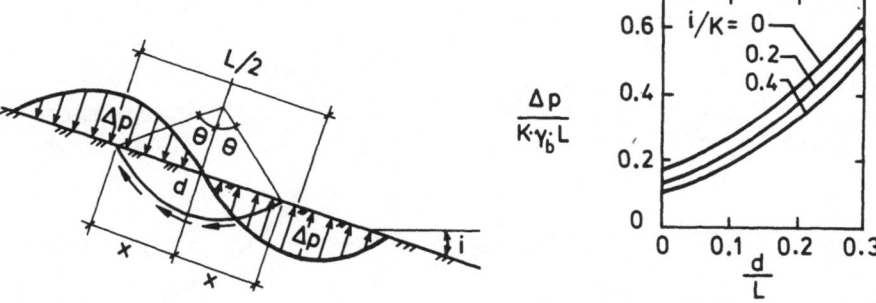

a) Geometry of failure surface

b) Limiting equilibrium solution (F = 1.0)

for $K = \dfrac{s_u}{\gamma_b \cdot z} = $ constant

Fig. 3. Limiting equilibrium analysis of wave loading using circular failure arcs (Henkel, 1970)

to wave loading also included earthquake forces and excess pore
pressures. An example of such an analysis is shown in Fig. 4.

In both the above mentioned cases, the bottom wave pressures were
taken from linear wave theory. In very soft and compressible seafloor deposits, the vertical displacements due to wave loading
might influence the bottom pressures. Doyle (1973) presented results of model tank tests which indicated that this could be a
reality.

Displacement type analyses can be an important tool in evaluating
wave loading effects on submarine slopes. Wright (1976) developed
a static finite element model which included wave forces and gravity stresses, and a fairly general hyperbolic stress-strain relationship. In analyses of South Pass Block 70 failure, Fig. 5, he
showed that lateral downslope displacements are more than twice
the corresponding deflections upslope. This is a direct result of
the downward component of gravity stresses. He also showed that
the deterioration of stress-strain parameters due to cyclic
loading led to drastic increase in displacements.

The limitation of this static displacement model is that it does
not consider the accumulation of deformations due to passage of a
number of waves. This problem has been approached by Schapery and
Dunlap (1978), modelling the soil as a linearly viscoelastic
material. Their analyses also included the effect of energy ad-
sorption of the seafloor on the wave characteristics.

That liquefaction of granular submarine soils due to earthquakes
can result in "flow-slides" of vast dimensions is a well-known
phenomenon. As summarized by Seed (1979), liquefaction analyses may
be based either on:

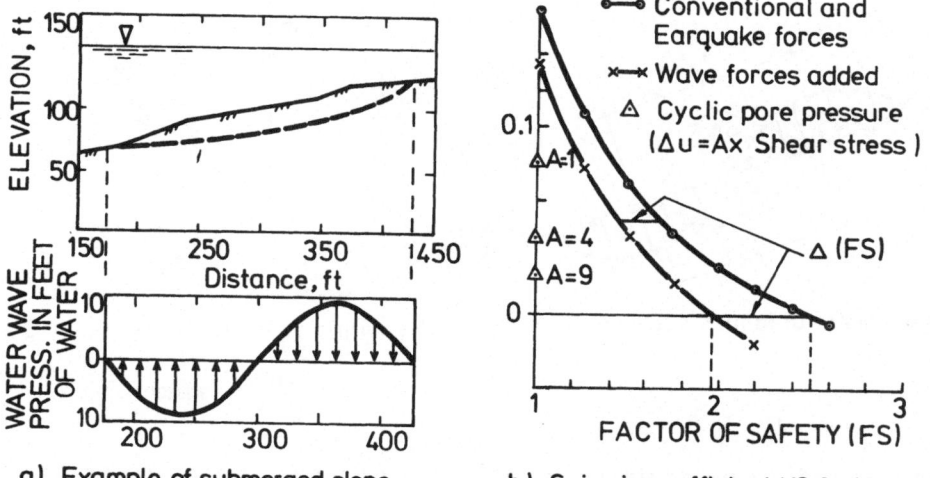

a) Example of submerged slope

b) Seismic coefficient VS factor of FS

Fig. 4.   Example problem for seafloor stability under seismic and wave loading. (Finn and Lee, 1979)

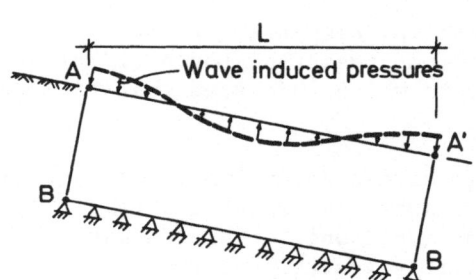

a) Schematic representation of the finite element model

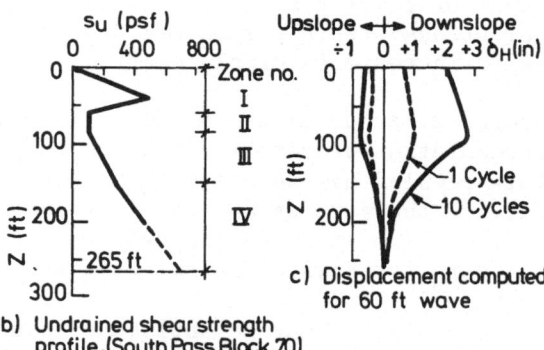

b) Undrained shear strength profile. (South Pass Block 70)

c) Displacement computed for 60 ft wave

Fig. 5.   F.E.M. analysis for wave induced movements (Wright 1976)

1) Observed performance of sand deposits in previous earthquakes
   (Evaluation of SPT, Rel. density, grain size)
2) Determination of stress conditions causing liquefaction in the
   laboratory

Although the state-of-the art within this area has advanced much
over the last years, the predictive capabilities of these methods
are still rather poor and strongly debated.

The possibility of wave-induced liquefaction has been looked at by
numerous investigators. Cluckey et al., (1980) compared wave induced
cyclic stresses with cyclic stresses causing liquefaction in the
laboratory on a medium dense sandy-silt sediment from the Yukon
Prodelta. They concluded that 6 m storm waves could cause liquefaction to 3,5 metres depth. Nataraja et al (1980) presented a
simplified liquefaction analysis of submarine soils based upon
comparison of wave-induced cyclic stresses with cyclic resistance
of soils based on earthquake experiences. Such approaches must be
used with great caution, mainly for the following reasons:

a) Earthquake loads vary much more rapidly and last for a shorter
   period than do wave loading. This will have a strong influence
   on pore pressure dissipation effects

b) Submarine soils down to water depths of hundreds of meters have
   undergone a very long history of repeated wave-loading, which
   might make them less succeptible to wave (or even earthquake)
   induced liquefaction

c) The difference in stress conditions during wave-loading and
   earthquakes and the possible influence of that is not fully
   resolved

Dynamic models for evaluating seismic loading effects have to the
authors knowledge, not been used so far for evaluating submarine
slope stability. However, the available models and programs should
be readily adaptable. The advantage of a dynamic analysis is that
it will give the time-varying stresses and deformations induced by
an earthquake, taking inertial effects properly into account. Nonlinear soil properties, pore pressures and deformations generated
by cyclic loading and pore pressure dissipation after dynamic
loading may also be included.

On the basis of this brief review of analytical methods that are
available for evaluating submarine slope stability, one might conclude that these methods have their limitations and that there is
room for theoretical improvements. Still, the authors feel that
there are many more uncertainties associated with the determination
of the necessary input parameters, particularly relating to soil
behaviour. This is even more of a problem than on land because
soil sampling and in-sity testing is difficult and very costly in

the marine environment. Another important aspect is the limited
evaluated experience with submarine slope stability analysis. There
are very few submarine slides which have been investigated in sufficient detail to allow definite conclusions to be drawn regarding
failure mechanism and geotechnical characteristics of the failed
sediments. Future work should therefore concentrate on developing
better and more suitable sampling and in-situ testing techniques,
and on making detailed investigations and analyses of existing
slides.

POST-FAILURE MASS MOVEMENTS

So far, relatively little attention has been given within soil mechanics research to the post-failure movements of masses involved
in submarine as well as subareal slope failures. In many offshore
engineering problems this is an increasingly important aspect of
the design considerations. When placing a structure on the sea
bottom it is not only necessary to evaluate the stability situation
at the specific site itself, but also to look at the possibility of
the structure being "hit" by slide debris from failures upslope.
Such an evaluation must include answers to a number of very difficult questions, some of which are:

1) What is the extent of a possible slide, and how does the slide
   develop with time?

2) What mechanisms govern the movements of the debris downslope,
   and what will be the velocity, height, density and runout
   distance of the masses?

3) What forces might the moving masses exert on obstructions
   (structures) of different kind as it moves downward, and to
   what extent will it erode sediments as it moves along?

With regard to the first question, it can be exceedingly difficult
to predict the extent of a slide and its development with time.
This is especially true for very loose sand and silt deposits which
have a metastable structure, and collapse or undergo "spontaneous
liquefaction" when they are disturbed by earthquakes or are otherwise brought to failure by an initial slide. It is in these materials that the well known flow slides of vast dimensions can develop
(examples are given later).

Masses involved in a slide or slump might behave in a variety of
ways after failure, ranging from slow creep movements, to rapid
debris flows, fluidized sediment flows and dilute turbidity
currents. Middelton and Hampton (1976) have discussed the development of different types of flows, and have suggested a classification as shown in Fig. 6. A major question in this context is what
are the soil characteristics of the intact material as well as the

environmental conditions that governs its transition into different
types of flowing material.

In dense medium to coarse sands and relatively competent non-
sensitive clays, masses involved in a slide will normally move as
more or less rigid material with fairly low velocity, and movements
will cease relatively rapidly when the slope angle is reduced beyond
some critical level.

Soft sensitive clays and loose fine sands and silts might on the
other hand turn into more or less liquid material and flow downslope with considerable velocity and run great distances on very
gentle slopes. As it moves downslope, water might be mixed into
the flowing mass, making it even more liquid in behaviour. Some
material might also in the process be "torn loose" from the main
body of the flow and turn into very dilute turbidity currents. This
was for instance observed in model tank studies by Van der Knaap
and Eijpe (1968). The authors feel, however, that unless the flowing mass meets some obstructions, or there are very abrupt changes
in slope geometry, creating "hydraulic jumps", only minor parts of
the main body of a flowing mass will normally turn into very low
density turbidity currents. This question must, however, be strongly dependant on the grain size distribution of the masses involved.
Grain size must also be the main factor that governs resedimenta-
tion, reconsolidation and cessation of flow.

Two principal types of models which can be used to describe the
mechanics of a flowing mass are discussed below.

a) Density (turbidity) current flow models

Experimental and analytical studies of the hydrodynamics of density
currents have been presented by for instance Kuenen (1952),

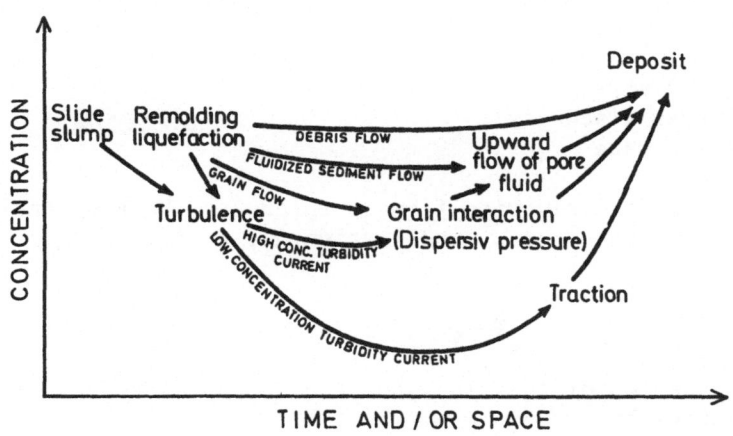

Fig. 6.   Evolution of "Flow slide" (Middelton and Hampton, 1976)

Middelton (1966 a, b, 1967), Hampton (1972).

Middelton and Hampton (1976) and Pantin (1979). Such models are primarily adaptible to low density flows where the soil particles are carried in water by suspension (i.e. there is little or no interparticle contact). For flow of a density suspension on relative flat slopes, the suggested equations for the average velocity of the head of the flow are of the form:

$$\bar{u} = k \ (g \ \frac{\rho_2 - \rho_1}{\rho_1} \ x \ h)^{1/2} \quad \text{where}$$

h = height of flow at head
ρ_2 = density of flow
ρ_1 = density of overlying fluid
k = constant

Suggested values for k lie in the range 0,7 - 1,3. It should be emphasized, however, that there are only limited experimental evidence on the validity of this model available, and most experiments have been carried out on materials with density less than 1,1 g/cm^3.

b) Plastic and viscous flow models

Such models have been applied to a variety of problems ranging from slow creep deformations and "mud flows" to more rapid fluid type of mass movements.

Johnson (1970) presented solutions to the latter type, describing the flow characteristics of the mass as a Bingham fluid. Herein the shear stress, τ_s, resisting displacement is related to the velocity, $\dot{w}$, of movements through:

$$\tau_s = k + \eta \ \frac{d\dot{w}}{d_y} \quad \text{where}$$

k = yield strength
η = linear viscosity
$\frac{d\dot{w}}{d_y}$ = shear rate

For a uniform, infinitely wide flow of thickness, H, and density, γ, down a slope with constant inclination, δ, a rigid plug of thickness:

$$T_c = \frac{K}{\gamma \sin \delta}$$

will form on the top of the flow, and the steady state velocity of the plug is given by:

$$\dot{w}_{plug} = \frac{1}{\eta} \ \frac{H^2 \gamma \sin \delta}{2} + \frac{k^2}{2 \ \gamma \sin \delta} - kH$$

Johnson (1970) also presented solutions for the complete velocity profile, and for flow in channels.

The limitations of Johnson's solutions with regard to application to actual submarine problems are:

- Hydrodynamic forces on the top boundary are not included, which might be important especially if relatively large velocities are generated.

- Considerable time is needed to accelerate the flow to steady state. If the slope geometry is not constant over very large distances, uniform steady-state flow velocities might never be reached.

The authors are currently expanding on Johnson's solutions to include these factors, However, the major limitations of this model is that there are very little evaluated experience relating to the determination of yield strength and linear viscosity of a remoulded, flowing sediment. Furthermore, these properties are not likely to be constants, but will probably vary with time as the sediment flows downslope.

Concluding this discussion, the authors feel that presently, the post-failure movements of masses involved in slides or slumps are poorly understood, and that there is a great need for theoretical and experimental work within this area.

EXAMPLES OF COASTAL SLIDES IN NORWAY

As previously mentioned, an initial instability in loose uniform sand and silt deposits might turn into a flow slide of vast dimensions. A number of slides of this nature have occurred in Norwegian fjords, Table 1, as described by for instance Terzaghi (1956), Andresen and Bjerrum (1967) and Bjerrum (1971), Karlsrud (1979). All of these slides occurred in recent post-glacial deposits near river outlets. All the slides have been located within a relatively concentrated area in middle Norway. To illustrate the nature and mechanisms of these flow slides, a description of the slide events in Orkdalsfjord, May 2nd, 1930, is reveiwed.

The inner part of Orkdalsfjord is approximately 2 km wide and 8 m deep, but widens out and meets the Trondheimsfjord further out, Fig. 7. At the bottom of the fjord the river Orkla has its outlet, and has since the last glaciation, about 10.000 years ago, filled the fjord with loosely packed sand, silt and clay deposits. For the first 5 kilometers from the river outlet, the fjord bottom has a slope of 5 to 10%, but gradually flattens out to a few percent or less as one approaches Trondheimsfjord, 25 km away, where the water depths reach 5-600 m. Bjerrum (1971) has described the sliding as follows:

Table 1. Summary of flow slides in Norwegian fjords

PLACE	INITIAL SLIDE
Orkdalsfiord, 1930.	Recent fill (low tide)
Trondheimsfiord, Brattöra,1888.	?? (low tide)
Trondheimsfiord, Illsvika, 1950.	During filling
Hommelvika , 1942.	During filling (low tide)
Follafiord, Kongsmo , 1952.	??
Finnvika , 1940.	During filling (low tide)
Sandnessjöen, 1967.	Blasting
Gullsmedviken, 1964-78.	Errosion from outlet pipe

"The slide occurred in the early morning at an
exceptionally low tide and was observed at the following
points (Fig. 2):
(1)   At Storaunet about 1000 to 1500 m^3 of a minor
filling which had been recently placed along the
shore, disappeared at 7:48 a.m. in a slide – slide A –
that stretched 500 m along the edge of the fjord.
(2)   A few minutes later a 600 to 700 m long slide at
Orkanger, approximately 2 km away, destroyed some piers
and harbour works – slide B.
(3)   On the opposite side of the fjord a slide took
place at Furenstranden at 7:55 to 8:00 a.m. – slide C.
(4)   Three kilometres from Storaunet the Ofstad –
Sandløkken telephone cable crossed the fjord at a water
depth of a maximum of 300 to 350 m. At 7:55 a.m. the
cable was broken, indicating that the slide had propagated from Storaunet with a speed of about 25 km/hour.
(5)   Approximately 18 km further along the fjord, the
Vorpenesset – Stadsbygden cable crossed at a water
depth of 500 m. Failure of this cable at 9:40 a.m.
corresponded to an average speed of propagation of
10 km/hour since the occurrence of the initial slide.

From the observed changes in water depth it has been
calculated that about 25 million m^3 of soil had been
moved by the three first-mentioned slides (Vogt, 1943).
Borings carried out near the areas where the two major
slides A and B occurred indicate the existence of
large deposits of a very loose and soft non plastic
silt, having a water content of about 33 percent –
corresponding to a porosity of about 49 percent.

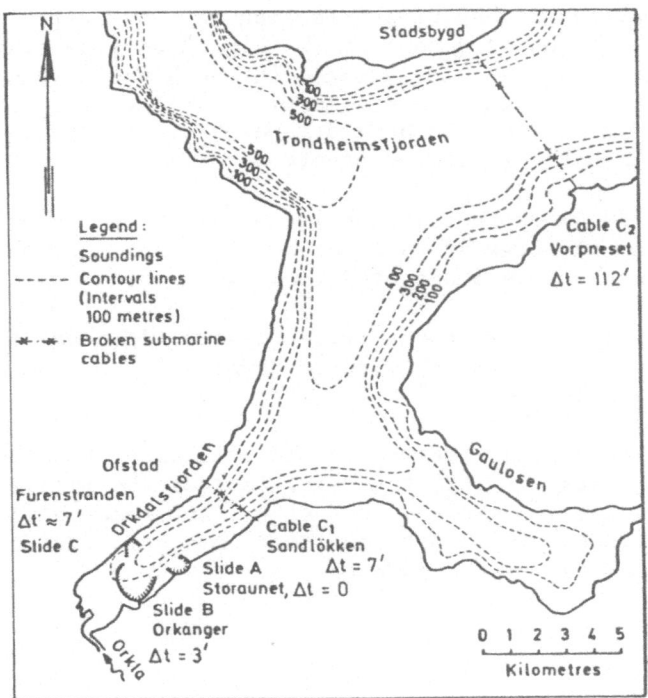

Fig. 7. Map of slides in the Orkdalsfjord (Terzaghi 1956)

From these events it seems clear that the masses involved in the
first slide A, must have <u>eroded</u> deeply into the bottom sediments
as they descended down the slope and met the fjord bottom. Erosion
must have been particularily severe when the masses with great velocity reached the bottom and was forced to change direction almost 90 degrees to proceed out the fjord. The scars left by this
erosion must have caused a new series of retrogressive flow slides
which eventually reached into and released Slide B. Erosion by
these masses as they descended out the fjord finally triggered
Slide C. Thus, after the first initial slide, erosion and retrogressive sliding must have given a more or less continous supply
of material to the flow which finally reached all the way out to
Trondheimsfjord. This general mechanism of erosion and retrogressive
sliding has been observed in all the Norwegian slides shown in
Table 1.

As summarized in Table 1, the initial slides which triggered these
large flow slides were in 4 cases apparently primarily caused by
man-made fills on the shore-line, and the time of the initial slide
frequently corresponded with low tide. Two of the slides started
at large water depths for no apparent reason. Earthquakes, which
have been reported to trigger spontaneous liquefaction and flow
slides in other parts of the world, were not recorded in any of the

Norwegian cases. The slide in Sandnessjøen was, however, triggered
by minor blastings inside a ship wreck. Fig. 8 outlines the situation. The boat "Sleperen" was engaged in the blasting and demoli-
tion of a wreck at 8 m water depth. During the day of June 1, 1967,
20 charges of 2,5 kg each had been detonated within the wreck, last
time at 1540. Then the following events occurred:

- At 1545 - 1550 "Sleperen" was pulled outwards by the anchor. At
  the same time the crew saw a flood wave of considerable height
  (4-7 m) appear, which propogated towards Vågen.

- At ~ 1550, the electric power line crossing Vågen and supplying
  electric power to the wreck yard at Høvding 350 m from
  "Sleperen" was broken.

- At ~ 1600 cracks were observed in the fill along the shore line
  at Høvding's yard. At 1650 the first part of the fill and a
  wharf at Høvding disasseared in a slide, but some retrogressive
  sliding continued after that, in which some buildings and other
  harbour structures also disappeared.

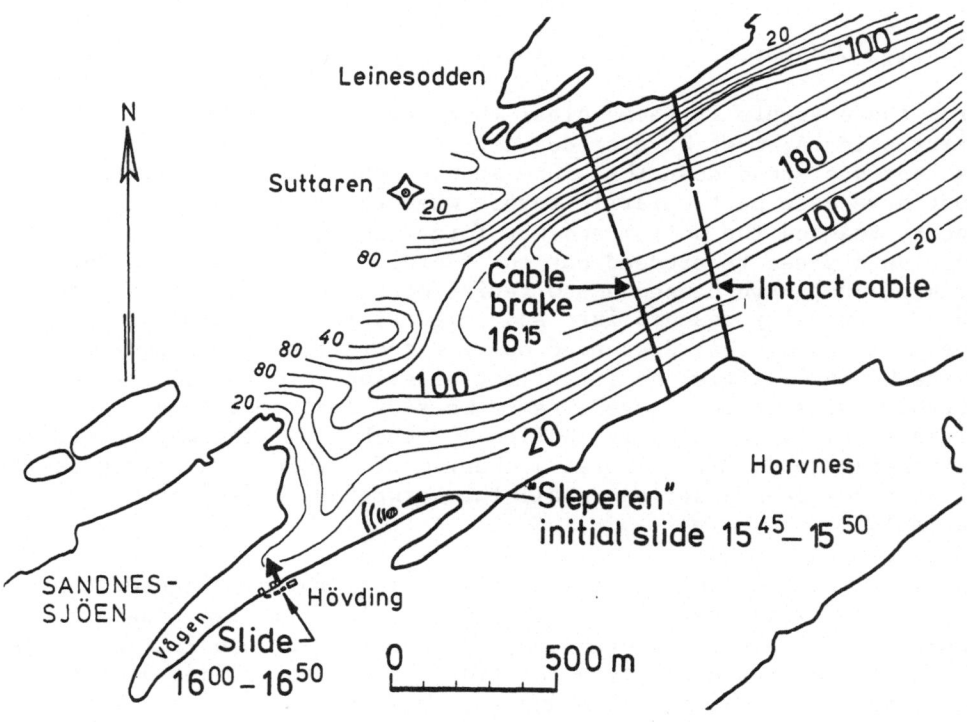

Fig. 8.   Slide in Sandnessjøen, 1967 (Karlsrud, 1979)

- At 1615, the telephone cable crossing the fjord 1,5 km N-E from
  Sleperen was cut. This indicates a speed at "propagation" from
  the initial slide of ~ 3 km/hour. Another cable, ~ 300 m further
  along was not damaged, indicating that the slide masses came to
  a rest before that.

Most of the masses which were involved in the slide at Høvding's
yard were redeposited in Vågen. The relatively slow propagation of
this part of the slide also indicates that the sand deposits in
this area were denser and not prone to collapse and flow.

A number of coastal slides in soft clays have been reported by
Karlsrud (1979). These were almost exclusively associated with man-
made fills on the shore-line. Recent submarine post-glacial clay
deposits, will normally have an undrained shear strength which increases linearly with depth (if they are not under-consolidated),
and which is close to zero at the mudline. (There might be some
very small strength due to true cohesion). Fig. 9 shows some typical shear strength profiles from
borings within an area along the
Drammensfjord. Notice here that
the borings in shallow water within the "beach zone" indicate some
type of a stronger "crust". This
is probably due to the depositional environment and breaking waves
in the "beach zone". Dessication
as such can not explain this
"crust" because low tide in the
area rarely goes deeper than
0,5 m below mean sea water level
(M.S.L.). The lack of an upper
stronger crust in the clay deposits beyond the beach zone means
that these deposits can sustain
very small additional loads be-

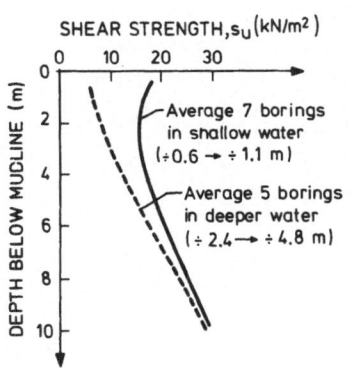

Fig. 9. Undrained vane shear
         strength of marine clay
         deposit in Drammensfjord

fore failing. This lack of a strong crust, which one always finds
in the subareal environment, is the main reason for slides in the
coastal clay deposits.

An example of such a slide is given in Fig. 10. A small man-made
fill rising 1,5 km above M.S.L. was to be placed within the "beach
zone". It was supposed to reach out to "B", ~ 10 m in front of the
break point, beyond which the sea bottom sloped downwards at an
inclination of ~ 16,1°. The minimum safety factor should then be
1,5. Filling was, however, continued beyond point "B". When it
reached point "A", just at the break point, the outer part of the
fill disappeared in the sea. There were some retrogressive sliding,
but the slide quickly came to a halt. There was no mystery about
the failure, as the theoretical safety factor was close to 1,0

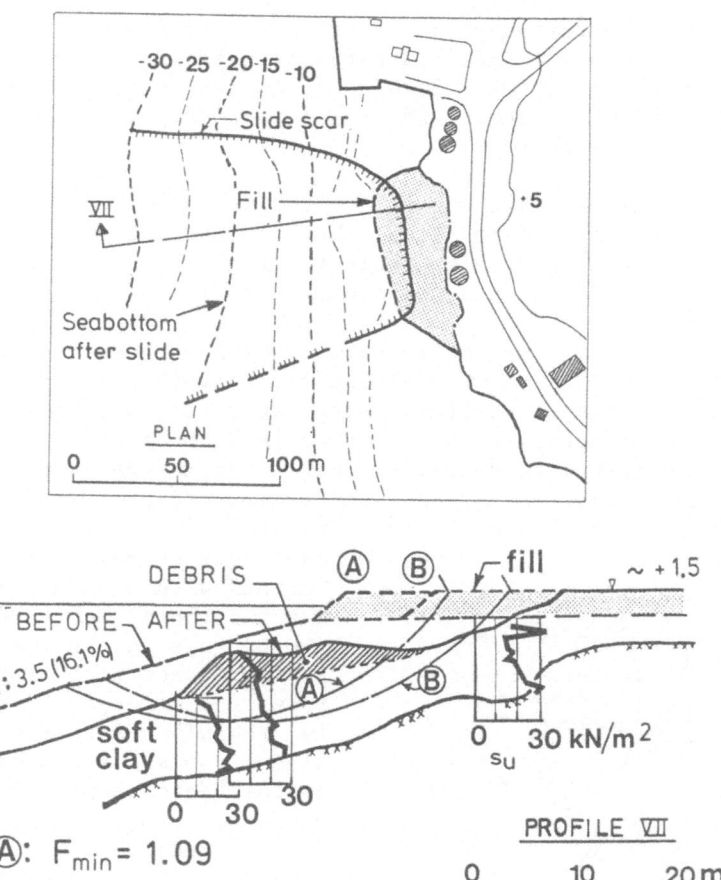

Fig. 10. Slide induced by filling along Drammensfjord, 1974

when the fill reached "A". Soundings after the slide revealed that
at least 200 m out from the edge of the fill, the upper 3-5 m of
the soft clay sediments down the slope had been swept along in the
slide, and must have continued to flow downslope for considerable
distances. This has also been the case in the other submarine
slides in clay deposits presented by Karlsrud (1979). Although not
directly comparable to true flow slides in sands and silts, it is
worthwhile noticing that slides in soft clay deposits also can involve relatively large volumes of material in addition to what is
involved in an initial slide, and that the debris can move great
distances downslope.

CONCLUSIONS

Analytic methods for engineering analysis of submarine slopes, although not perfected, are quite well developed. The major uncertainty in the use of these methods is in the determination of the soil input parameters. On the other hand, relatively little attention has been given within soil mechanics research to the post-failure movements of masses involved in submarine as well as sub-aerial structures. Major questions remain as to the extent, velocity, height and density of submarine slides after they develop and what forces the moving masses exert on submarine structures as they move down slope.

Case studies, such as those described above, can be instructive in numerous ways. First, these studies help to identify slide prone deposits. Secondly, they provide insight on the mechanisms by which submarine slides are initiated and then propagated. This insight is essential for the development and evaluation of the theoretical models used in engineering analysis. Finally, if sufficiently well documented and analyzed, case studies will provide meaningful data on the input geometries and soil parameters for engineering analysis.

However, for maximum benefit, the following information is necessary:

- age of slide
- geologic conditions at time of the slide
- geotechnical parameters of the slide mass
- extent of slide pit
- thickness, run out distances and velocities of the slide material
- geometry of slide path

Other papers at this workshop describe the many improvements in geophysical exploration techniques and the recent discoveries of many submarine slides. However, very few of these slides are sufficiently well documented, as described above, to be very instructive. There is great need for further research in this area.

REFERENCES

Andersen, K.H., Brown, S.F., Foss, I., Pool, J.H., and Rosenbrand, W.F. (1976), "Effect of Cyclic Loading on Clay Behaviour". Proceedings, Conference on Design and Construction of Offshore Structures. Institution of Civil Engineers, London, England, 1976, pp. 75-79.

Andresen, A. and L.Bjerrum (1967)
        Slides in subaqueous slopes in loose sand and silt.
        In. Marine Geotechnique. Ed. by A.F. Richards. Urbana,
        University of Illinois Press, pp. 221-239. Also publ.
        in: Norwegian Geotechnical Institute, Publ., 81.

Bennet, R.H. (1977)
        "Pore-water pressure measurement: Mississippi Delta
        submarine sediments".
        Marine Geotechnology, Vol. 2, pp. 177-189.

Bjerrum, L. (1971)
        Subaqueous slope failures in Norwegian fjords.
        Norwegian Geotechnical Institute. Publ. 88, pp. 1-8.

Cluckey, E., D.A. Cacchione and C.H.Nelson (1980)
        Liquefaction potential of the Yukon prodelta, Bearing Sea.
        Offshore Technology Conference, 12. Houston 1980. Pro-
        ceedings, Vol. 2, pp. 315-325.

Doyle, E.H. (1973)
        Soil wave tank studies of marine soil instability.
        Offshore Technology Conference, 5. Houston 1974.
        Preprints, Vol. 2, pp. 753-766.

Esrig, M.I. and R.C.Kirby (1977)
        Implications of gas content for prediction of stability
        of submarine slopes.
        Marine Geotechnology, Vol. 2, pp. 81-100.

Finn, W.D.L. and M.K.W. Lee (1979)
        Sea floor stability under seismic and wave loading. 25 p.
        American Society of Civil Engineers. National Convention,
        Boston 1979. Soil dynamics in the marine environment.

Garrison, L.E. (1977)
        "The SEASWAB experiment".
        Marine Geotechnology, Vol. 2, pp. 117-122.

Gibson, R.E. (1958)
        The progress of consolidation in a clay layer increasing
        in thickness with time.
        Géotechnique, Vol. 8, pp. 171-182.

Hampton, M.A. (1972)
        The role of subaqueous debris flow in generating turbidity
        currents. Journal of Sedimentary Petrology, Vol. 42,
        No. 4, pp. 775-993.

Henkel, D.J. (1970)
        The role of waves in causing submarine landslides.
        Géotechnicque, Vol. 20, No. 1, pp. 75-80.

Hirst, T.J. and A.F.Richards (1977)
        "In-situ pore pressure measurements in Mississippi Delta
        front sediments". Marine Geotechnology, Vol. 2, pp. 191-205.

Johnson, A.M. (1970)
        Physical processes in geology.
        San Francisco, Freeman. 557 p.

Karlsrud, K. (1979)
        Undersjøiske utglidninger og skred, 30 p.
        Norske sivilingeniørers forening. Skredfare og areal-
        planlegging; vurdering av faregrad og sikringstiltak;
        kurs.
        Lofthus i Hardanger 1979.

Kuenen, Ph.H. (1964)
        Deep sea sands and ancient turbidites.
        In. Turbidities. Ed. by: A.H.Bouma and A.Brouwever.
        Amsterdam, Elsevier.

Madsen, O.S. (1978)
        Wave induced pore pressures and effective stresses in
        a porous bed. Géotechnique, Vol. 28, No. 4, pp. 377-393.

Middelton, G.V. (1966 a)
        Experiments on density and turbidity currents, I.
        Motion of the head.
        Canadian Journal of Earth Sciences, Vol. 3, pp. 523-546.

Middelton, G.V. (1966 b)
        Experiments on density and turbidity currents, II.
        Uniform flow of density currents.
        Canadian Journal of Earth Sciences, Vol. 3, pp. 627-637.

Middelton, G.V. (1967)
        Experiments on density and turbidity currents, III.
        Deposition of the sediments.
        Canadian Journal of Earth Sciences, Vol. 4, pp. 475-505.

Middelton, G.V. and M.A.Hampton (1976)
        Subaqueous sediment transport and deposision by sediment
        gravity flows. Chapter 11 in:
        Marine sediment transport and environmental management.
        Ed. by: Stanley and Swift. American Geological Inst./
        Wiley.

Morgenstern, N.R. (1967)
        Submarine slumping and the initiation of turbidity
        currents. Marine Geotechnique. Ed. by: A.F.Richards.
        University of Illinois Press, pp. 189-220.

Nataraja, M.S., H. Singh and D.Maloney (1980)
        Ocean wave-induced liquefaction analysis: a simplified
        procedure. International Symposium on Soils under
        Cyclic and Transient Loading, Swansea 1980. Proceedings,
        Vol. 2, pp. 509-516.

Pantin, H.M. (1979)
        Interaction between velocity and effective density in
        turbidity flow; phase-plane analysis, with criteria for
        autosuspension. Marine Geology, Vol. 31, No. 1/2, pp. 59-100.

Sangrey, D.A., E.C.Cluckey and B.F.Molnia (1976)
        Geotechnical engineering analysis of underconsolidated
        sediments from Alaskan coastal waters.
        Offshore Technology Conference, 11. Houston 1979.
        Proceedings, Vol. 1, pp. 677-682.

Schapery, R.A. and W.A.Dunlap (1978)
        Prediction of storm-induced sea bottom movement and
        platform forces. Offshore Technology Conference, 10.
        Houston 1978. Proceedings, Vol. 3, pp. 1789-1796.

Seed, H.B. (1979)
        Soil liquefaction and cyclic mobility evaluation for
        level ground during earthquakes.
        American Society of Civil Engineers. Proceedings, Vol. 105,
        No. GT 2, pp. 201-255.

Sills, G.C. (1980)
        Private communication

Terzaghi, K. (1956)
        Varieties of submarine slope failures. 41 p.
        Texas Conference on Soil Mechanics and Foundation
        Engineering, 8. Austin 1956. Proceedings.
        Also publ. in: Norwegian Geotechnical Institute.
        Publication, 25.

Van der Knaap, W. and R.Eijpe (1968)
        Some experiments on the genesis of turbidity currents.
        Sedimentology, Vol. 11, pp. 115-124.

Wright, S.G. (1976)
        Analyses for wave induced sea-floor movements.
        Offshore Technology Conference, 8.Houston 1978.
        Preprints, Vol. 1, pp. 41-52.

Yamamoto, T. (1978)
        Sea bed instability from waves.
        Offshore Technology Conference, 10. Houston 1978.
        Proceedings, Vol. 3, pp. 1819-1828.

ON AN EXPLANATION OF MARINE FLOW SLIDES IN SAND

H.L. Koning

senior research engineer
DELFT SOIL MECHANICS LABORATORY
Delft, the Netherlands

ABSTRACT

Sometimes due to a failure in a marine sand deposit some million m^3 of sand are removed. These flow slides occur in sand-masses with a high porosity. When sheared such high porosity sand tends to a volume decrease and this tendency causes an increase of the pore water pressure. Depending on circumstances this increase of the pore water pressure may be so great that the sand liquefies. So it is important to study the relevant properties of the sand and also the porosity in situ.

Some remarks will be made on the possible causes of increasing shear stresses in marine sandmasses. A pair of flow slides which occurred in the Netherlands was investigated rather intensively and some results of these investigations will be shown.

1. Introduction

In the Netherlands many flow slides in sand have occurred in the province of Zeeland, situated in the south-western part of the country. These flow slides ara characterized by the after the slide remaining very flat slopes with angles ranging between 2 and 6 degrees. Mostly only a part of the riverbank flows away, but sometimes the levee fails also and the land behind it is flooded. The main reason that these flow slides occur in sand is deemed to be the high porosity of the sand.

A sandmass consists of a number of grains and voids. The ratio of the volume of these voids to total volume of sand including voids is called porosity. The magnitude of this porosity can be illustrated by considering an unreal sample of sand

83

consisting of uniform spheres. In the cubic packing - in which
the centres of the spheres are situated at the angular points
of cubes - the porosity has its highest value and amounts to
47.6%. The densest state of equal-sized spheres is attained
with a hexagonal array, the porosity is equal to 26.0% then.
So for this unreal type of sand the range between maximum and
minimum porosity should be 21.6%. However for the in reality
encountered sands this range amounts to about 12%.

When shear stresses are exerted on a sandmass with a high
porosity the grains tend to take a denser state, consequently
a decrease of volume occurs. If the voids are filled with air
the sandmass can undergo this deformation almost without constraint. But if the pores are filled with a liquid a reduction
in volume must result in the outflow of an equivalent volume of
the liquid. Excess pressures in the liquid are created then to
make this outflow possible. These excess pressures can reach
very high values. For instance when in a water saturated sandmass
the porosity decreases from 40% to 39.5%, the excess pore water
pressure could theoretically amount to about 41 MN/m^2, all
exterior circumstances remaining constant and no water flowing
out. However it is almost sure that these high value cannot be
attained. Through these excess pore water pressures the effective
stresses between the individual grains are reduced and as the
friction in sand is proportionate to the effective stresses,
the frictional resistance of the sand consequently also decreases.
This means that the deformation can go on faster and likewise
do the excess pore water pressures. Depending on circumstances
the increase in the water pressures may be so great that the
effective stresses become zero, so that the frictional resistance
is lost and the sand behaves like a heavy fluid: the sand liquefies. Usually this state will be entered long before the above
theoretically estimated value of the excess pore water pressure
is attained. If the surroundings allow the sand to flow away a
flow slide occurs.

The opposite happens if shear stresses are generated in a
water saturated sandmass with a low porosity. In that case an
increase in volume is caused, for which a quantity of water must
flow in. Then an under-pressure in the pore water occurs and
this causes the effective stresses between the individual grains
to increase. So the frictional resistance of the sand becomes
greater and a more stable situation is obtained. However, this
situation is only a temporary one since the under-pressures in
the pore water decrease to zero according as water flows in.

It will be clear that a transitional case can occur in
which shear stresses do not give rise to changes in volume.
The corresponding density is called the critical density and is
usually stated as a porosity.

The phenomenon that shear stresses exerted on a granular
mass also bring about changes in volume is called dilatancy
(Reynolds, 1885). Casagrande (1936) drew attention to this

phenomenon again and demonstrated its importance to the stability of slopes.

It follows from the above that for an evaluation of the stability of marine slopes in sand knowledge of the porosity in situ and of the critical density is essential. There are other reasons why the porosity in situ should be known, since this porosity is of influence on other properties of the sand, like angle of internal friction, compressibility and permeability.

2. The critical density test

A critical density test can be executed on a dry sample or on a water-saturated sample. In both cases a dry cylindrical sample is built up inside a rubber membrane. The sample is surrounded by water, which in turn is bounded by a cell.

In the dry test the sample is placed under an ordered all-around pressure and the porosity is determined. The sample is now vertically deformed at a rate of 1 mm/min while the required force is measured. At the Delft Soil Mechanics Laboratory the horizontal stress is adjusted in such a way that the octahedral stress remains constant. The changes in volume of the sample are mainly caused by shear stresses then. During the test the state of stress, the relative vertical deformation and the relative volume deformation of the sample are recorded. The test is performed at different porosities to determine the transition from volume decrease to volume increase.

In the wet critical density test the sample is saturated and placed under an ordered all-around pressure; after that the porosity is determined. The sample is now vertically loaded with a gradually increasing stress. Again the horizontal stress is adjusted to keep the octahedral stress constant. During the test the state of stress, the vertical deformation and the excess pore water pressure in the sample are recorded. In a loose sample the pore water pressures increase and the sample collapses in a fraction of a second. In a dense sample negative pore water pressures develop. Also this test is performed at different porosities to determine the above mentioned transition. More details about these critical density tests are given by Lindenberg and Koning (1981).

The aforementioned tests are quasi-static ones. If dynamic forces come up for discussion - like with earthquakes or wave attack - cyclic tests have to be performed. These tests are executed on saturated samples. Usually the horizontal stress remains constant, while the vertical stress is changing cyclically. During the test the stresses, the vertical deformation of and the excess pore water pressures in the sample are recorded. In a loose sample the pore water pressures increase with every cycle until liquefaction occurs. In dense samples often a slight pore water pressure increase is measured which remains practically constant, while the vertical deformations

mostly are negligible. For densities in between the pore water
pressures increase significantly, but not to the state of
liquefaction while important vertical deformations occur. In
addition to the porosity the ratio of the shear stress to the
normal stress proved to be an important parameter.

3. The density measurement in situ

Obviously it is important to know the porosity of sandlayers in the field. Because of this the Delft Soil Mechanics
Laboratory developed an apparatus for determining the density
of sand in situ as a function of depth. The apparatus is fully
adapted to that used for executing static cone penetration tests.
Hence simultaneously with a density measurement the cone
resistance and the local skin friction are obtained.

The density measurement of sand in situ is based on the
fact that the sand grains themselved consist of electrically
non-conducting minerals. On the other hand the pore water is
electrically conducting,especially if it contains dissolved salts.
The more voids are present in a sandmass, the lower is the
electrical resistance of the total mass of soil (grains and
water). However,the resistance is determined not only by the
amount of water, but also by the resistance of the water itself.
It now can be shown that the ratio of the specific electrical
resistivity of the pore water to the specific electrical resistivity of the total mass of soil is a criterion of the
porosity.

The specific electrical resistivity of the soil is determined by means of a "soil probe", consisting of a sounding tube
supplied with four insulated electrodes. The probe is pushed
into the ground by means of a sounding apparatus and at every
20 cm difference in depth a reading is made. For this purpose
a voltage difference is applied to the two outer electrodes and
the specific resistivity of the soil is measured with the aid
of the two inner electrodes.

For the measurement of the specific electrical resistivity
of the pore water a "water probe" is used. This probe too
consists of a sounding tube, now supplied with a filter and a
measuring cell inside. At any desired depth - normally every
20 cm - some pore water can be sucked through the filter into
the measuring cell and its resistivity can be measured.

The ratio of these two resistivities is a criterion of the
porosity. The theoretical determination of the connection between
this ratio and the porosity is unfeasible, so a calibration curve
is established in the laboratory by means of tests on samples
taken from the investigated sandlayers.

4. Triggering mechanisms

A flow slide does not necessarily occur when a slope cuts

through a watersaturated loosely packed sandlayer. A triggering
mechanism is needed to increase the shear stresses in the sandmass. Then due to the decrease in volume excess pore water
pressures are generated leading to liquefaction. A number of
triggering mechanisms will be summarized:

a. scour. Through scour locally steeper slopes occur,
resulting in an increase of the shear stresses and ultimately
liquefaction may occur. A quantity of sand flows away, so that
a higher part of the slope looses its support. This in turn
leads to an increase in shear stresses, followed by liquefaction,
and so on. In this retrogressive flow slides large quantities of
sand can liquefy and flow away (Koppejan et al, 1948).

b. dredging. Dredging creates locally steeper slopes and a
retrogressive slide as described above can occur.

c. sedimentation. Through sedimentation locally steeper
slopes can develop which induces flow slide.

d. deposition. Deposition of material by man can bring about
locally steeper slopes or can create an overload on top of the
slope. In both cases the shear stresses are increased which can
lead to liquefaction.

e. seepage. Seepage forces can locally endanger the stability
of the slope. This mechanism often has been found in combination
with scour.

f. earthquakes. Earthquakes cause a cyclic loading of the
sand and consequently can generate liquefaction.

g. wave action. Through wave action seepage forces are introduced which can lead to limit equilibrium.

h. other reasons. There may be several other reasons which
occur less frequently, like explosions, vibrations, a ship
colliding with a slope.

In the following sections some results will be shown of
investigations executed in connection with a pair of flow slides
which occurred in the Netherlands.

5. Flow slide Roggenplaat

The Roggenplaat is a sandplate in the Eastern Scheldt basin.
In consequence of the closing works in this basin large quantities of sand have to be worked up. Part of this sand is carried
off by the stream and settled elsewhere. Also the Roggenplaat
received a certain mass of sand as followed from the results of
consecutive recordings of the bottomdepth. This sedimentation
caused a steepening of the slope and narrowed the stream channel.
The bank across the plate was protected and could not be eroded,
so to maintain the necessary wet profile the gully had to be
deepened too. Obviously the combination of these two phenomena
triggered between April 5 and 7, 1973 a flow slide. In this slide
a total mass of about 1.3 million m^3 of sand was removed. On
request of the Department of Public Works of the Netherlands
field and laboratory investigations were executed by the Delft

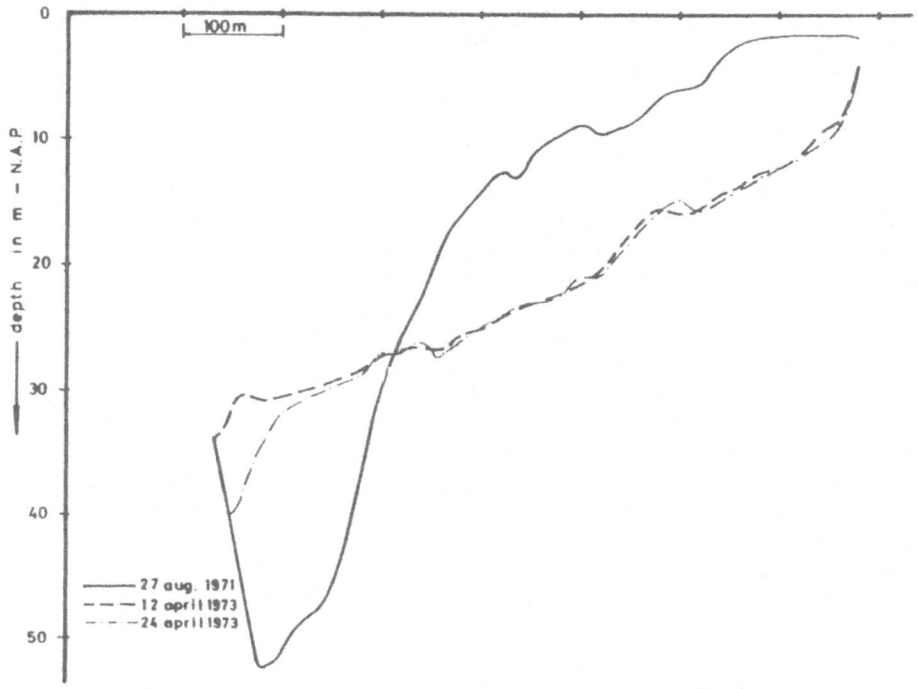

Fig. 1 Cross section flow slide Roggenplaat

Soil Mechanics Laboratory.

In fig. 1 a cross section along the centreline of the flow slide is shown. It can be seen that the gradient of the slope some time before the flow slide was about 1 vert. to 3 hor., i.e. an angle of about 18.5°. The gradient immediately before the flowslide is not known however. After the flow slide this gradient was decreased to 1 : 25 (about 2°), locally even 1 : 35 (about 1.5°). The depth of the stream channel decreased from over 50 m - N.A.P. to about 30 m - N.A.P. Also it can be seen that immediately after the flow slide scour starts again; after about 18 days some 10 m have been eroded already.

At several locations inside and around the area influenced by the flow slide, density measurements and borings were executed. The borings were laid open in the laboratory and samples were borrowed from them for further investigation. The determination of the maximum and minimum porosities indicated that the maximum porosities range between 46 and 48%, and the minimum porosities between 34 and 36%. Further a number of critical density tests was executed. As an example the results of such a test on sample 29 are shown in figure 2. In this test the octahedral stress was equal to 50 kN/m² and the critical density proved to be about 38.5%. In a test on the same sample with an

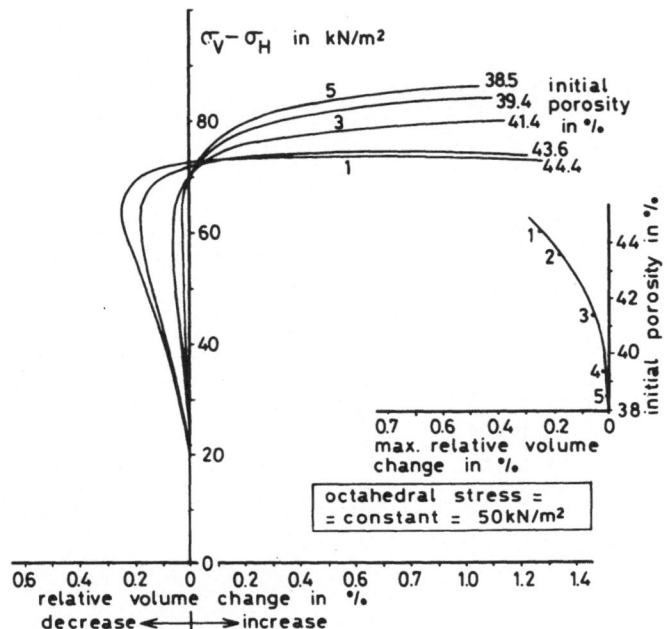

Fig. 2 Results of critical density test on sample 29

octahedral stress of 13 kN/m^2 the critical density amounted to
about 44%, so in that case there was a marked influence of the
stress level at which the test was executed. However, it must
be added that this influence is not always as strong as in this
example.

The results of the density measurements and of the borings,
made in the centreline of the flow slide are shown in figures
3 and 4. In figure 4 also the locations of the investigated
samples is pointed out, just as the results of the determinations
of maximum, minimum and critical densities. In comparison with
the maximum and minimum densities only the in the stream channel
resettled sand is medium packed, all other sandlayers are relatively loosely packed. Owing to the results of the critical
density tests some surface layers are dense when compared with
the results of critical density tests at the relevant state of
stress. However it will be shown that these layers most probably
are resettled layers. From comparison of the density in situ of
the remaining layers with the results of the critical density
tests it follows that these layers all are loosely packed.

The dash and dot line in figures 3 and 4 separates two
layers with different properties. In the upper layer the cone
resistances and the local skin frictions (not shown in figure 3)
are lower than in the layers underneath, while the porosities
are higher. Moreover the grainsize distribution curves of the
samples located above this borderline show similar characteristics
as the grainsize distribution curves of the samples borrowed from

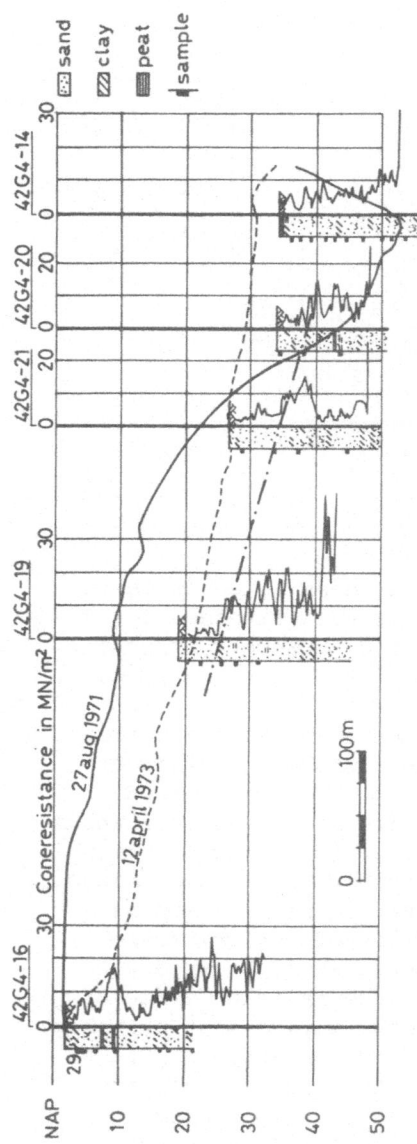

Fig. 3 Results of static cone penetration tests Roggenplaat

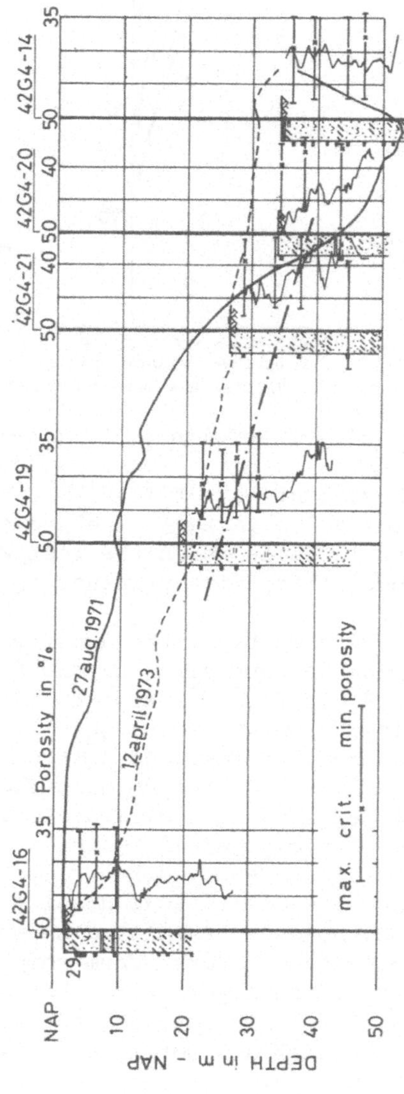

Fig. 4 Results of density measurements in situ Roggenplaat

the in the stream channel resettled sand. So it is very probable
that the sand above the dash and dot line is resettled sand,
coming from sandlayers at a higher level, and that the flow slide
has been initiated at a depth of 38 m - N.A.P.

6. Flow slide Vinkeveen lake

The municipality of Amsterdam needed sand for the extension
of the city. A borrowpit was found at the Vinkeveen lake, situated inside the province of Utrecht. The bottom of the lake consisted of peat, laying on pleistocene sands. In the first instance these sands were dredged to a depth of 30 m - N.A.P., since
1972 the dredging was carried out to a depth of 50 m - N.A.P.
with the help of the suction dredger Weesperkarspel. The water-
level in the lake was kept at a level of 2 m - N.A.P.

Round about the lake a number of islands had to be made and
to be kept up. The purpose of these islands was protection of the
banks of the lake and to serve recreation. In August 1975 dredging

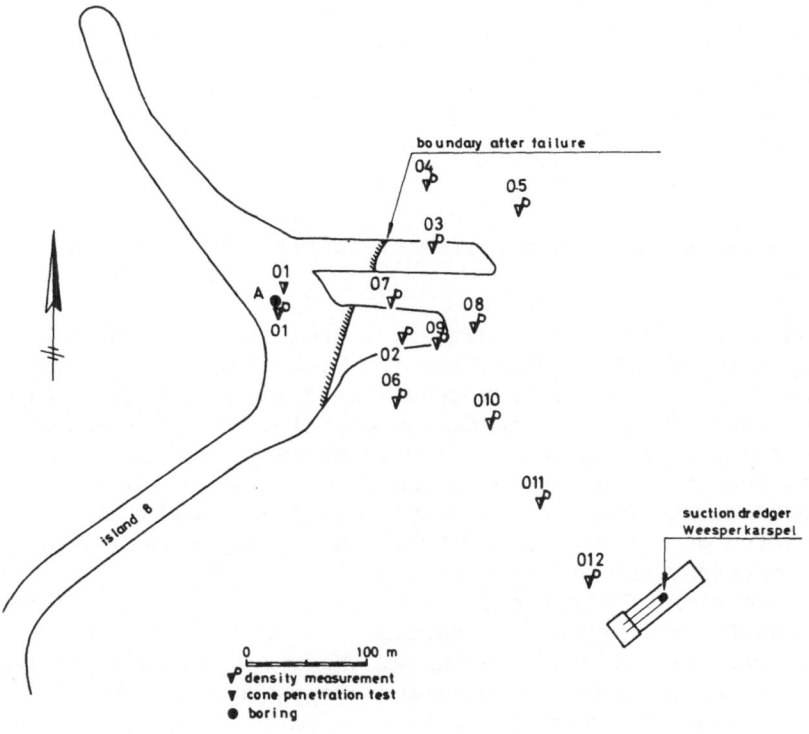

Fig. 5 Plan of site investigations Vinkeveen Lake

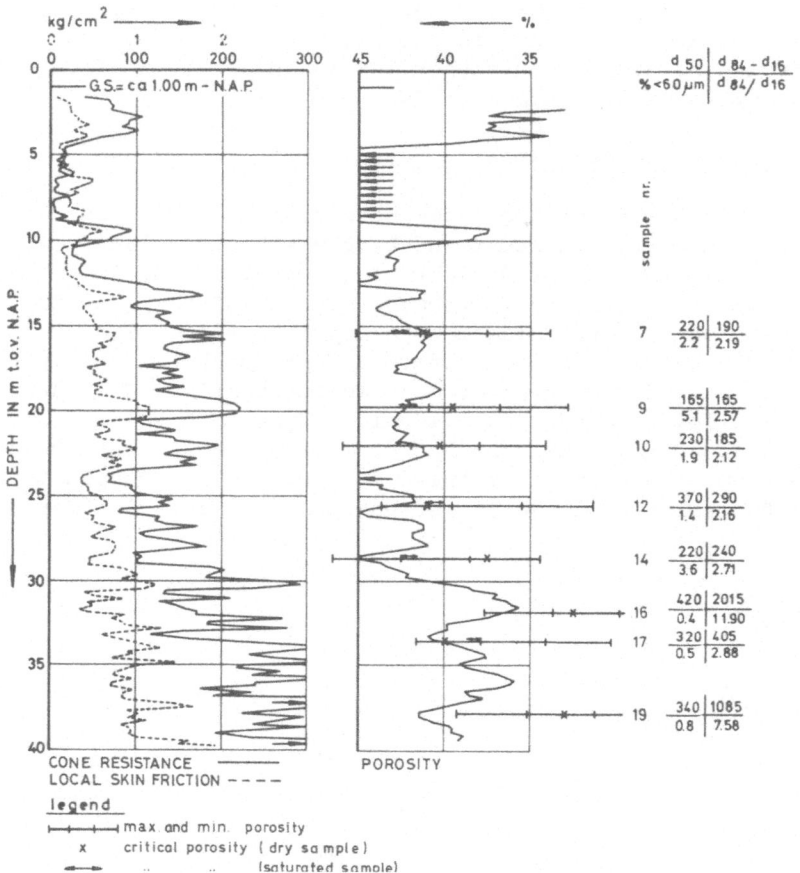

Fig. 6 Results of density measurement in situ 01, executed on
        island 8, Vinkeveen Lake

continued at a distance of about 275 m from island no. 8, with
the local bottomdepth at that time of about 41 m - N.A.P. During
recording of the bottomdepth between the dredger and the island
anomalous records were obtained. The dredging was stopped but
nevertheless part of island no. 8 disappeared. Obviously a flow
slide had occurred, triggered by the dredging works. To verify
this hypothesis a number of density measurements and a boring
were executed on request of the Public Works Departments of the
City of Amsterdam and of the Province of Utrecht.

        The location of the field investigations is shown in figure
5. The density measurements 02 through 012 were located above
water and were executed from a barge. The results of the density
measurement carried out on the island are shown in figure 6. This
figure also shows the results of the determinations of the maximum, minimum and critical densities and some data subtracted
from the grainsize distribution curves. The island consists of

sand to a depth of about 4.5 m - N.A.P., followed by the original
peatlayer to a depth of about 8.5 m - N.A.P. In the sand beneath
the peat two layers can be distinguished to the attained depth.
Between 8.5 and 30 m - N.A.P. lower cone resistances and local
skin frictions were measured than in the deeper layers. The mean
porosity in the sand between 8.5 and 30 m - N.A.P. is about 43%
and in the sandlayer between 30 and 40 m - N.A.P. about 38%.
From the grainsize distribution curves it followed that the upper
layer consists of a finer sand than the lower layer. In accordance
with this the maximum and minimum porosities in the upper layer
amount to about 45 and 35% and in the lower layer to approximately
40 and 30%. Comparison with the data of the density in situ given
above shows that the sand is in a loose state. This is confirmed
by the results of the critical density tests.

   The results of the density measurements 01 through 010 are
summarized in figure 7. It can be observed that especially in
locations 02, 06, 08, 09 and 010 very high porosities are ascer-
tained in the layers between 18.5 and 32 m - N.A.P. Considering
the results of the critical density tests it follows that these
layers are very loosely packed. Attention is called to the fact

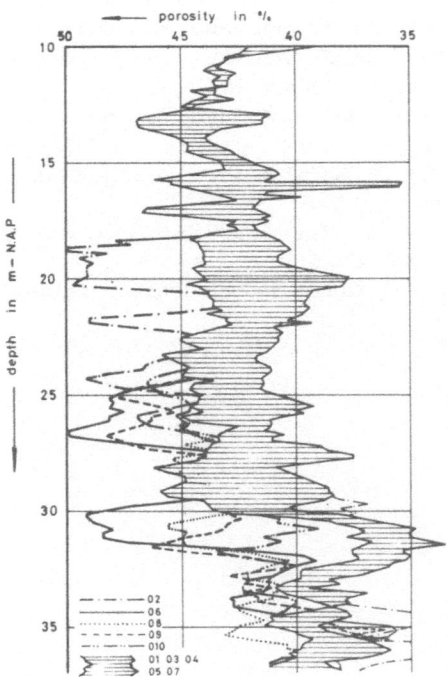

Fig. 7 Summary of results of density measurements in situ 01
    through 010, Vinkeveen Lake

that these locations are situated south-east of the island in the
direction to the dredger.

    Due to the applied method of dredging locally a steeper part
of a slope is created, which moves upward to the bottom of the
lake. This steeper part runs into the very loose sandlayers and
causes liquefaction. So the conclusion of the investigations was
that a flow slide occurred in the observed very loose sandlayers
and that this flow slide was triggered by dredging. A total mass
of about 420.000 m^3 of sand flowed away, a fair big mass in view
of the local circumstances. However this can be explained by an
available storage capacity at a distance of some 450 m. This
storage consisted of a local depth of over 50 m - N.A.P., which
after the flow slide had a depth of 40 to 45 m - N.A.P.

7. Acknowledgements

    The kind permission of the Departments of Public Works of
the Netherlands, of the province of Utrecht and of the city of
Amsterdam to publish the results of these investigations is
gratefully acknowledged.

REFERENCES

Casagrande, A. (1936), Characteristics of cohesionless soils
affecting the stability of slopes and earth fills.
        Contributions to Soil Mechanics, 1925-1940, Boston Society
        of Civil Engineers, October 1940.
Koppejan, A.W., Van Wamelen, B.M. and Weinberg, L.J.H. (1948),
Coastal flow slides in the Dutch province of Zeeland.
        Proc. 2nd Int. Conf. Soil Mech. Found.Engng., Rotterdam,
        Vol. 5, p. 89-96.
Lindenberg, J. and Koning, H.L. (1981), Critical density of sand
(to be published in Geotechnique).
Reynolds, O. (1885), The Dilating of Media composed of Rigid
Particles in Contact.
        Philios. Mag. S5, Vol. 20, No. 127, p. 469 - 481.

SUBMARINE SLUMPING AND MASS MOVEMENTS

ON THE CONTINENTAL SLOPE OF ISRAEL

Gideon Almagor[1] and Gdalyah Wiseman[2]

[1] Marine Geology Division, Geological Survey of Israel
Jerusalem, Israel

[2] Faculty of Civil Engineering, Technion - Israel
Technological Institute, Haifa, Israel

ABSTRACT

The continental margin of Israel has the shape of a lens with
a foreset structure, which narrows from 40-50 km off northern Sinai
to 5 km off southern Lebanon. The lens was formed by accumulation,
since the Pliocene time, of mainly fine clastics derived from the
Nile and transported by the counterclockwise currents of the southeastern Mediterranean. After initial deposition, the detritus was
redistributed over the continental slope and the adjacent deep sea
by slumping. The slumped materials were probably largely transported
downslope in the form of mudflows and debris flows, and occasionally
by sliding of large blocks of sediments. The continental slope
steepens from south (average $3-4^\circ$, maximum $5-7^\circ$) to north (average
$6-8^\circ$, maximum $14-18^\circ$).

A geotechnical study was carried out on a large number of
undisturbed core samples to provide a basis for the quantitative
analysis of the slumping. Angles of internal friction of $24-25^\circ$
measured by drained direct shear tests specify the maximum possible
inclination of a stable infinite slope. The steepest slump-scar wall
slopes are about 20° and indicate that a drained slumping mechanism
is unlikely, and that these slopes have long-term stability. Conso-
lidated-undrained triaxial compression tests and laboratory vane
tests yielded angles of internal friction of $15-17^\circ$ and $c_u/\bar{p}_o$ values
of 0.22-0.91. An analysis of the force equilibrium within the slope
sediments, based on these results, lead to the conclusion that
horizontal earthquake-induced accelerations, as little as 5-6% of
gravity, are sufficient to cause slope failures. Collapse resulting

from liquefaction is unlikely, as the sediments are normally consolidated silty clays with intermediate sensitivity of 2-4.

Mass creep phenomena of apparently stable sediments ($c_u/\bar{p}_o$ =
0.25) are ubiquitous over the sub-horizontal shelf edge and uppermost slope (0.5-2.5^o), where they form undulating ground. Analysis
of the static strength properties of the sediments, together with
the stability of the slope, suggests that these phenomena result from
long-term deterioration in shear strength of the sediments due to
repeated loading effects. In view of the great water depth (above
80 m) and the mild oceanographic conditions in the region, this
weakening could be caused by accumulating effects of earthquakes. This
is presently being investigated by studies of the cyclic load
properties of the sediments.

INTRODUCTION

Seismic surveys in the southeastern Mediterranean revealed that
slumping and other mass movement processes in the continental margins
are rather ubiquitous. The slumping phenomena recorded range from
gigantic, deep-seated compound, rotational slumps to block and slab
sliding, debris flows and mudflows, and mass creep. Whereas the early
studies of the submarine slope stability were merely academic, aiming
to define the role of slumping in the deposition and distribution
patterns of the sediments in the southeastern Mediterranean (Almagor,
1964; 1967), recent studies have become of great practical value
concerning possible earthquake-triggered hazards, especially since
the mid-70's, when assessment of the suitability of the Nitzanim site
for construction of a nuclear power plant was undertaken (Garfunkel
et al., 1979). Interest in potential natural hazards, that may be
caused by active tectonism in the offshore of Israel, is steadily
growing, as the coast of Israel is its most populated area where
large shore installations, such as power stations, heavy industrial
plants and harbours, are located.

This paper summarizes the submarine slope stability studies of
the continental slope that were carried out by the authors since 1972.
The analyses herein presented are based on the static load properties
(consolidation and shear strength characteristics) of the slope
sediments. The geological interpretation and the results of the
slope stability analyses were published in several papers (Almagor.
1978; 1979; 1980; Almagor and Wiseman, 1977; Almagor and Garfunkel,
1979).

Outline of a programme of slope stability analysis based on
cyclic load properties of the continental shelf and upper slope
sediments are also presented. This programme, which is now being
carried out together with Dr. S. Frydman of the Faculty of Civil
Engineering, the Technion - Israel Technological Institute, aims to
study the cumulative earthquake effects on the stability of the

slope sediments.

THE CONTINENTAL MARGIN OF ISRAEL

The continental margin of northern Sinai and Israel forms the eastern extension of the huge Nile delta and its submarine cone (Fig. 1). Its shape is of a thick lens with a foreset structure that narrows from southeast to north. The continental shelf narrows from 60 km off northern Sinai to 20 km off central Israel to 5 km off northern Israel and Lebanon. The continental slope extends to a depth of 900-1,000 m, and narrows from 50 km in the south to 10 km in the north. The average gradient of the slope is 3° in the south, with maximal inclinations of $5-6^{\circ}$ at a depth of 400-450 m, and $6-8^{\circ}$ in the north, with maximal inclinations of $14-18^{\circ}$ at depths of 400-450 m and 700-800 m. Off southern Israel, a shallow (50-70 m) depression, 15 to 40 km wide, marks the base of the continental slope (Fig. 2). It widens northward until its identity is lost off central Israel.

The continental margin is made up of large volumes of Nile-derived sediments, that have been deposited continuously since early Pliocene times, over the entire southeastern Mediterranean (Nachmias, 1969; Venkatarathnam and Ryan, 1971; Nir and Nathan, 1972). These sediments were transported by the counterclockwise southeastern Mediterranean contour currents along the northern Sinai-Israeli coasts, and redistributed seaward by periodic downslope slumping processes, thus forming the basically smooth and rounded configuration of the present continental margin (Fig. 2). This is well attested to by the numerous slump scars, both buried and recent (Figs. 2, 3, 4), the numerous large debris flows on and within the continental slope (Figs. 2, 5), and by the sedimentary micro-structures observed in the sediment cores (Einsele and Werner, 1968; Almagor, 1976; 1978; Maldonado and Stanley, 1976).

The continental slope is marked with numerous slump scars in its central and lower sections (Fig. 6). The scars have widths of up to 3 km, downslope lengths of up to 4 km, and by comparison with the adjacent topography, subseafloor depths of up to 45 m (Figs. 3, 4). Their shapes suggest several successions of slumping activity. The recent slump scars appear "fresher": the subseafloor depths are large (40-45 m), the walls are steep (20°), and they have well-defined boundaries and frequently cut older scars. They are located at water depths of 400-450 m, where the continental slope attains its greatest angle. The "freshness" of these slump scars is further corroborated by recent fault planes on the slope just above them. The latter dip 50° from the vertical at or immediately below the seafloor, becoming flatter with depth, possibly dying out in the bedding (Fig. 4). These faults probably indicate disruption of the force equilibrium and the release on stresses in the sedimentary mass upslope. The older slump scars are shallower (15-30 m), and have less clearly defined boundaries. They are often only partly preserved, and are located

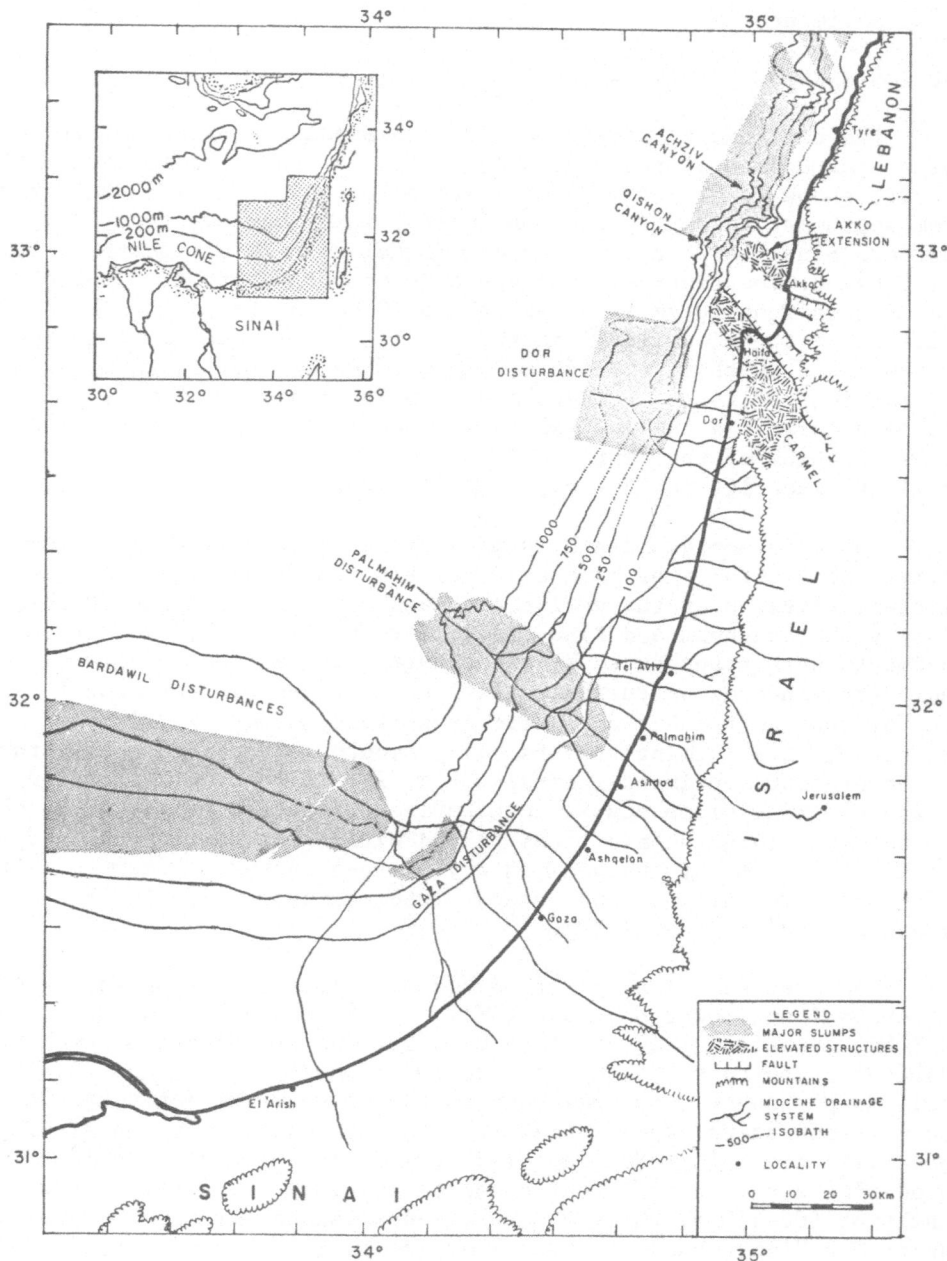

Figure 1.   Main physiographic and structural elements in the
continental margin of Israel (from Almagor and Garfunkel,
1979).

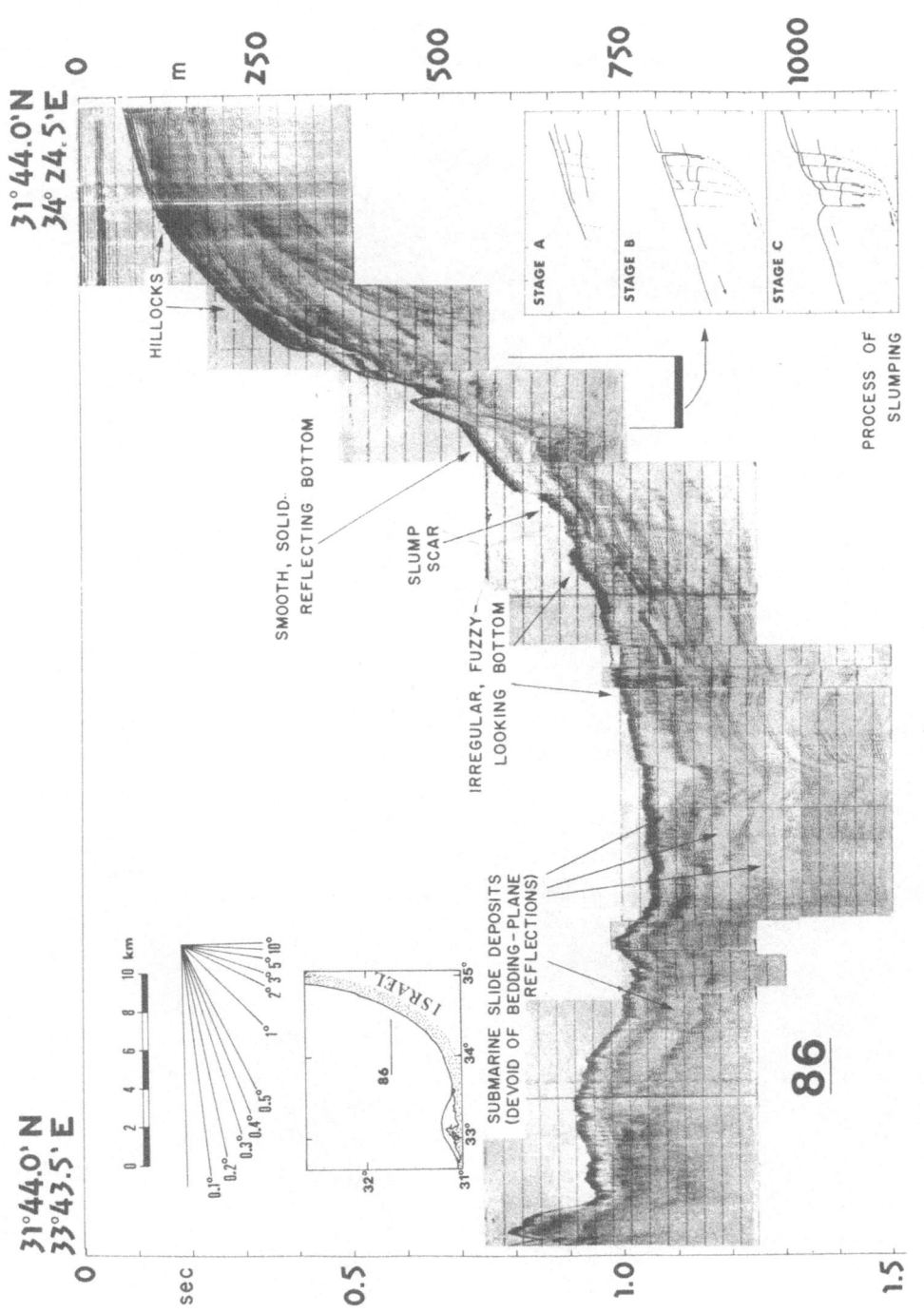

Figure 2. A seismic profile perpendicular to shore off southern Israel. Vertical exaggeration = 46x (modified after Neev et al., 1976).

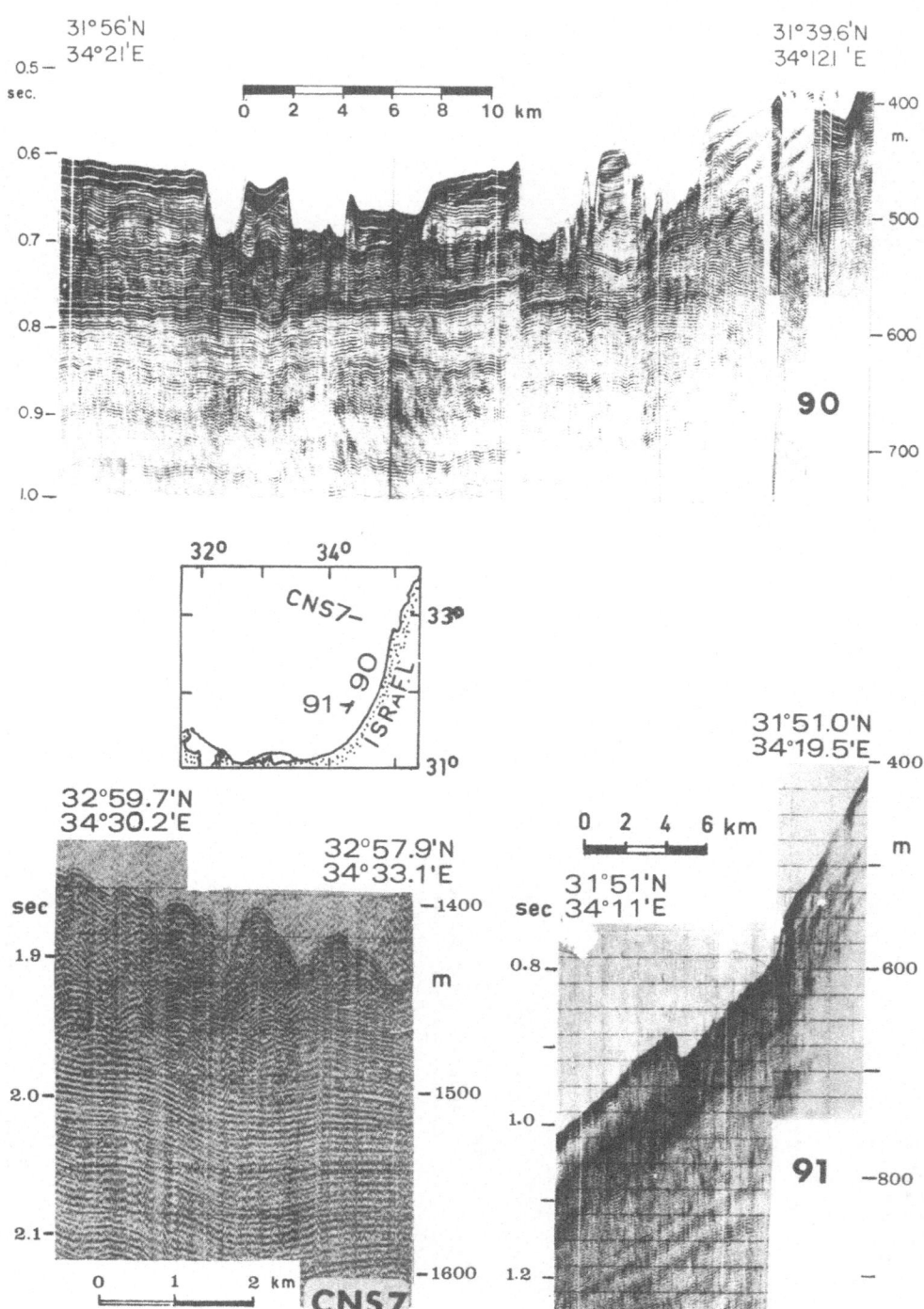

Figure 3.   Slump scars and debris. Vertical exaggeration = 46x.

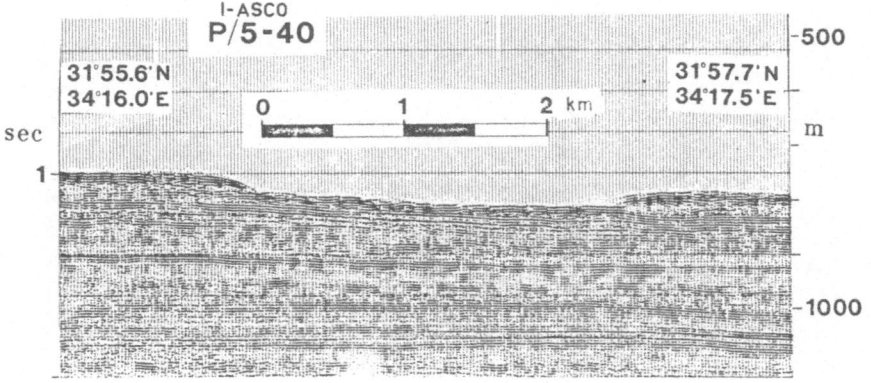

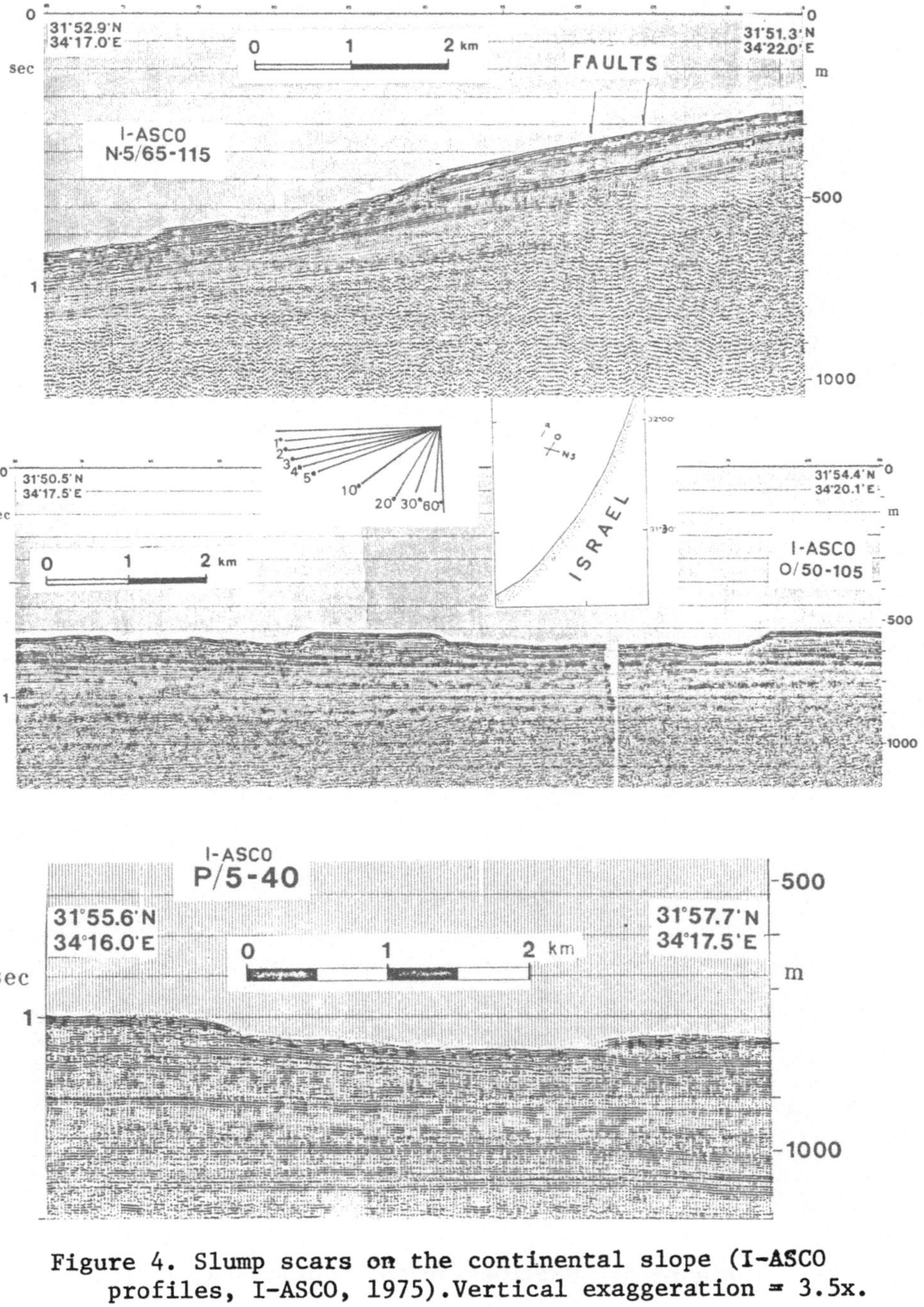

Figure 4. Slump scars on the continental slope (I-ASCO
profiles, I-ASCO, 1975).Vertical exaggeration = 3.5x.

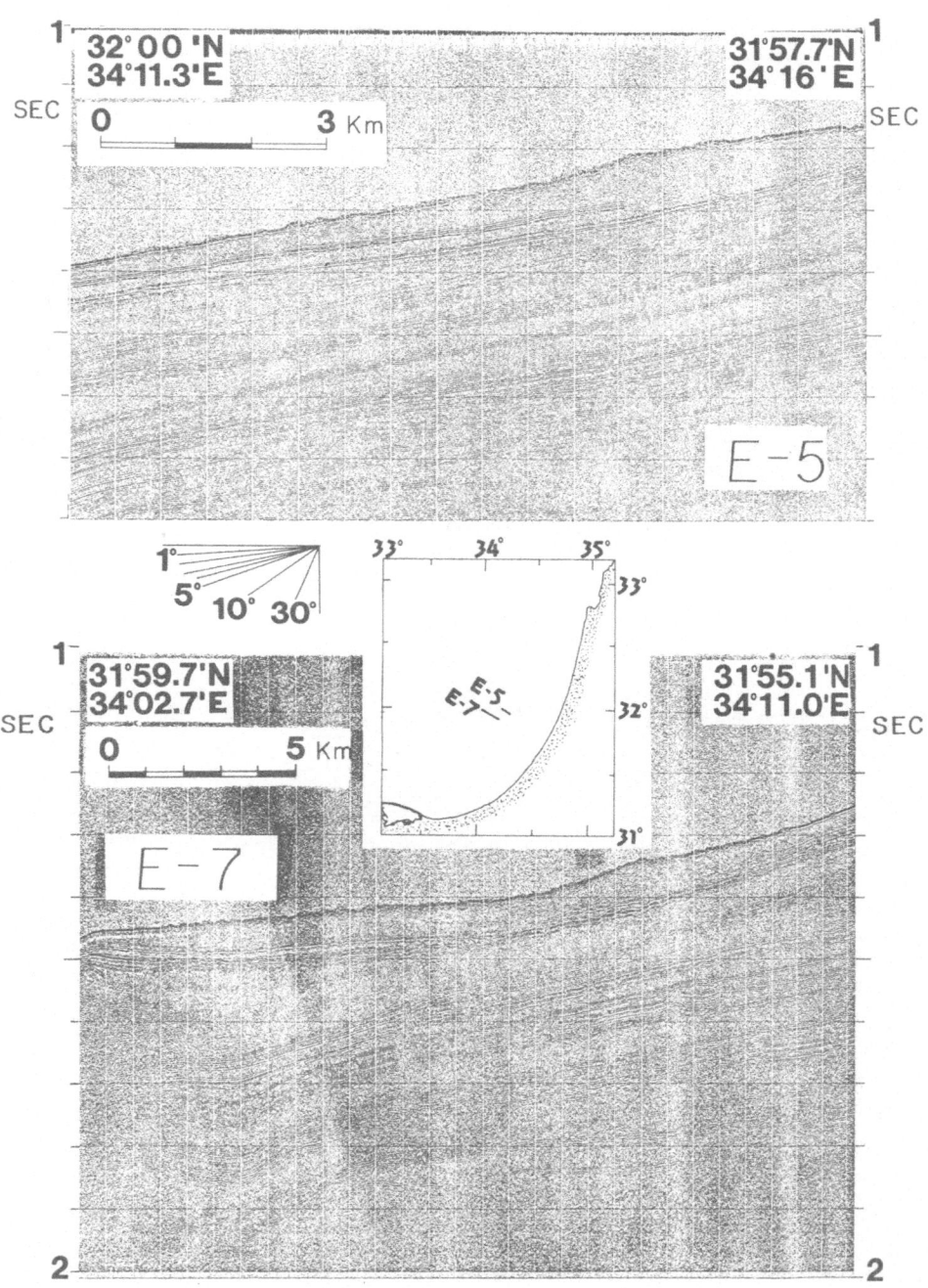

Figure 5.   Mudflows in the continental slope.
Vertical exaggeration = 3.5x.

in the lower sections of the continental slope (Fig. 6).

The slumping materials were largely transformed into suspensions that flowed downslope in the form of debris flows. The debris flows are readily identified on the seismic profiles by the acoustically fuzzy appearance of the bottom beneath the slump scars when on the seafloor, and by their transparent, devoid-of-bedding-plane reflections, which are interbedded in-between well stratified reflectors when buried (Figs. 3, 5). The debris flows, which are found along wide portions of the lower slope, fill the shallow depression at the base of the slope, and extend into the deepsea proper beyond (Figs. 2, 5). Core samples obtained from the debris flows yielded large amounts of silty materials, and sizable chunks of yellowish continental loess with carbonate nodules. Radiographs revealed graded microbedding, which indicate deposition from flows.

Regression of the sea during Pleistocene glacial times, and ingressions during interglacial time intervals resulted in the accumulation of quartz sands on the present-day coastal plain, continental shelf, and the uppermost continental slope (down to 300-350 m depth at present). Longitudinal sand dunes demarcate the ancient shorelines of the regressing sea. Fossilization of these dunes during periods of subaerial exposure periods formed a physiographic pattern of subparallel longitudinal carbonate-cemented quartz sandstones separated by longitudinal depressions, filled with continental deposits (loess, red sandy soils, marsh deposits, and so forth) (Itzkhaki, 1961; Issar, 1968; Neev et al., 1966; 1976). In the marine regime to the west, continuous sedimentation of terrigenous silty clays took place. The last (Holocene) ingression that started some 13,000 years ago (Milliman and Emery, 1968) covered most of the area. The sedimentation of silty clays that followed, nearly totally covered the present-day smooth topography of the continental terrace. The rate of sedimentation has not exeeded 1 m in 1,000 years (Reiss et al., 1971; Nir, 1973; Horowitz, 1974).

Mass creep and small rotational slumping are widespread along the entire shelf edge at 80 to 150 m water depth (about 4 km wide strip, $0.5-1°$ inclination), and the uppermost continental slope, at water depths of 200-325 m (4-5 km wide shore-parallel strip, $1-2.5°$ inclination)(Figs. 6, 7). These strips are characterized by undulating topography, made up of a system of low amplitude, elongated hillocks that are separated by elongated depressions, and by numerous shallow horizontal clefts. This topography is in outstanding contrast to the very smooth, rounded topography of the shelf and upper slope of Israel. The height of these hillocks range between 5 and 15 m, and their widths range between 100 and 300 m, and occasionally 400 m. The clefts are 150-200 m wide and reach 20-25 m depth. The horizontal dimensions of both the hillocks and the clefts are less than 1 km long. Owing to its low relief this hillocky topography is only partially expressed in the bathymetric chart (Fig. 6).

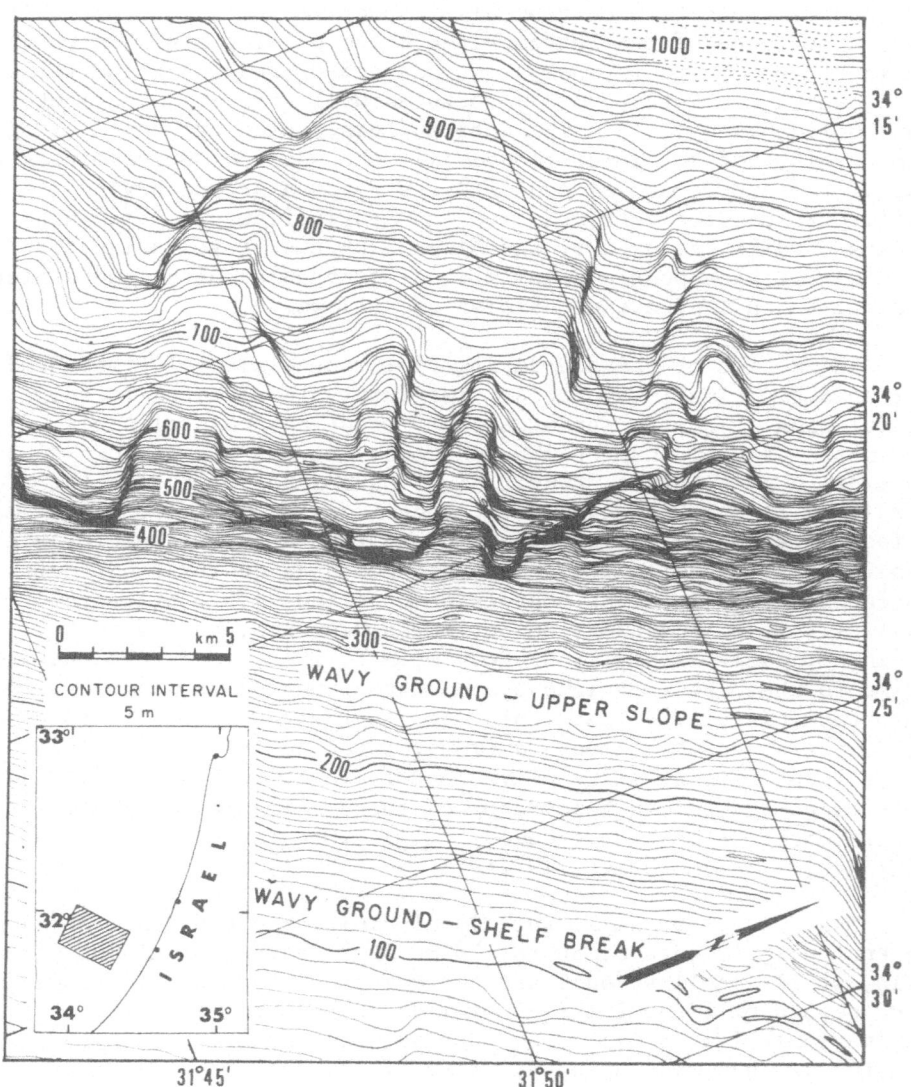

Figure 6. Area affected by slumping and creep of surficial sediments
    on the continental slope off southern Israel (from Almagor and
    Wiseman, 1977).

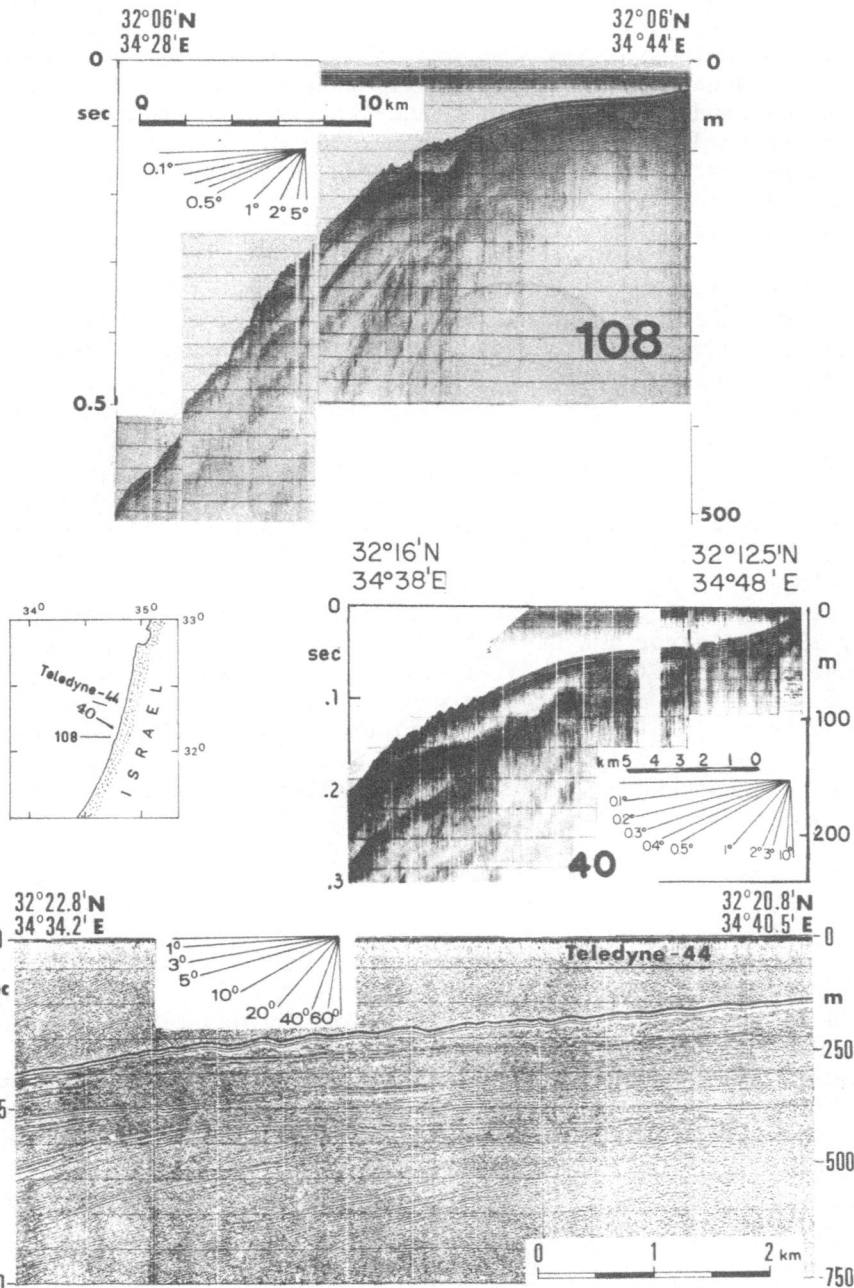

Figure 7. Mass creep and small rotational slumping at the shelf edge
and upper continental slope. Subbottom reflectors are not affected
by crumpling of the surficial sediments. Vertical exaggeration = 46x
in profiles 40 and 108, and 4x in profile Teledyne-44.

These hillocks and clefts are surficial features, as the underlying reflectors of the muddy sediments remain horizontal (Fig. 7). The seismic profiles also show that the hillocky bottom abuts against the underlying, occasionally outcropping rocky ridges in the shelf edge and the upper slope, and that the clefts are located next to the seaward facing-slope of these ridges. These suggest mass creep downslope that is blocked by the ridges, and small scale rotational downslope with the seaward-facing slopes of the rocky ridges acting as failure planes (Fig. 7).

The Pliocene-Quaternary sedimentary lens attains a thickness of 1-2 km under the continental slope and outer shelf, and thins gradually seaward and landward. It overlies seaward-thickening evaporitic series, that fill the lower portions of erosion drainage channels developed during the desiccation of the Mediterranean, and the ensuing exposure period of its margins, that took place during the Messinian (7 to 5 million years before present) (Hsü et al., 1973). The thick sedimentary overload of the present-day continental slope atop the relatively thick, mobile evaporites embedded within these ancient drainage channels caused instability that resulted in downslope flow of the evaporites, and subsequent rotational slumping and tilting of huge slabs of the overlying sediments (Garfunkel et al., 1979; Almagor and Garfunkel, 1979; Almagor, 1980). These large downslope slumps (termed "disturbances" by Garfunkel et al., 1979) vastly disturbed and distorted appreciable portions of the originally smooth northern Sinai-Israeli continental margin (Figs. 1, 8). No quantitative analysis of these slump features was attempted, since essential geotechnical data are not available; therefore, they will not be discussed in this paper.

GEOTECHNICAL TESTING

A full-fledged geotechnical testing programme was carried out on 54 large diameter (8 cm), relatively long (3-5 m) undisturbed core samples from the continental margin of Israel (from water depths of 30 to 1,000 m) to provide a basis for the quantitative analysis of the slumping phenomena. Special coring procedures and laboratory treatment were used to minimize the degree of disturbance (Almagor and Wiseman, 1977); therefore, the geotechnical properties measured are considered representative of the natural conditions within the subbottom sedimentary column. The results were described in detail by Almagor and Wiseman (1977) and Almagor (1978), and therefore will be only briefly summarized here.

Mineralogically the sediments are characterized by their high content of montmorillonite (60-80%), some kaolinite (20-40%), low content of illite (less than 15%), lack of chlorite, and low carbonate content (0-9%). Their mineralogy is characteristic of the Nile-transported sediments of the southeastern Mediterranean Sea and proves their Nile origin (McCoy, 1974; Nir and Nathan, 1972; Venka-

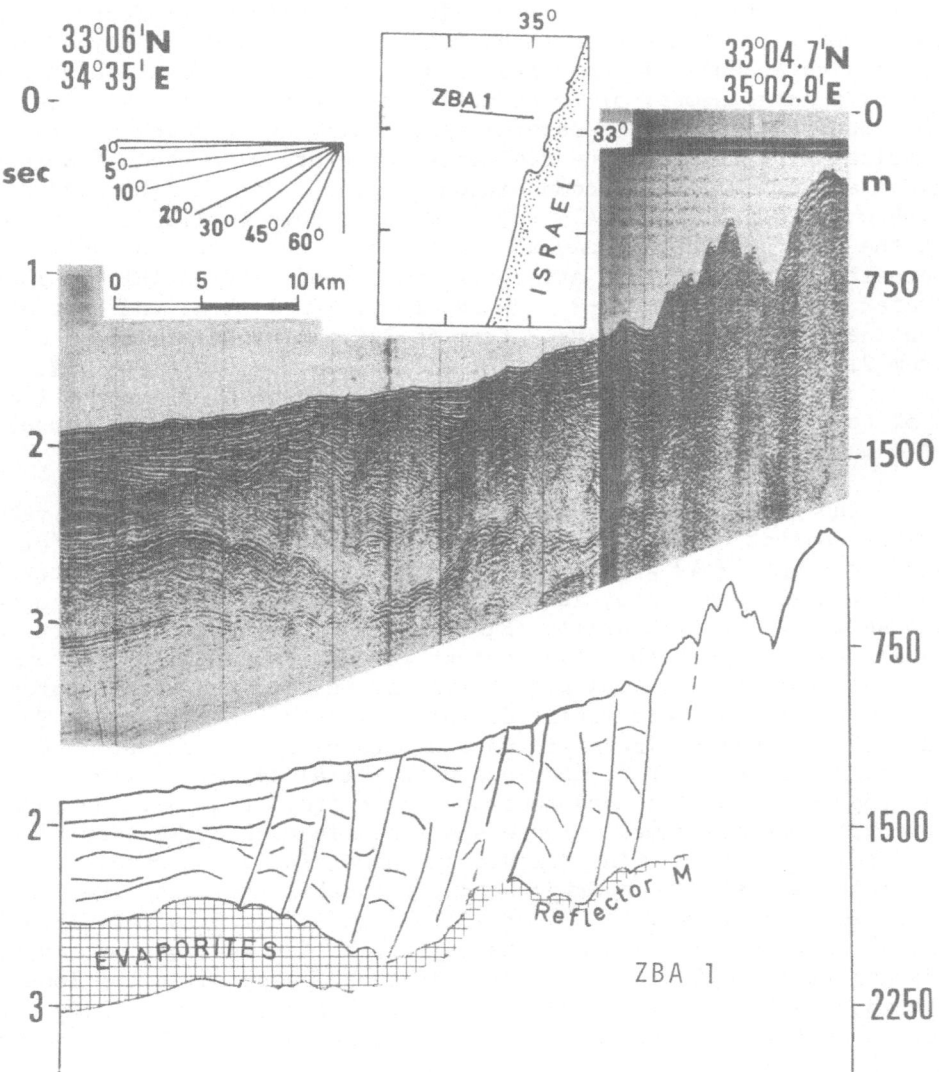

Figure 8. Deep-seated rotational slumping of huge Pliocene-
Quaternary sediment slabs over downslope-flowing mobile
Messinian evaporites. Vertical exaggeration = 12.5x
(profile ZBA-1 modified from Ben-Avraham, 1978)

tarathnam, 1971). Grain specific gravity values of the sediments
range from 2.69 to 2.82 with an average of 2.78. These values are
within the range characteristic of marine clayey and silty-clayey
sediments (Boswell, 1961; Richards, 1962). The clay-size fraction
(less than 2 μm) of the sediments increases with water depth from
about 30% at 30 m depth to about 60% at 350-400 m depth beyond which
it remains essentially constant. However, the clay fraction at the
top of a large number of cores obtained from the deeper sections of
the continental slope is surprisingly small, not exceeding 20%. The
physical properties of the sediments correlate with their mineralogical and granulometric composition (Fig. 9). Unit weight decreases
with water depth from 1.41-1.61 g/cm^3 nearshore to 1.25-1.37 g/cm^3
at the base of the continental slope, while the water content,
porosity and void ratio, Atterberg limits and compressibility increase
(water content - from 75-95% nearshore to 128-143% at the base of the
continental slope; porosity - from 68-72% to 78-80%; void ratio -
from 2.1-2.6 to 3.5-4.0; liquid limit - from 60-80% to 85-115%;
plasticity index - from 35-50% to 65-80%; compression index - from
0.52 to 1.30). The shear strength of the sediments also increases
with water depth.

The geotechnical properties that were measured (average values:
water content - 106%; porosity - 74.5%; unit weight - 1.45 g/cm^3;
shear strength - 42 g/cm^2) are characteristic of the southeastern
Mediterranean sediments, and are similar to those obtained by others
for sediments of Nile origin (Einsele, 1967; Keller and Lambert,
1972). In each core the clay content remains essentially constant,
the unit weight increases with burial depth, and the water content
(and porosity and void ratio) decreases with depth of burial. The
rate of unit weight increase and the rate of water content (and
porosity and void ratio) decrease with burial depth are steep within
the top 40-50 cm, and more moderate thereafter (Fig. 9). The shear
strength of the sediments increases linearly with burial depth (Fig.
9). Because of their fine texture and mineralogical composition, the
sediments are almost impervious: the permeability values that were
measured range from $1.4 \cdot 10^{-7}$ to $25 \cdot 10^{-7}$ cm/sec ($1.5 \cdot 10^{-4}$ to $25.9 \cdot 10^{-4}$
darcy). Permeability values measured in the top sections of the cores
are generally 5-10 times greater than those measured in the lower
sections, 3-4 m below the seafloor. The change of permeability
versus burial depth reflects the consolidation of the accumulating
sediment column.

The shear strength of the sediments was measured by means of
drained direct shear tests, undrained laboratory miniature vane tests
and triaxial consolidated undrained compression tests with pore-water
pressure measurements. Angles of internal friction measured by direct
shearing under drained conditions are $\bar{\phi}_d$ = 24-25°, designating the
maximum possible angle of a stable infinite slope. The angles of
internal friction measured by undrained triaxial compression tests
are ϕ_{cu} = 15-17°. The relation of undrained shear strength of the

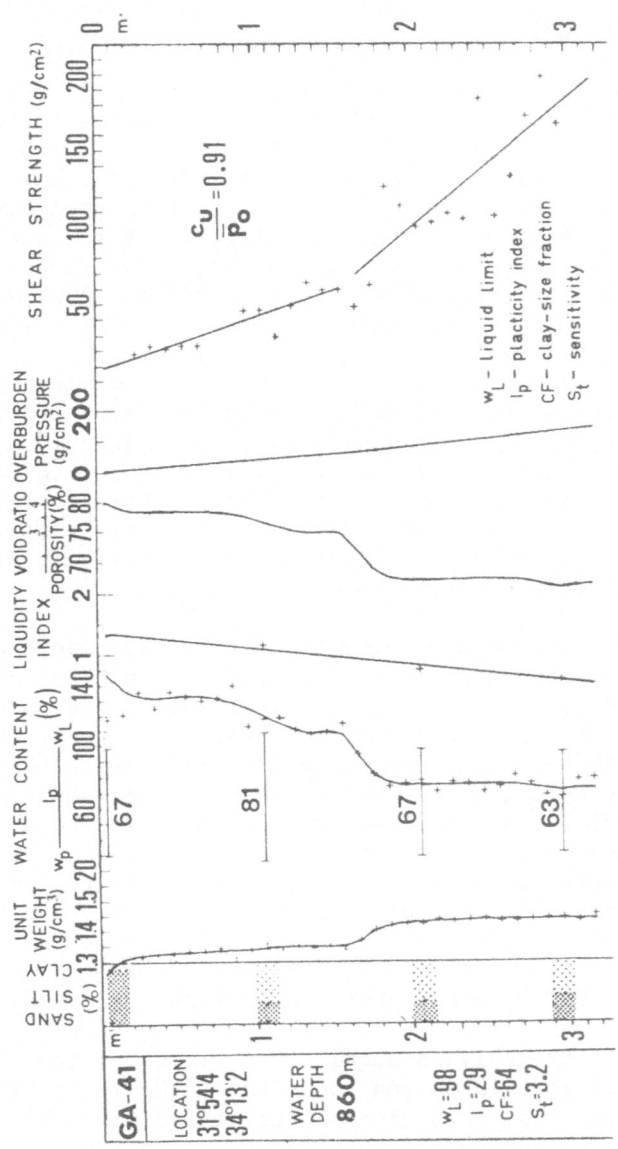

Figure 9. Geotechnical data, Core CA-41. Overconsolidated sediments, buried some 1.6 m below the seabottom, are well identified by their geotechnical properties (from Almagor, 1978).

sediments to the effective overburden pressure $c_u/\bar{p}_o$, derived from
the vane tests, range from 0.24 nearshore to 0.76–0.91 at the base
of the slope (Fig. 10). The lowermost limit of the $c_u/\bar{p}_o$ values
plotted versus water depth were used in the slope stability analyses,
as failure is likely to occur in the sedimentary layers whose shearing
resistance is minimal.

The measured small preconsolidation pressures (i.e., the highest
pressure to which the sediment samples were subjected in their geological past) that are correlatable to the effective overburden
pressures in the sediments, the regular increase of the sediment
shear strength with burial depth which starts from practically zero
at the sea-bottom (Fig. 9), and the sensitivity of the sediments
(i.e., the ratio of the shear strength of an undisturbed sediment
sample to its shear strength after remoulding) of 2–5 suggest that
the sediments are normally consolidated despite their relatively high
deposition rate. This was confirmed by calculations of the degree of
consolidation of the sediments in the continental margin of Israel
and by calculations of the hydraulic gradients that exist in the
sediments. The degree of consolidation is practically 100%, and the
hydraulic gradients are rather low, ranging from 0.02 to 0.07. As
the sediments are normally consolidated, it is therefore possible to
extrapolate reliably properties measured in the core samples from the
uppermost 5 m of the sedimentary column to even greater depths
provided the sediments are homogeneous.

Deviations from the general pattern presented above are common
in a number of core sections (Fig. 9). Stiff silty clays of identical
mineralogical and textural composition were sampled in three cores.
Their high values of preconsolidation pressure (0.470–0.600 kg/cm^2)
and shear strength (110–200 g/cm^2) clearly indicate that they are
overconsolidated. As the recent and subrecent slope sediments have
never been subaerially exposed, which might have led to their desiccation, their overconsolidated state must be related to some sediment removal process, such as slumping. It is estimated that these
sediments supported a sedimentary column that was 12–18 m thick
(Almagor, 1976).

SLUMPING - ANALYSIS OF STATIC LOAD PROPERTIES

Slope stability analysis requires knowledge of the slope
topographic profile,the shape and location of the major slip surface,
the water-pressure conditions at the time of failure, the sediment
strength parameters, and the sediment density. In the case of submarine slumps, especially those that occur in deep water, the
necessary information is seldom available. Consequently, it is
necessary to extrapolate the seabottom and nearbottom sediment
geotechnical properties measured either in place or evaluated by
laboratory tests on sediment samples.

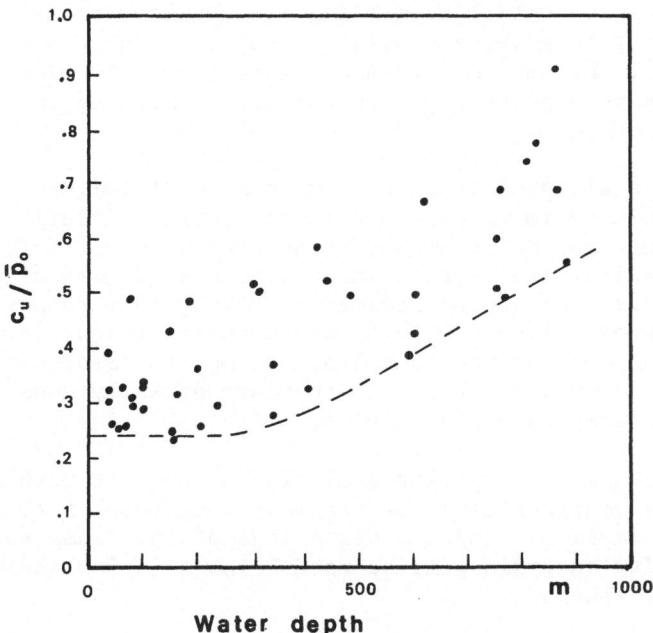

Figure 10. Relation $c_u/\bar{p}_o$ values to distance from shore and water depth. Dashed lines mark the lowest $c_u/\bar{p}_o$ boundary.

The following discussion is adopted from Almagor and Wiseman (1977). It is based on Morgenstern's (1967) work, in which possible mechanisms of submarine slumping were defined and analysed.

Stability of the sediments on a given slope basically depends on the shear strength of the sediments and the rate of increase of this strength with burial depth. A slope fails when the average shearing stress along the potential surface of sliding becomes equal to the average shearing resistance along this surface.

The shear strength of the sediments depends on the conditions of drainage during the shear and on the time required for draining. It is therefore essential to distinguish between *drained slumping*, which is a long-term process owing to the time lag between the rapid loading (i.e., the sedimentary accumulation and the slow dissipation of the excess pore pressure), and *undrained slumping*, which is a short-term process.

Collapse slumping is a third mechanism of failure, associated with metastable sediments where failure occurs initially under drained conditions, but the deformations associated with the failure bring about a large increase in pore-water pressure, drastically diminishing the shear resistance of the sediments, making them temporarily assume the character of a liquid with a low viscosity (liquefaction) and flow downslope. Collapse of the continental slope sediments in the investigated area is most unlikely, as they are normally consolidated silty clays with intermediate sensitivity.

Infinite slope stability analysis is the best method for analyzing the continental slope sediments, because of their parallel bedding, gentle slope, and the dimensions of the slump scars. Finite slope stability analysis will be used below only for analyzing the scarps of the slumps.

Drained Slumping

The stability of a given slope, whose angle is α is dependent on the equilibrium between the weight of the sedimentary column that rests on the surface of failure $\bar{W} = \gamma bh$, the normal effective force $\bar{P}$ on the surface of failure, and the shear resistance s of the sediments along the surface of failure $L = b/\cos\alpha$ (Fig. 11). During drained slumping, excess pore pressure does not develop; therefore, the normal effective force resulting from the weight of the sedimentary column is

$$\bar{P} = \bar{W}\cos\alpha = \bar{\gamma}hL\cos^2\alpha \tag{1}$$

the shear force is

$$T = \bar{W}\sin\alpha = \bar{\gamma}hL\sin\alpha\cos\alpha \tag{2}$$

and failure occurs when the shear force equals the shear resistance

$$\bar{\gamma}hL\sin\alpha\cos\alpha = \bar{c}L + \bar{\gamma}hL\cos^2\alpha\tan\bar{\phi} \tag{3}$$

The critical height of the sedimentary column h_c (i.e., the maximal thickness of a sedimentary column that can be supported before sliding of the sedimentary mass is initiated) is therefore defined by

$$h_c = \bar{c}\{\bar{\gamma}\cos^2\alpha(\tan\alpha - \tan\bar{\phi})\}^{-1} \tag{4}$$

From this equation, it is seen that even small values of cohesion $\bar{c}$ of the sediments are sufficient to support a thick sedimentary column. The maximum possible inclination of a stable slope is defined by

$$\tan\alpha = \tan\bar{\phi} + (\bar{c}/\bar{\gamma}h)\sec^2\alpha \tag{5}$$

The limited thickness to which a sedimentary layer can accumulate on a slope when the slope angle α is greater than the angle of internal friction $\bar{\phi}$ of the sediment is depicted in Figure 12. Line BCE depicts the maximal shear resistance of the sediments (the Mohr rupture envelope), and line OAE depicts the normal stress $\bar{\sigma}$ acting on them. As long as $\bar{\sigma}$ is within the area bounded by the line OE, the shear stress is smaller than the shear resistance of the sediments, $\tau < s$, and the slope is stable. For example, the normal stress OA generates the shear stress DA, which is smaller than the shear resistance of the sediments DC. At the critical depth h_c the normal stress OE effects the shear stress EF, which equals the shear resistance of the sediments, and failure occurs.

If the sediments are normally consolidated clays and $\bar{c} = 0$, failure occurs when the angle of internal friction of the sediments $\bar{\phi}$ is equal to the angle of the slope α:

$$\tan\alpha = \tan\bar{\phi} \tag{6}$$

As long as the angle of internal friction $\bar{\phi}$ is greater than the angle of the slope α, sediment accumulation does not disturb the stability of the slope. In such cases, lines BC and OA in Figure 12 will never cross. In the investigated area, the angles of internal friction under drained conditions range $\bar{\phi}_d = 24-25^\circ$, and are much larger than tha steepest angles of the continental slope, $\alpha = 6^\circ$, even if it is considered that the the $\bar{\phi}_d$ values evaluated in much longer time shear tests were smaller than those measured. The measured $\bar{\phi}_d$ values are also larger than the angles of the slopes of the scarps of the slump scars in the investigated area ($\alpha = 20^\circ$). These friction angles indicate that the drained condition for slumping mechanism in the investigated area is unlikely.

Undrained Slumping

Undrained shear failure causing slope instability can be effected by rapid changes in slope geometry (undercutting or over-steepening of the slope), fluctuations in pore-water pressure or by

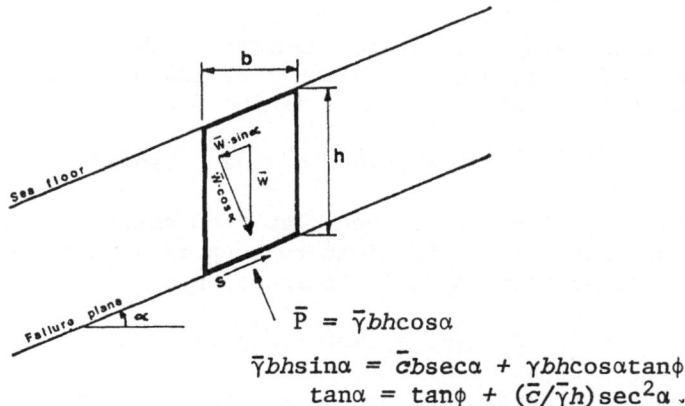

$$\bar{P} = \bar{\gamma}bh\cos\alpha$$
$$\bar{\gamma}bh\sin\alpha = \bar{c}b\sec\alpha + \gamma bh\cos\alpha\tan\phi$$
$$\tan\alpha = \tan\phi + (\bar{c}/\bar{\gamma}h)\sec^2\alpha .$$

Fig. 11. Equilibrium of infinite slope under drained conditions.

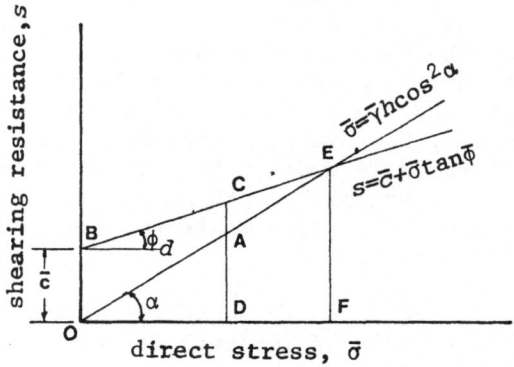

Fig. 12. Mohr diagram illustrating the limited thickness to which a sedimentary layer can accumulate on a slope when the slope angle α is greater than the internal friction angle $\bar{\phi}_d$ of the sediment.

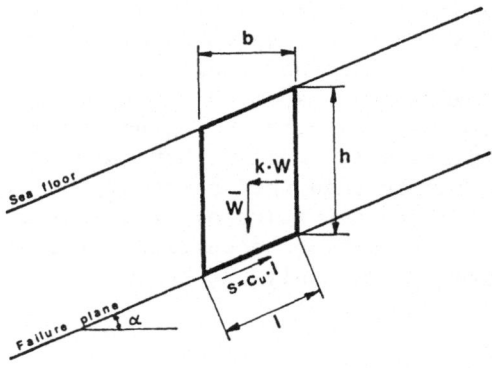

Fig. 13. Equilibrium of infinite slope under undrained conditions.

accelerations associated with earthquakes and blasts.

Increase of the sedimentary load on sediment layers in which drainage conditions are poor may generate excess pore pressure u in the sediments. When excess pore pressures are present in the sediments, the shear strength at a particular depth is reduced, and the depth to a potential shearing plane is similarly reduced. The slope is stable as long as the total angle of internal friction ϕ is greater than the inclination α of the infinite slope (assuming that the pore pressures developed during the laboratory shear tests are identical to the *in situ* pore pressures developed in the sediments). As the depositional rate is increased for a given sediment, pore pressures may be generated in the sediments at much greater a rate than they can dissipate, and failure may occur even at very small slope angles.

The criterion for the stability of the slope is expressed by the force equilibrium between the shear force $\bar{W}\sin\alpha$, generated by the weight of the sedimentary column, and the shear resistance of the sediments $c_u L$ along the potential failure plane (Fig. 13):

$$c_u L = \bar{W}\sin\alpha \tag{7}$$

Resolving this equation for the slope angle

$$1/2 \ \sin 2\alpha = (c_u/\bar{\gamma}h) \simeq (c_u/\bar{p}_o) \tag{8}$$

This mechanism is likely to be significant at deltaic deposition rates, which are required to produce a substantial degree of under-consolidation in clayey sediments. Such submarine slumps were indeed described and analyzed in the deltas of the Mississippi (Shepard, 1955; McClelland, 1956; Terzaghi, 1956; Fisk and McClelland, 1959), the Frazer (Mathews and Shepard, 1962; Terzaghi, 1962), and the Magdalena (Heezen, 1956) rivers. However, in the investigated area, the total angles of internal friction that were measured are ϕ = 15-17°, and are much greater than the steepest slope in the continental slope in the investigated area (α = 6°). The $c_u/\bar{\gamma}h$ values evaluated from Eq. (8) for α = 6° are four to eight times larger than the $c_u/\bar{p}_o$ values evaluated for the sediments in the investigated area (Fig. 10).

Excess pore pressures that can lead to failure may be generated by the passage of large ocean waves over soft, underconsolidated sediments (Henkel, 1970), and by the exposure of the seabottom that follows rapid drawdown of the sea level in areas subjected to high tides (Terzaghi, 1956). The water depth (400 m and more), the small amplitude of the tide, and the normal consolidation of the sediments in the investigated area exclude these mechanisms as the causes for slumping.

Indications of possible undrained slumping were found only in several places:

1. The bottom sediments of core GA-38 collected from a slope steeper than 20° are overconsolidated ($\bar{p}_c = 0.600$ kg/cm^2). This preconsolidation pressure could be indicative of slumping of an approximately 12 m thick layer of silty-clayey materials.
2. Buried overconsolidated sediments were found in several sediment cores, as shown in Figure 9. They indicate undrained slumping that exposed them several thousand years ago (considering a rate of deposition of less than 1 m in 1,000 years) after which they were re-covered by modern sediments.
3. Sediments similar to those described above were cored in several places inside the submarine canyon off Ahziv, northern Israel, where slope angles steeper than 20° were measured (Nir, 1973). Although these samples were not subjected to geotechnical testing they also seemed to be overconsolidated. Oversteepening of the canyon slopes due to either sediment accumulation or undercutting could have effected the slumping in the Ahziv canyon.

Earthquakes as a Cause for Undrained Slumping

Forces generated by the earthquake-induced periodical displacements within the sediment may lead to slope failure. The effect of an earthquake in the analysis of undrained slumping is based on the following: (1) the abruptness of an earthquake shock renders impossible the drainage of pore water from the sediments, even if the sediments are very permeable; (2) an earthquke shock introduces a horizontal body force k, expressed as some percentage of gravity; and (3) the vertical acceleration induced by an earthquake is small compared with the horizontal acceleration, and can therefore be neglected. {This assumption lacks a sound basis: recent acceleration measurements during earthquakes show that the vertical accelerations may be large, sometimes even larger than the horizontal accelerations (Benfer, 1974).}

The horizontal acceleration acts on the whole sedimentary mass, the granular material, and the pore water; therefore, the force equilibrium along the potential failure plane in the infinite slope is given by

$$c_u L = \bar{W}\sin\alpha + kW\cos\alpha \qquad (9)$$

where $W = \gamma bh$ designate the weight of the sediment slice, and $\bar{W} = \bar{\gamma}bh$ designates the submerged weight of the slice (Fig. 13). By arranging Eq. (9):

$$c_u/\bar{\gamma}h = {}^1/_2\sin2\alpha + k(\gamma/\bar{\gamma})\cos^2\alpha \qquad (10)$$

If average values of γ and $\bar{\gamma}$ are used, the ratios $\gamma/\bar{\gamma}$ and $c_u/\bar{\gamma}h$ are constant. For the sediments in the investigated area $\gamma/\bar{\gamma} \approx 2.75$ and $c_u/\bar{\gamma}h \approx c_u/\bar{p}_o$, therefore, and substituting these equivalents in Eq. (10), the relationship between the slope angle α and the undrained strength $c_u/\bar{p}_o$ at limiting equilibrium and subject to an earthquke acceleration k percent of gravity for undrained slumping of the slope sediments in the investigated area, is given by

$$c_u/\bar{p}_o = 1/2 \sin\alpha + 2.75k\cos^2\alpha \qquad (11)$$

Figure 14 illustrates this relationship for the range of $\alpha = 0\text{-}20^\circ$, $c_u/\bar{p}_o = 0.20\text{-}0.50$ and $k = 0\text{-}16\%$ of gravity.

Considering first the stability of the continental slope free of earthquke loading, it is observed from Figure 14 that (1) the continental slope cannot stand an inclination greater than $15\text{-}25^\circ$ (at water depths of 400-1,000 m, respectively), and that (2) undrained slumping on very gentle gradients (less than 2°) can only occur in very soft materials ($c_u/\bar{p}_o < 0.1$), as was first stated by Morgenstern (1967).

From Figure 14 it is concluded that even small earthquake-induced accelerations, $k = 5\text{-}6\%$ of gravity, are sufficient to effect undrained slumping in the steepest slopes in the investigated area ($\alpha = 6^\circ$ at 400-450 m depth, $c_u/\bar{p}_o = 0.24$). The recent "fresh"-looking slump scars are indeed located at these depths (Fig. 6). Upslope, the gradient is gentler, containing no slump scars of this type. The existence of slump scars in the lower portion of the continental slope is hard to understand, as the slope of the lower portion is also gentle ($\alpha = 1\text{-}3^\circ$), whereas the undrained shear strength of the sediments have the highest observed values ($c_u/\bar{p}_o = 0.35\text{-}0.75$). Force equilibrium analysis clearly indicates that much larger horizontal accelerations ($k = 12\text{-}16\%$ of gravity) can cause these slumps. This problem will be discussed subsequently.

An attempt to calculate the width of slump scars generated by the earthquake-induced horizontal accelerations proved to be inconclusive. The force equilibrium of the sliding slab is

$$blh\bar{\gamma}_{av}\sin\alpha + blh\gamma_{av}k = bl(c_u/\bar{p}_o)(\bar{\gamma}_{av}h) + 2lh(c_u/\bar{p}_o)(\bar{\gamma}_{av}h/2) \qquad (12)$$

where $blh\sin\alpha$ is the downslope weight component along the failure plane (the floor of the slump scar), $blh\gamma_{av}k$ is the lateral force generated by the earthquake horizontal acceleration, $bl(c_u/\bar{p}_o)(\bar{\gamma}_{av}h)$ is the shearing resistance along the failure plane, and $hl(c_u/\bar{p}_o)(\bar{\gamma}_{av}h/2)$ is the shearing resistance along one of the lateral failure planes (the slump scar wall). Arrangement of Eq. 12 leads to

$$\frac{\text{width of the slump}}{\text{height of the slump}} = \frac{b}{h} = \frac{(c_u/\bar{p}_o)}{k(\gamma/\bar{\gamma}) + \sin\alpha - (c_u/\bar{p}_o)} \qquad (13)$$

since $\sin\alpha$ and k are very small compared to $\gamma/\bar{\gamma}$ and $c_u/\bar{p}_o$, variations in the slope angles and the earthquake horizontal accelerations have practically no expression in relation to the slump dimensions.

From the above analysis, which is based on a constant $\gamma/\bar{\gamma}$ ratio throughout the sedimentary column, calculations of depth to the

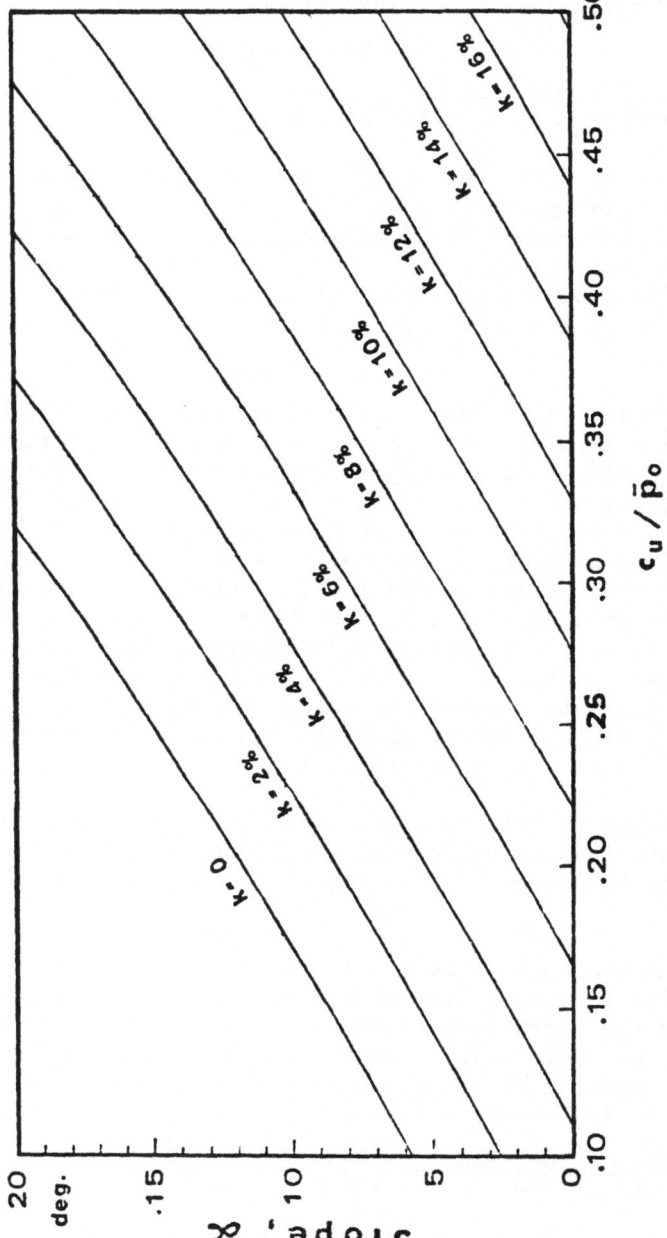

Figure 14. Relation of slope angle to undrained strength for an infinite slope at
limiting equilibrium and subject to an earthquake acceleration
(from Almagor and Wiseman, 1977).

failure plane or changes of k for factor of safety $F=1$ with burial
depth cannot be performed. To do so, changes of γ and $\bar{p}_O$ with depth
in the sediments were extrapolated to a depth of 50 m (the deepest
slump scars are 45 m), based on γ and $\bar{\gamma}$ measurements in the cores
and on the consolidation tests. For values $c_u/\bar{p}_O = 0.25$ (at 400–450
m water depth, where the deepest slump scars are located), it now
becomes possible to calculate the c_u and k values needed to cause
failure, and present the relationship of the factor of safety F to
burial depth for a constant k. The results show that in order to cause
a failure at 45 m depth, the horizontal acceleration must be twice
what is needed to cause a failure near the surface (Fig. 15).

   Morgenstern (1967) accumulated data on submarine slumps caused
by earthquakes and was able to provide a correlation between submarine slumping and earthquakes of magnitudes larger than 6.5. The
available data on earthquakes in the southeastern Mediterranean Sea
indicate that seismic activity is presently concentrated in a broad
zone off the coasts of Israel and Lebanon (Arieh, 1967)(Fig. 16) and
that the earthqukes of large magnitudes occured in the area in the
historical past (Willis, 1928; Amiran-Kallner, 1950; Shalem, 1951).

Stability of the Scarps of the Slump Scars

   The scarps of the slump scars in the investigated area have
steep inclinations, sometimes greater than 20°, and lengths of several
hundred meters (Fig. 6). Their stability should be examined by finite
slope analysis, owing to their small areal extent and high inclinations. The "method of slices" is the most common method used to
predict and analyze failures in finite slopes along circular surfaces
(Bishop, 1955), and for noncircular failure surfaces (Morgenstern
and Price, 1965). The factor of safety calculated for the steepest
scarps that were located ($\alpha = 20^\circ$) by use of this method is $F=1.735$
(Fig. 17).

   Using the theory of plasticity, Booker and Davis (1972)
presented a solution for the stability of submerged slopes made of
nonhomogeneous clayey sediments. The stability of the scarps of the
slump scars was calculated by this method. The rate of increase of
the strength with depth $c_u/\bar{p}_O$ at equilibrium ($F = 1$) for the slope
of the slump scar scarp ($\alpha = 20^\circ$) is readily evaluated from the
graph formulated by

$$\rho/\gamma = (c_u/\bar{p}_O)_{equilibrium} = 0.14 \qquad (14)$$

where $\rho = (\gamma c_u)/\bar{p}_O$. Since $c_u/\bar{p}_O = 0.25$ throughout the sedimentary
column in the slump area, the factor of safety is

$$F = \frac{(c_u/\bar{p}_O)_{measured}}{(c_u/\bar{p}_O)_{equilibrium}} = \frac{0.25}{0.14} = 1.786 \qquad (15)$$

This value is similar to the F value evaluated using the "method of
slices".

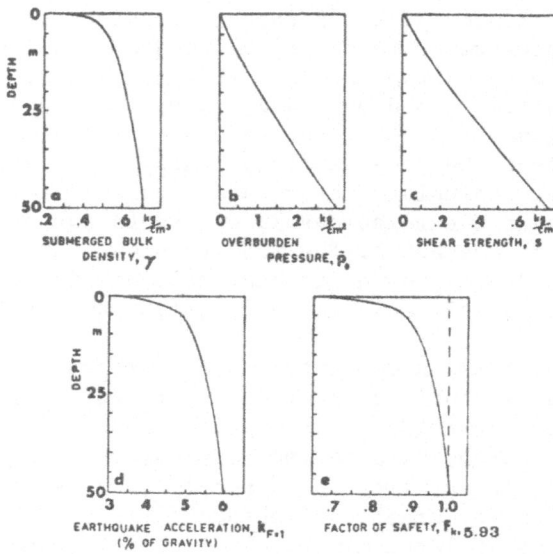

Fig. 15   Relation of (a) submerged bulk density, (b) overburden
           pressure, (c) shear strength, (d) earthquake horixontal
           acceleration needed to cause failure, and (e) the factor
           of safety of seafloor sediments to 50 m depth in the sed-
           imentary column. Typical properties of marine silty clays
           at water depth of 400 m were used as starting values in
           constructing these graphs. The construction of these graphs
           is discussed by Hamilton (1959)(from Almagor and
           Wiseman, 1977).

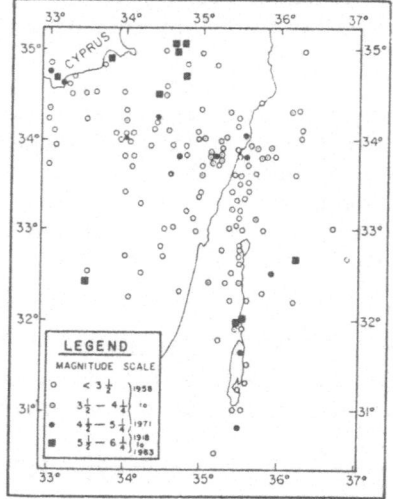

Fig. 16   Earthquake epicenters map. (from Arieh, 1967).

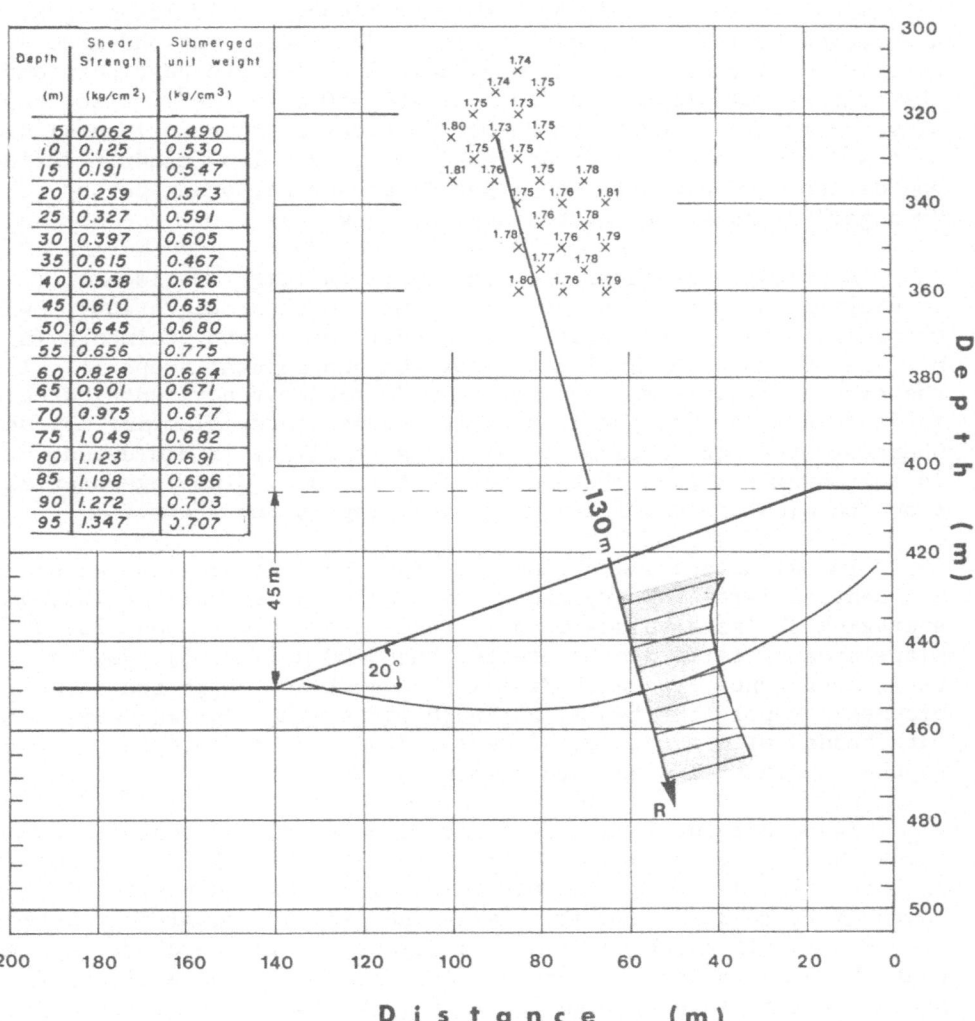

Figure 17. Results of undrained slope stability analysis (finite slope, $\phi = 0$, c and γ increasing with depth) (from Almagor and Wiseman, 1977).

Discussion and Conclusions

The existence of slump scars in the lower portions of the continental slope, where the slope angle is $\alpha = 1\text{-}3^{\circ}$ and the undrained shear strength of the sediments is $c_u/\bar{p}_o = 0.35\text{-}0.91$, is not readily explained, as large horizontal accelerations, $k = 12\text{-}16\%$ of gravity, are needed to initiate slumping there. Such accelerations should have been expected to completely "demolish" the steepest portions of the continental slope at a depth of 400-450 m ($\alpha = 6^{\circ}$). Also, the slump scars that occupy the deeper portions of the continental slope have an "older look": they are shallower, have less clearly defined boundaries, and are often only partly preserved, often cut by "younger" slump scars located upslope (Fig. 6).

A possible explananation incorporates larger horizontal accelarations ($k = 12\text{-}16\%$ of the gravitational acceleration). Owing to the wide range of geotechnical properties of the sediments ($c_u/\bar{p}_o = 0.20\text{-}0.91$) and the inclination of the continental slope ($\alpha = 1\text{-}6^{\circ}$) the same earthquake may generate a wide range of horizontal accelerations in different portions of the slope. Thus, slumping may occur wherever the equilibrium of the slope stability is disrupted, even in the lower portions of the continental slope. Slumping materials from the midslope disfigure the lower slope slump scars.

The slump scars are common on the continental slope of southern and central Israel, where the slope attains a general inclination averaging 3° (peak values of 6°). There they occupy about 30% of the slope area at water depths greater than 400 m (Fig. 6). However, their occurrence gradually diminishes northward where the slope becomes steeper ($\alpha = 6\text{-}8^{\circ}$), although large size slumped debris and interbedded wide stretches of debris flows in the deep sea adjacent to the continental slope are common.

The continental margin of northern Israel and southern Lebanon forms the outermost zone of influence of the Nile River. It consists of finer Nile-derived sediments compared to those deposited in the south (Nir, 1973). Presumably, also the rate of deposition there is slower than off northern Sinai and Israel. It is, therefore, assumed that these sediments are stronger and form more stable slopes that better resist slumping when low earthquake-induced horizontal accelerations are applied. However, this assumption is unsubstantiated, as no geotechnical data pertaining to these sediments exist.

MASS CREEP - ANALYSIS OF CYCLIC LOAD PROPERTIES

As is discussed above, the shelf edge and upper slope sediments are normally consolidated silty clays with normal sensitivity of 2-5 and are sufficiently strong ($c_u/\bar{p}_o = 0.24$, Fig. 10) to sustain the slopes at the shelf edge and upper slope ($\alpha = 0.5\text{-}2.5^{\circ}$), even if subjected to the effect of earthquake accelerations greater than

those detrimental to the steepest slopes ($\alpha = 5-7^{\circ}$) of the middle
continental slope (Fig. 14). Yet, mass creep phenomena (Fig. 7) are
ubiquitous along the entire Israeli shelf edge zone. The term *creep*,
or *plastic flow*, is herein defined as continuous yielding of the
soil particles under applied undrained stress.

It is suggested that these creep phenomena reflect long-term
deterioration in shear strength of the sediments due to repeated
loading effects. Frequent loading reversals can occur rapidly during
earthquakes, or more slowly when caused by wave loading. The effects
of repeated loading depend mainly on the cyclic stress level,their
number, frequency and duration, and on the sediment types. Application of cyclic loading on normally consolidated and slightly overconsolidated clays, where drainage is poor,(and on metastable and
confined loose sands) will lead to a build-up of pore-water pressure,
increased strain and subsequent decrease of shear strength of the
sediment with time (Fig. 18). In addition to this long-term deterioration in shear strength, which may develop over a period of hundreds
and thousands of years, each individual earthquake may cause a
further, momentary decrease in shear strength due to a build-up of
fluid pressure in the sediment pores during action of the quake. If
the excess pore-water pressure reduces the effective normal stress
to a sufficiently low level, an effective stress failure (liquefaction) will develop, which may result in mass movement downslope.

In view of the great water depth of the sediments involved
(above 80 m) and the mild oceanographic conditions in the region,
the weakening of the sediments in question (outwardly expressed as
creep failures) is probably caused by cumulative effects of earthquakes, which abundantly occur in the region (Fig. 16).

As with metal fatigue, it is assumed that the deterioration in
shear strength of the sediments is not time history dependent.Thiers
and Seed (1969) showed that the deterioration in stress-strain
properties is due entirely to the amount of cyclic strain, whether
developed by a few strong or many small stress pulses. Therefore,
as the basin-filling sediments of the continental shelf and upper
slope (Fig. 7) represent a complete undisturbed Holocene sequence,
cyclic shear testing of long core samples could yield results
pertaining to the complete earthquake history of this sedimentary
column since deposition, if the basic postulate of the phenomena
being independent of time history is correct.

From Figure 14 it is seen that only a minor addition of an
earthquke-induced horizontal acceleration k of gravity is needed to
initiate undrained mass movement of the shelf edge and upper slope
sediments if a static analysis is applied. This means that even a
small amount of weakening of the sediments by the low magnitude
earthquakes that generally occur in the region (Fig. 16) is sufficient to cause mass creep.

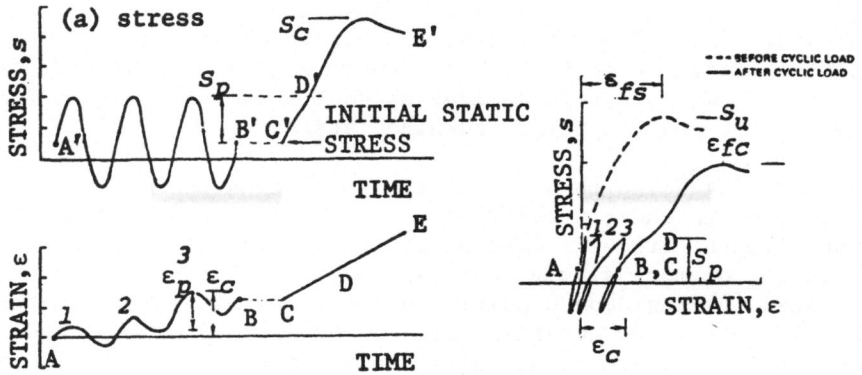

Figure 18. Illustrative cyclic-ststic test record. AB and A'B'
equal the cyclic stress-strain path; BC and B'C' equal the rest and
adjustment period following cyclic leading; CDE and C'D'E' equal
the one-directional static loading test; and the dashed curve refers
to one-directional static loading test on an identical sample with
not previous cyclic loading (form Lee and Focht, 1976 b).

S_u – Static strength in an undrained test
    performed on an undisturbed sample
    prior to any cyclic loading.

S_c – Static strength in an undrained test
    following cyclic loading.

S_p – Cyclic strength amplitude.

ε_{fs} – Strain to failure in an undrained
    test performed on an undisturbed
    sample prior to any cyclic loading.

ε_{fc} – Strain to failure in a static test
    following cyclic loading.

ε_p – Cyclic strain amplitude.

ε_c – Accumulative strain.

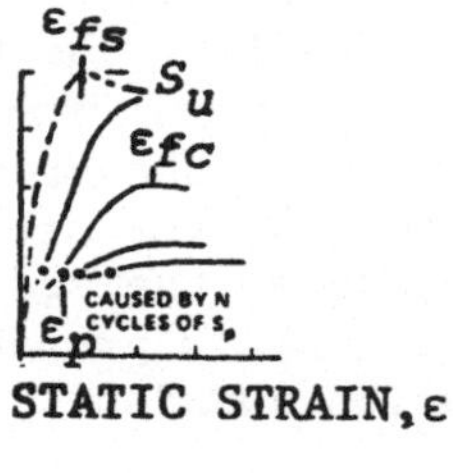

Figure 19. Illustration of static strength after cyclic loading.
from Lee and Focht, 1976).

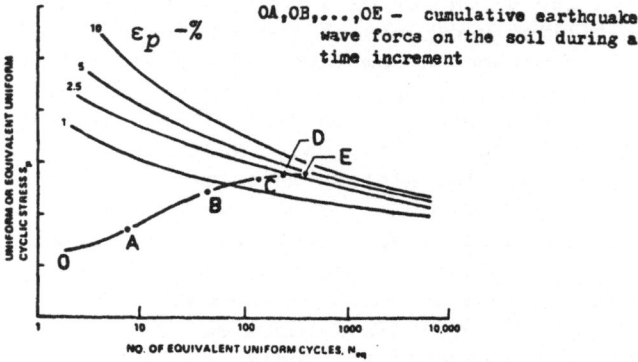

Figure 20. Effect of cyclic storm loading on clay soil.
(from Lee and Focht, 1976 b).

The methodology applicable was formulated by Andersen (1976) and Lee and Focht (1975; 1976a; 1976b), though for ocean waves rather tnan for earthquake effects. The amount of cyclic softening, as expressed by the decrease of tne static stress requirea to cause failure, is accentuated by increasing prior cyclic loading (Fig. 19). From cyclic tests, sucn as illustrated in Figure 18, series of cyclic strain, strength and pore-water pressure curves, that are characteristic to tne sediment in question, can be developed (Figs. 19, 20); the tailure criterion being the amount of pulsating strain amplitude corresponding to a particular curve. As the effect of pulsating stress cycles in a group of any intensity is assumed to be independent of when it is applied within this group, it is possible to use various assumed earthquake inputs to calculate the cyclic loading effects with time (or number of cycles) on the seaiments (Fig. 20). From historical records, which in the Middle East contain earthquake data that date back some 2,000 years (Willis, 1928; Amiran-Kallner, 1950), a fair approximation of the cumulative earthquake cyclic loading effect can be constructed, and roughly extrapolated to the entire Holocene period. Conversion methods of the randomly irregular earthquake loads, as recorded by use of accelerometers, to equivalent uniform cycles tnat can be used in the analysis are described by Andersen (1976) and Lee ana Focht (1975; 1976b). For the slope stability analysis it may be convenient to use the cyclic earthquake loading data in terms of the reciprocal of the factor of safety

$$1/F = \tau_{max}/S_u \qquad\qquad (16)$$

where τ_{max} is the maximum shear stress in the slope sediment produced by the largest pulse in the designed quake, and S_u is the static strength in an undrained test performed on an undisturbed sample prior to any cyclic loading.

REFERENCES

Almagor, G., 1964, Studies of sediments in core samples collected from the shelf and slope of Tel Aviv - Palmahim coast, Geological Survey oi Israel Report QGR/64/2 (in Hebrew, English abstract), 80 p.

Almagor, G., 1967, Interpretation of strength and consolidation data from some bottom cores off Tel Aviv - Palmahim coast, Israel, in Richards, A.F. (ed.), Marine Geotechnique, University of Illinois Press, Urbana, Ill., p. 131-353.

Almagor, G., 19/6, Physical properties, consolidation process and and slumping in recent marine sediments in the Mediterranean continental slope of southern Israel, Geological Survey of Israel Report MG/76/4 (in Hebrew, English summary), 132 p.

Almagor, G., 1978, Geotechnical properties of the sediments of the continental margin of Israel, Journal of Sedimentary Petrology, v. 48, no. 4, p. 1267-1274.

Almagor, G., 1979, Relict sandstones of Pleistocene age on the continental shelf of northern Sinai and Israel, Israel Journal of Earth-Sciences, v. 28, p. 70-76.

Almagor, G., 1980, Halokinetic deep-seated slumping on the Mediter-
        ranean slope of northern Sinai and southern Israel, Marine
        Geotechnology, v. 4, no. 1, p. 83-105.

Almagor, G., and Garfunkel, Z., 1979, Submarine slumping in the
        continental margin of Israel and northern Sinai, The American
        Association of Petroleum Geologists Bulletin, v. 63, no. 3,
        p. 324-340.

Almagor, G., and Wiseman, G., 1977, Analysis of submarine slumping
        on the continental slope off the southern coast of Israel,
        Marine Geotechnology, v. 2, p. 349-388.

Amiran-Kallner, D. H., 1950, A revised earthquake catalogue of
        Palestine. I., Israel Exploration Journal, v. 1, p. 223-246.

Andersen, K. H., 1976, Behaviour of clay subjected to undrained
        cyclic loading, Norwegian Geotechnical Institute Publication
        No. 114, p. 33-44.

Arieh. E., 1967. Seismicity of Israel and adjacent areas, Geological
        Survey of Israel Bulletin No. 43, p. 1-14.

Benfer, N. A. (ed.), 1974, San Fernando, California earthquake of
        February 9, 1971. vol. 1: Effects on Building Structures,
        parts A and B, Washington D.C., National Oceanic and Atmos-
        pheric Administration, 841 p.

Bishop, A. W., 1955, The use of slip circle in the stability analysis
        of slopes, Géotechnique, v. 5, p. 7-17.

Ben-Avraham, Z., 1978, The structure and tectonic setting of the
        Levant continental margin - eastern Mediterranean, Tectono-
        physics, v. 46, p. 313-331.

Boswell, P. G. H., 1961, Muddy Sediments, W. Heffer & Sons, Ltd.,
        Cambridge, England, 140 p.

Booker, J. R., and Davis, E. H., 1972, A note on a plasticity solution
        to the stability of slopes in inhomogeneous clays, Géotechnique
        v. 22, p. 509-513.

Einsele, G., 1967, Sedimentary processes and physical properties of
        cores from the Red Sea, Gulf of Aden and off the Nile Delta,
        in Richards, A. F. (ed.), Marine Geotechnique, University of
        Illinois Press, Urbana, Illinois, p. 154-169.

Einsele, G., and Werner, F., 1968, Zusammensetzung, gefüge und mecha-
        nische eingeschaften rezenter sedimente von Nile Delta, Roter
        Meer und Golf von Aden, 'Meteor', v. 3, no. 1, p.21-42.

Fisk, N. H., and McClelland,B., 1959, Geology of the continental
        shelf of Louisiana, its influence on offshore foundation
        design, Geological Society of America Bulletin, v. 70, p.
        1369-1394.

Garfunkel, Z., Arad, A., and Almagor, G., 1979, The Palmahim Distur-
        bance and its regional setting, Geological Survey of Israel
        Bulletin No. 72, 56 p.

Hamilton, E. L., Thickness and consolidation of deep-sea sediments,
        Geological Society of America Bulletin, v. 70, p. 1399-1424.

Heezen, B. C., 1956, Corrientes de-turbidez del Rio Magdalena, Boletin
        de la Sociedad Geografica de Colombia No. 51-52, p. 134-143.

Henkel, J. D., 1970, The role of waves in causing submarine landslides Géotechnique, v. 20, p.75-80.

Horowitz, A., 1974, Preliminary palynological indications as to the climate of Israel during the last 6,000 years. Paleorient, v. 2, no. 2, p. 407-414.

Hsü, K. J., Cita, M. B., and Ryan, W. B. F., 1973, The origin of the Mediterranean evaporites, in Ryan, W. B. F., Hsü, K. J. et al., Initial Reports of the Deep Sea Drilling Project, U.S. Government Printing Office, Washington, D.C., v. 13, p. 1203-1231.

I-ASCO, 1975, Report on high resolution geophysical survey offshore Israel submittedto Israel Electric Corporation (unpublished), 27 p.

Issar, A., 1968, Geology of the central coastal plain of Israel, Israel Journal of Earth-Sciences, v. 17, p. 16-29.

Itzkhaki, Y., 1961, Contributions to the study of the Pleistocene in the coastal plain of Israel: Pleistocene shore-lines in the coastal plain of Israel, Geological Survey of Israel Bulletin No. 32, p. 1-9.

Keller, G. H., and Lambert, D. N., 1972, Geotechnical properties of submarine sediments, Mediterranean Sea, in D. J. Stanley (ed.), The Mediterranean Sea, Hutchinson and Ross, Inc., Stroudsburg, Pennsylvania, p. 401-415.

Lee, K. L., and Focht, J. A.,Jr., 1975, Liquefaction potential at the Ekofisk tank in the North Sea, Journal of the Geotechnical Engineering Division, American Society of Civil Engineers, v. 101, no. GT1, p. 1-18.

Lee, K. L., and Focht, J. A., Jr., 1976a, Strength of clay subjected to cyclic loading, Marine Geotechnology, v. 1, no. 3, p. 165-168.

Lee, K. L., and Focht, J. A., Jr., 1976b, Cyclic testing of soil for ocean wave loading problems, Marine Geotechnology, v. 1, no. 4, p. 305-325.

Maldonado, A., andStanley, D.J., 1976, The Nile Cone: submarine fan development by cyclic sedimentation, Marine Geology, v.20, p. 27-40.

Mathews, W. H., and Shepard, F. C., 1962, Sedimentation of Frazer River Delta, British Columbia, American Association of Petroleum Geologists Bulletin, v. 46, p. 1416-1443.

McClelland, B., 1956, Engineering properties of soils on the continental shelf of the Gulf of Mexico, Proceedings of the 8th Texas Conference on Soil Mechanics and Foundation Engineering, University of Texas, Bureau of Engineering Research, Special Publication 29, 28 p.

McCoy, F. W., 1974, Late Quaternary sedimentation in the eastern Mediterranean Sea, unpublished Ph. D. thesis, Harvard University, Mass.,USA, 132 p.

Milliman, J. D., and K. O. Emery, 1968, Sea level during the past 35,000 years, Science, v. 162, p. 1121-1123.

Morgenstern, N.R., 1967, Submarine slumping and the initiation of turbidity currents, in Richards, A. F. (ed.), Marine Geotech-

nique, University of Illinois Press, Urbana, Illinois, p.
189–220.

Morgenstern, N. R., and Price, V. E., 1965, The analysis of stability
by general slip surface, Géotechnique, v. 15, p. 79–93.

Nachmias, J., 1969, Source rocks of the Saqiye Group sediments in
the coastal plain of Israel – a heavy mineral study, Israel
Journal of Earth-Sciences, v. 10, p. 1–16.

Neev, D., Almagor, G., Arad, A., Ginzburg, A., and Hall, J. K., 1976,
The geology of the southeastern Mediterranean Sea, Geological
Survey of Israel Bulletin No. 68, 51 p.

Neev, D., Edgerton, H. E., Almagor, G., and Bakler, N., 1966, Preliminary results of some continuous seismic profiles in the
Mediterranean shelf of Israel, Israel Journal of Earth-Sciences
v. 15, p. 170–178.

Nir, Y., 1973, Geological history of the recent and subrecent sediments of the Israel Mediterranean shelf and slope, Geological
Survey of Israel Report MG/73/2, 179 p.

Nir, Y., and Nathan, Y., 1972, Mineral clay assemblages in recent
sediments of the Levantine basin, Mediterranean Sea, Bulletin
Groupe Française des Argiles, v. 24, p. 187–195.

Reiss, Z., Merling-Reiss, P., and Moshkovitz, S., 1971, Quaternary
planktonic foraminifera and nannoplankton from the Mediterranean continental shelf and slope of Israel, Israel Journal
of Earth-Sciences, v. 20, p. 141–147.

Richards, A.F., 1962, Investigation of deep-sea sediment cores, II.
Mass physical properties, U.S. Hydrographic Office Technical
Report 106, 146 p.

Shalem, N., 1951, La seismicité au Levant, Research Council of Israel
Bulletin, v. 2, p. 1–16.

Shepard, F.P., 1955, Delta-front valleys bordering the Mississippi
distributaries, Geological Society of America Bulletin, v.
66, p. 1489–1498.

Terzaghi, K., 1956, Varieties of submarine slope failures, Proceedings of the 8th Texas Conference on Soil Mechanics and
Foundation Engineering, University of Texas, Bureau of Engineering Research, Special Publication 29, 40 p.

Terzaghi, K., 1962, Discussion on sedimentation of Frazer River
delta, British Columbia, American Association of Petroleum
Geologists Bulletin, v. 46, p. 1438–1443.

Thiers, G. R., and Seed, H. B., 1969, Strength and stress-strain
characteristics of clays subjected to seismic loading conditions, Proceedings of the Symposium on Vibration Effects of
Earthquakes on Soils and Foundations, American Society for
Testing and Materials, Special Technical Publication 450, p.
3–56.

Venkatarathnam, K., and Ryan, W.B.F., 1971, Dispersal patterns of
clay minerals in the sediments of the eastern Mediterranean,
Marine Geology, v. 11, p. 261–282.

Willis, B., 1928, Earthquakes in the Holy Land, Bulletin of the
Seismological Society of America, v. 18, no. 2, p. 73–102.

PIEZOMETER PROBES FOR ASSESSING EFFECTIVE STRESS AND STABILITY IN SUBMARINE SEDIMENTS

Richard H. Bennett, John T. Burns, Thomas L. Clarke, J. Richard Faris*, Evan B. Forde, and Adrian F. Richards**

NOAA, Atlantic Oceanographic and Meteorological Laboratories, Miami, Florida 33149

*128 Holladay Avenue, San Francisco, CA 94110

**Marine Geotechnical Laboratory, Lehigh University, Bethlehem, PA 18015

ABSTRACT

Multisensor piezometer probes were deployed at four different sites in the Mississippi Delta in water depths ranging from 13.5 to 43.6 m with sensor penetration depths of up to 15.6 meters. Absolute and differential pressure sensors were used to measure pore water pressure and excess pressures, respectively. The free water column pressure was measured with absolute pressure sensors. Pore pressures induced by probe insertion were determined as well as ambient excess pore pressures following the time-dependent decay of induced pressures. Significant differences in the pore pressures and related geotechnical properties were found between East Bay and Main Pass sediments. Generally higher probe insertion pressures and lower ambient excess pore pressures were characteristic of Main Pass compared to East Bay. Probe insertion pressures (U_i) were found to correlate well with the undrained shear strength (S_u) of the sediments, indicating reasonably good agreement with the predicted relation: $U_i = 6S_u$ as suggested by an earlier study[42]. Using this relationship undrained shear strengths were calculated and compared with measured values.

Surface wave activity due to tides and short-period surface waves were recorded by the piezometers. Bottom pressure changes

due to tidal activity were found to be transmitted through the
sediments to depths of at least 15.6 m. Short-period (approximately 6.5 sec) surface waves were found to transmit pressure
pulses through the sediments to depths of between 6.5 and about
12.5 meters with severe attenuation of the pressure at 12.6 m and
greater. Input of excess pore pressure data to a computer program to assess seafloor stability at a selected site resulted in
significant reductions (up to 20%) in the factors of safety even
without including surface wave effects. Further field experiments and analysis of existing data are necessary to evaluate the
possible time-dependent changes in effective stress with depth
below the mudline as a function of surface wave activity. The
previous field experiments have demonstrated the feasibility of
making pore pressure measurements with multisensor probes and
have shown the criticality of pore pressure measurements to the
study of seafloor stability and to offshore geotechnical site
evaluations.

INTRODUCTION

Increasing seafloor utilization for commercial, military and
governmental activities has stimulated both engineering and
scientific investigations of submarine deposits. Detailed
geological, geophysical and geotechnical studies[1-20] on continental margins during the last decade establish the importance of
mass movements and related processes in shaping the seafloor and
in transporting vast quantities of material. Seemingly more
areas than previously expected have either experienced submarine
slumping and mass movement or reveal geological and geotechnical
properties indicative of potentially unstable material on continental shelves, slopes and rises[21-26]. In light of the numerous
areas found to be unstable or potentially unstable, interest has
focused on understanding the history and development of seafloor
deposits. Although descriptive geological and geophysical
studies are necessary to assess areas of interest, geotechnical
techniques add a critical quantitative dimension to the analyses
and evaluation of seafloor stability studies. Theory and
modeling provide quantitative methods to assess seafloor stability[25,27-34]; however, "ground truth" and quality geotechnical
properties measurements are critical in providing confidence in
analyses. In order to better evaluate stability problems, scientists and engineers have recognized the importance of high-qual-
ity samples for testing and the need for in situ geotechnical
measurements[35,36].

The importance of sediment pore pressures, specifically
excess pressures, has been recognized for decades by soils engineers as a critical property affecting the stability of soil
deposits. The effect of surface wave activity during storm
periods has been considered a major factor in triggering slope

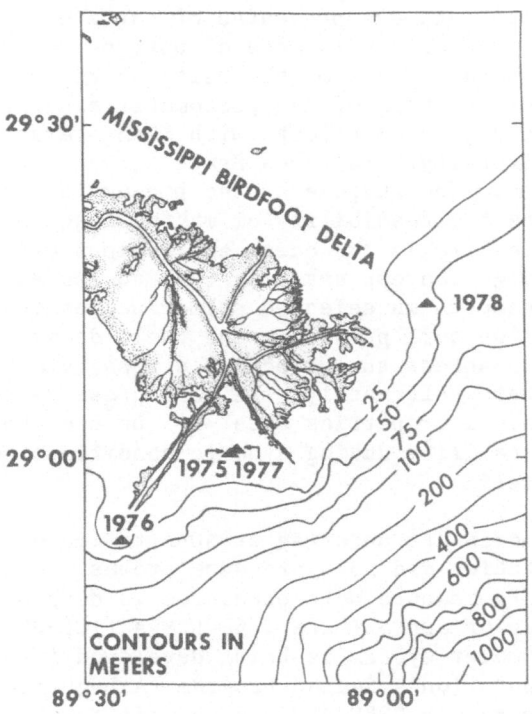

Fig. 1.   Piezometer probe sites, 1975-1978.

failures and mass movements in submarine sediments of the Mississippi Delta[1,2,28,37]. Despite the recognized importance of both the role of excess pore pressures and the effect of surface wave activity on the ultimate stability of Mississippi Delta sediments, essentially no field data were available to evaluate these aspects prior to 1975[38,39].

This paper summarizes the results of four in situ pore pressure experiments conducted in the Mississippi Delta by the National Oceanic and Atmospheric Administration (NOAA) between 1975 and 1978 (Figure 1). A brief description is given of the piezometer instrumentation, geotechnical properties at the probe sites, and limited data are presented of surface wave effects on pore pressures. Comparison is made of pore pressures measured at East Bay and Main Pass areas of the Delta where the longest-term data were collected. Most of the piezometer experiments were made possible through team efforts with NOAA, Texas A&M University, Lehigh University, Louisiana State University, and the U.S. Geological Survey. The purpose of the piezometer experiments was (1) to determine the feasibility of making long-term measurements of pore pressure in the ocean environment using prototype and improved design concept systems; (2) to assess ambient and dynamic pore pressures in selected submarine sediments; (3) to determine insertion pore pressures and their decay characteristics; and (4) to assess the effective stress (state of stress) at the probe sites. Ultimately, the pore pressure results and related geotechnical properties data will be used to evaluate the stability or instability during dynamic conditions for selected design wave criteria.

Current studies are underway at NOAA to improve the shallow-water instrumentation and pore pressure probes. A new system is being developed to measure pore pressures in deep. ocean basins at ambient pressures of approximately 68.9 MPa (10,000 psi) The deep-water piezometer system is being developed in cooperation with Sandia Corporation's Seabed Program (Albuquerque, New Mexico). During recent DSRV ALVIN dives off the U.S. Atlantic continental slope, a deep-water mini-piezometer was field tested in water depths of 1450 meters. The probe is a prototype to be used during Sandia's In Situ Heat Transfer Experiment (ISHTE). The shallow and deep-water piezometer capabilities will have direct applications to a variety of interest in engineering, offshore development, geology, and environmental studies.

GEOSTATIC AND EFFECTIVE STRESS: SIGNIFICANCE OF PORE PRESSURE

Pore pressure plays a critical role in determining the state of effective stress in sedimentary deposits. The effective stress determines the degree or state of consolidation and plays a critical role in determining the utlimate strength of a sedi-

mentary unit. An understanding of the state of stress and the
strength of submarine sediments is important in the analyses of
seafloor stability. Review of a few fundamental concepts will
clarify the importance and role of pore pressures in submarine
sediments.

Neglecting atmospheric pressure at sea level, the vertical
geostatic (total) stress (σV_Tg) at a selected depth in the sediment (Fig. 2) is given by:

free water column   sediment column

$$\sigma V_Tg = \sum_{x=0}^{x=d} \gamma_{sw} \Delta x \quad + \quad \sum_{x=d}^{x=z} \gamma_t \Delta x \qquad \text{eq. (1)}$$

γ_{sw} = Unit Weight of Seawater
γ_t = Saturated Unit Weight of Sediment

and the effective vertical stress ($\overline{\sigma}v_e$ is given by:

$$\overline{\sigma}v_e = \sigma V_Tg - Uw \qquad \text{eq. (2)}$$

where Uw is the total pore water pressure (stress) at a given
depth (z).

It is readily apparent that the stress due to the water
column above the seafloor is included in the terms σV_Tg and Uw
and thus the effective vertical geostatic stress ($\overline{\sigma}vg$) can be
written as follows assuming linear hydrostatic fluid stress in
the sediment (no excess pore pressure):

$$\overline{\sigma}vg = \sum_{x=d}^{x=z} \gamma_b \Delta x \qquad \text{eq. (3)}$$

where $\gamma_b = \gamma_t - \gamma_{sw}$ = buoyant unit weight

In sedimentary deposits where excess pore pressure exists,
the effective vertical stress ($\overline{\sigma}v_e$) is given by:

$$\overline{\sigma}v_e = \overline{\sigma}vg - U_e \qquad \text{eq. (4)}$$

where U_e = excess pore pressure at depth Z
U_e = Uw - Us
Uw = total pore water pressure
Us = hydrostatic pressure at depth Z.

Clearly, the excess pore pressure is the water pressure that is in excess of the linearly increasing water column pressure at any point (z) in the sediment (Fig. 2). Therefore, in determining the effective vertical stress at any depth (z) in the sediment column, either the total pore water pressure (Uw) can be subtracted from the vertical geostatic (total) stress (σ_{vTg}) as in eq. (2), or the excess pore pressure (U_e) can be subtracted from the vertical effective geostatic stress ($\overline{\sigma}vg$) in eq. (4) which assumes a linear hydrostatic increase in fluid stress.

The degree of underconsolidation can thus readily be seen to be directly proportional to the presence and amount of excess pore water pressure. An additional example of the importance of pore water pressures on the geotechnical properties of sediment is revealed by examining the parameters critical in determining the shear strength of sedimentary materials. Consider the shear strength of fine-grained sediments (muds) which depends on:

1. Cohesion = $\overline{c}$
2. Effective normal stress = $\overline{\sigma}v_e$
3. Angle of internal friction = $\overline{\phi}$

Therefore:

Shear Strength = S = $\overline{c} + \overline{\sigma}v_e \tan\overline{\phi}$
where again:

$$\overline{\sigma}v_e \quad = \quad \sigma_{vTg} \quad - \quad Uw$$

effective stress	geostatic (total) stress	total pore water stress

As the pore water pressure approaches the geostatic (total) stress the effective stress term approaches zero (0) and the sediment strength depends almost entirely upon its cohesion. For clean sands, that possess virtually no cohesion, the effective stress term is critical in determining the strength of the deposit and potential for liquefaction. When the pore water pressure in clean sands approaches the total geostatic stress, the deposits will have virtually no strength and the material will essentially fluidize. The state of stress and presence or absence of excess pore pressure are not only critical in determining the stability of a sedimentary deposit under static conditions, but also are important during dynamic environmental changes because low-strength underconsolidated deposits are more susceptible to slope failure than a normally consolidated deposit.

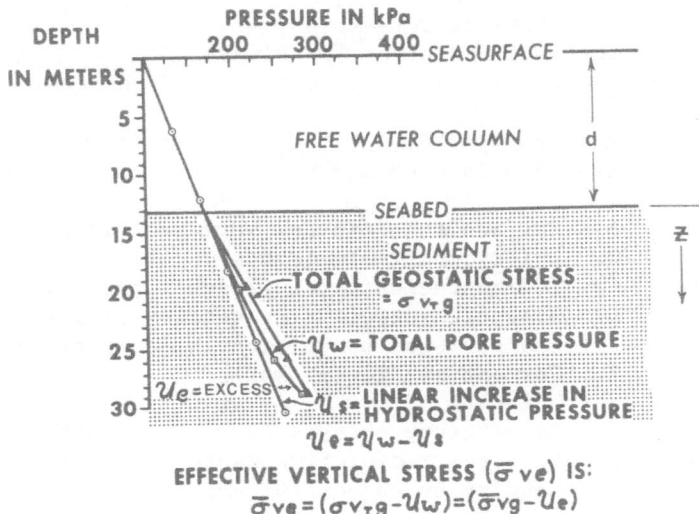

Fig. 2. Total geostatic stress, total pore pressure (measured during 1977), and hydrostatic pressure plotted against depth below the mudline. Note derivation of excess pore pressure.

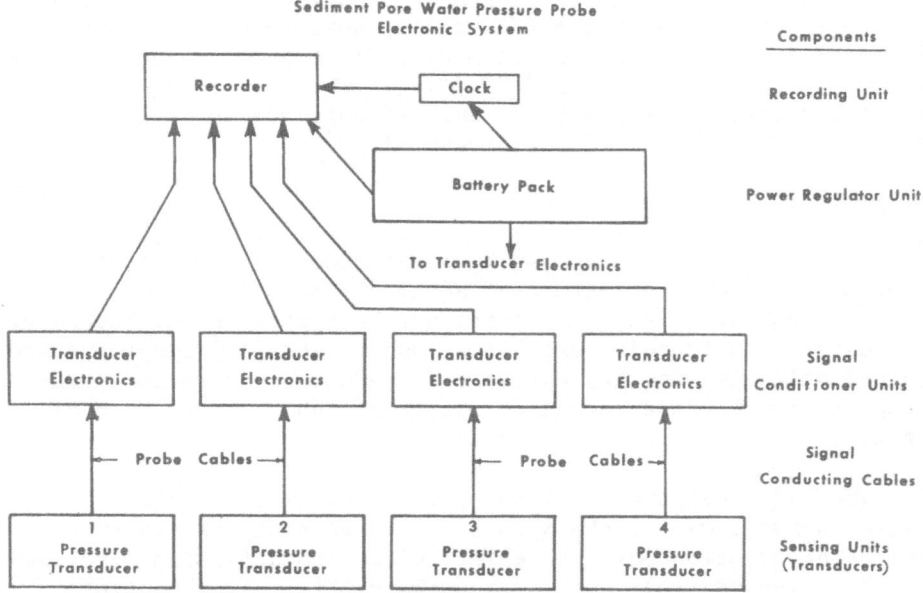

Fig. 3. Block diagram of major components of the shallow-water piezometer system.

Critical factors that determine the stability of a sedimentary deposit are:

1. Strength.
2. State of stress.
3. Seafloor slope (gradient).
4. Environmental conditions and changes.

Environmental changes and particularly dynamic effects can be either natural and/or man-induced and can include the following:

<table>
<tr><td rowspan="5">Natural
Factors</td><td>A.</td><td>Surface and internal wave activity.</td></tr>
<tr><td>B.</td><td>Seismic activity (shock waves).</td></tr>
<tr><td>C.</td><td>Sedimentation rates (loading rates).</td></tr>
<tr><td>D.</td><td>Biogenic activity: generally slow rates of changes (gas production, bioturbation).</td></tr>
<tr><td>E.</td><td>Current erosion of soil.</td></tr>
<tr><td rowspan="3">Man-Induced
Factors</td><td>A.</td><td>Temperature changes (Example: heat generated from radioactive waste disposal).</td></tr>
<tr><td>B.</td><td>Loading by objects/structures.</td></tr>
<tr><td>C.</td><td>Mining/dredging/dumping.</td></tr>
</table>

Thus, numerous natural and man-induced factors can affect the stability of sedimentary deposits and clearly, the strength and state of stress are critical factors, particularly when dynamic environmental factors are superimposed on the static properties. Theoretical studies have shown that pore pressure and states of stress change drastically as a function of surface wave activity during active storm periods in water depths of generally less than 122 meters[28,29,40,41]. This paper presents limited field data showing some dynamic effects due to surface wave activity (tidal and short-period waves).

INSTRUMENTATION

The piezometer system was designed specifically for the study of shallow-water submarine sediments on continental shelves. An essential requirement of the system was to record at high data rates (continuously) for long periods of time (days and months) while leaving the system unattended. Synchronous measurements of all pressure sensors were desired in order to evaluate wave effects on pore pressures.

The major components of the piezometer system include several subsystems: probe with pressure sensing transducers; underwater junction box for cables; signal-conditioning electronics; data recorder(s); and power package (Fig. 3). The probe shell enclosing the pressure transducers is a 0.10 m seamless steel pipe composed of several 3.05 m segments with "O" ring

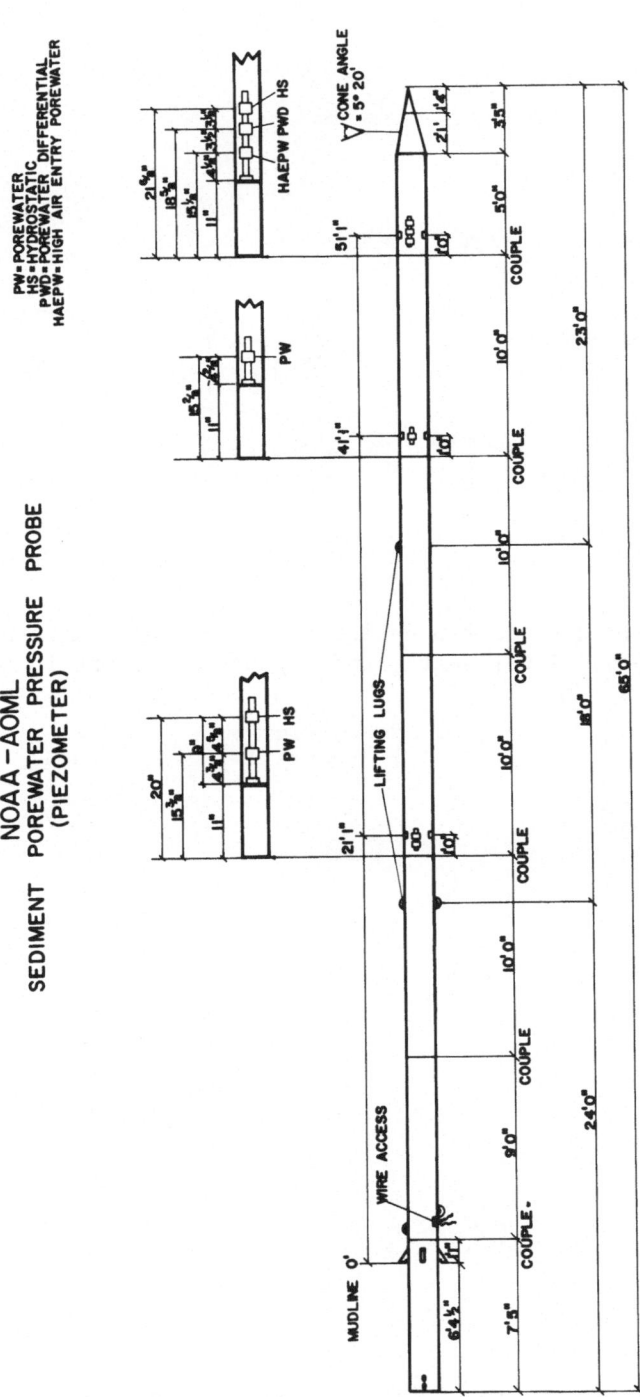

Fig. 4. Piezometer probe with locations of pressure sensors, pipe section lengths, and specifications of probe tip.

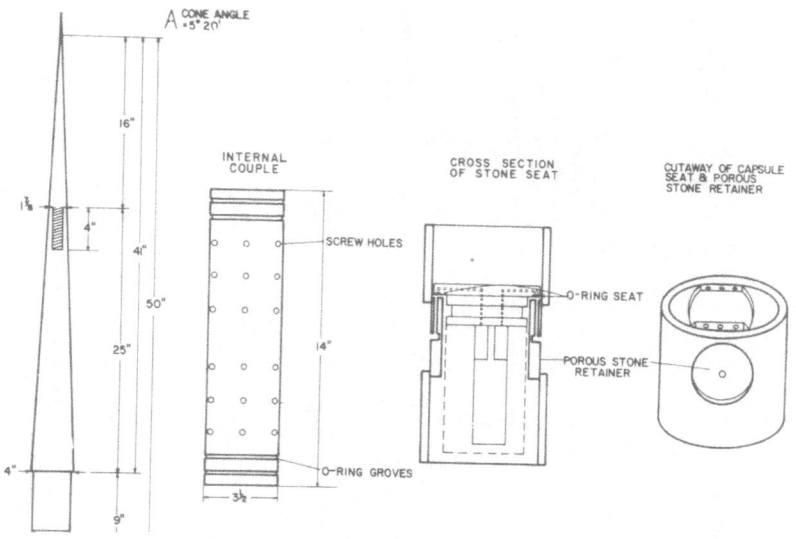

Fig. 5. Detailed probe tip design specifications, internal
couple, filter seats, and capsule placement in pipe.

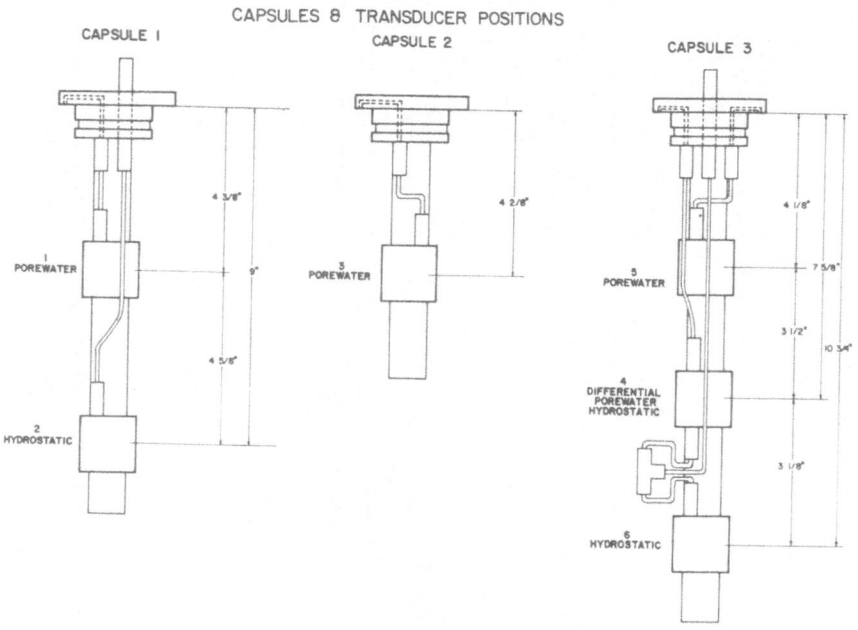

Fig. 6. Pressure sensor locations inside transducer capsules.

sealed couplings (Figs. 4, 5). This design allows selection of
the sensor and porous filters along the probe length.

Pressure sensors are of the variable-reluctance type transducers (four in the 1975 and 1976 deployments; six in the 1977
deployment; and seven in the 1978 deployment). Two absolute
pressure transducers are used to measure the "hydrostatic" or
free water column pressure inside the probe at selected depths.
Absolute pressure sensors also are used to measure pore pressure
at selected intervals and excess pore pressure is measured
directly with differential pressure sensors placed at selected
intervals along the probe length. Excess pore pressure can be
derived from the absolute difference between the pore water measurements and the free water column measurements. Transducers are
enclosed and sealed in oil-filled capsules (Figs. 5, 6), and they
are connected to porous filters on the exterior of the pipe.
Filters are coarse corundum or high-air-entry stones having
approximate porosity and permeability values of 45-50%, k =
0.1-0.3 mm/s and 35-38%, 1-3 nm/s, respectively. The rationale
for the use of different types of filters has been discussed
earlier[43].

The early prototype probe systems used a four-pen (ink)
analog recorder which proved difficult to maintain unattended
over long periods of time. The second-generation piezometer uses
three two-channel analog pen recorders as a back-up to a 14-chan-
nel analog magnetic tape recorder. The prototype piezometer
(Fig. 7) utilized a 45° single taper tip during the 1975 and 1976
deployments. This tip design was selected to minimize sediment
disturbance on the sensor side of the probe. The second-generation probe tip was designed to control sediment disturbance and
to produce a simulated plane-strain soil deformation during probe
insertion (Figs. 5, 8). The design specification for the tip was
provided by Dr. M. Esrig, and theoretically, with this design the
induced pore pressures during probe insertion can be related
directly to the undrained shear strength of the sediment[42,43].
Additional details of the piezometer instrumentation system have
been described in earlier papers[38,43,44].

SEDIMENTS

Sediment core samples were collected in Blocks 109 (1976),
28 (1977), and 73 (1978) near the piezometer sites and tests were
made to determine their geotechnical properties (Fig. 1). The
predominant sediment types from all three locations are silty
clays. Gas bubbles and laminations were apparent in cores from
each probe site. Distinctive silt lenses (characterized by their
dark color) were fairly common in cores from Block 109. The sand
content averaged: <2% with a maximum of 6% in Block 109; <1%
with a maximum of 1% in Block 28; and <5% with a measured maximum

TABLE I   GEOTECHNICAL SITES SUMMARY*

Site	Shear Strength (Field) kPa	Shear Strength (Lab) kPa	Water Content %	Liquid Limit %	Plastic Limit %	Wet Unit Weight Mg/m³	Specific Gravity	Sand Silt Clay %	Soil Description
Units:	kPa	kPa	%	%	%	Mg/m³	---	%	
1975 Block 28	1 to 10	---	85 to 109	91 (avg.)	32 (avg.)	1.43	2.70	1 Sa 29 Si 70 Cl	Generally the same soil type as Block 28 1977 Site (see below).
1976 Block 109	---	3.3 avg.	50 to 100	50 to 78	20-30	1.53 to 1.66	2.68	1 Sa 45 Si 54 Cl	Silty clay avg. <2% sand (5.9% sand max.), silt lenses and laminations, gas bubbles apparent.
1977 Block 28	2 to 5	~3.0	95 to 110	65 to 85	30 (avg.)	1.46 to 1.50	2.70	1 Sa 25 Si 79 Cl	Silty clay avg. <1% sand, worm burrows faint laminations, gas bubbles apparent.
1978 Block 73	6 to 23	6.8 (avg. from four cores)	25 to 120	60 to 90	40	1.56 (avg. from four cores)	2.70 to 2.78	10 Sa 25 Si 65 Cl	Silty clay avg. <5% sand (81% sand max.), sand lenses and laminations.

*Data compiled from several sources including boring and gravity core data from:

Mobil Oil Co., Shell Oil Co., Texas A&M Univ., U.S. Geol. Survey and National Oceanic and Atmospheric Administration (NOAA).

TABLE II PIEZOMETER OBSERVATIONS SUMMARY

Site	Latitude Longitude	Water Depth	Length of Data Recorded (hours)	Sensor Depth Below Mudline† (Meters)	Porous Filter		Sensor Type		Maximum Insertion Pressure kPa (psi)	Induced Insertion Pressure kPa (psi)	Ambient Excess Pore Pressure kPa (psi)	Tip Design
Units:					Corundum	High Air Entry	Absolute	Differential				
1975 Block 28	29°00′ N 89°15′ W	19 m	~650	8.4	X		X		50.3 (7.3)	38.6 (5.6)	11.7 (1.7)*	45° single taper
				15.2	X		X		99.2 (14.4)	55.1 (8.0)	44.1 (6.4)*	
1976 Block 109	28°52′ N 89°27′ W	36 m	~39	8.4	X		X		~46.2 (6.7)	---	---	45° single taper
				15.2	X		X		~93.0 (13.5)	---	---	
1977 Block 28	29°01′ N 89°15′ W	13 m	~2650	6.5	X		X		40.0 (5.8)	35.1 (5.1)	4.8 (0.7)	5°20′ cone taper
				12.6	X		X		77.2 (11.2)	57.9 (8.4)	19.3 (2.8)	
				15.6	X		X		100.6 (14.6)	55.1 (8.0)	45.5 (6.6)	
				15.6		X		X	86.8 (12.6)	74.4 (10.8)	12.4 (1.8)	
1978 Block 73	29°15′ N 88°55′ W	44 m	~530	6.5	X		X		59.9 (8.7)	51.0 (7.4)*	9.0 (1.3)*	5°20′ cone taper
				12.6		X	X		115.1 (16.7)	----(Anomalous Data)----	15.8 (2.3)	
				12.6		X		X	125.4 (18.2)	99.2 (14.4)	17.9 (2.6)	
				15.6	X			X	136.4 (19.8)	107.5 (15.6)	14.5 (2.1)	
										122.0 (17.7)		

*Estimates based on graphical extrapolation of decay curves.

†Estimates based on diver observations and mud remaining on weight stand.

of 81% associated with sand lenses in Block 73. The occurrence
of numerous sand layers in the sediments of Block 73 is worth
noting. For the cores analyzed, the wet unit weights were somewhat constant but increased slightly with depth. Miniature vane
shear strength measurements were performed in the field shortly
after the cores were collected in Blocks 28 and 73, and these
measurements were performed in the laboratory on cores collected
from Block 109. Noteworthy is that shear strengths from Block 73
are slightly greater than two times the shear strength of the
sediments in Blocks 28 and 109. The highest percentages of sand,
the highest shear strengths, and greatest ranges in water content
occur in the sediment cores from Block 73 as compared to the sediments from the other Blocks studied. These data and additional
geotechnical properties are summarized in Table 1, which also
includes selected data from available borings.

DISCUSSION OF PIEZOMETER DATA

General Considerations

    Pore pressure data obtained from piezometer sites in Blocks
28 (1977, East Bay) and 73 (1978, Main Pass) of the Mississippi
Delta are of primary interest in this paper for the following
reasons: (1) duration of the experiment was sufficient to
permit reasonable estimates of the ambient excess pore pressures,
(2) dissipation of excess pore pressures due to probe insertion
could be monitored without the influence of storms, (3) surface
wave activity (tides and/or short-period waves) were monitored at
these probe sites, (4) significant differences in the geology
between the probe sites and an interesting comparison of pore
pressure data and related geotechnical properties was possible,
and (5) improvements in the piezometer design and data recording
subunits provided more complete and reliable data than were
obtained with the prototype piezometer (1975, 1976). Pore pressure data obtained during 1975-76 with the prototype systems are
of secondary importance here because of the short-term deployment
in Block 109 (Southwest Pass area) and because of the severe
influence of Hurricane Eloise as it passed over the 1975 probe
site prior to the complete dissipation of the induced pore pressures. The passage of the storm complicated interpretation of
the original data, but the details of the experiment have been
described in earlier papers[38,44,45]. Insertion pressures at
Block 109 were observed with only minor changes in pore pressures
occurring during the 39-hour period (Table 2). Insertion pressures were obtained at Block 28 (1975), and the ambient excess
pore pressures were estimated based on graphical extrapolation of
the initial decay curves and the approximate times observed for
decay at the 1977 probe site. The 1977 probe site was only about

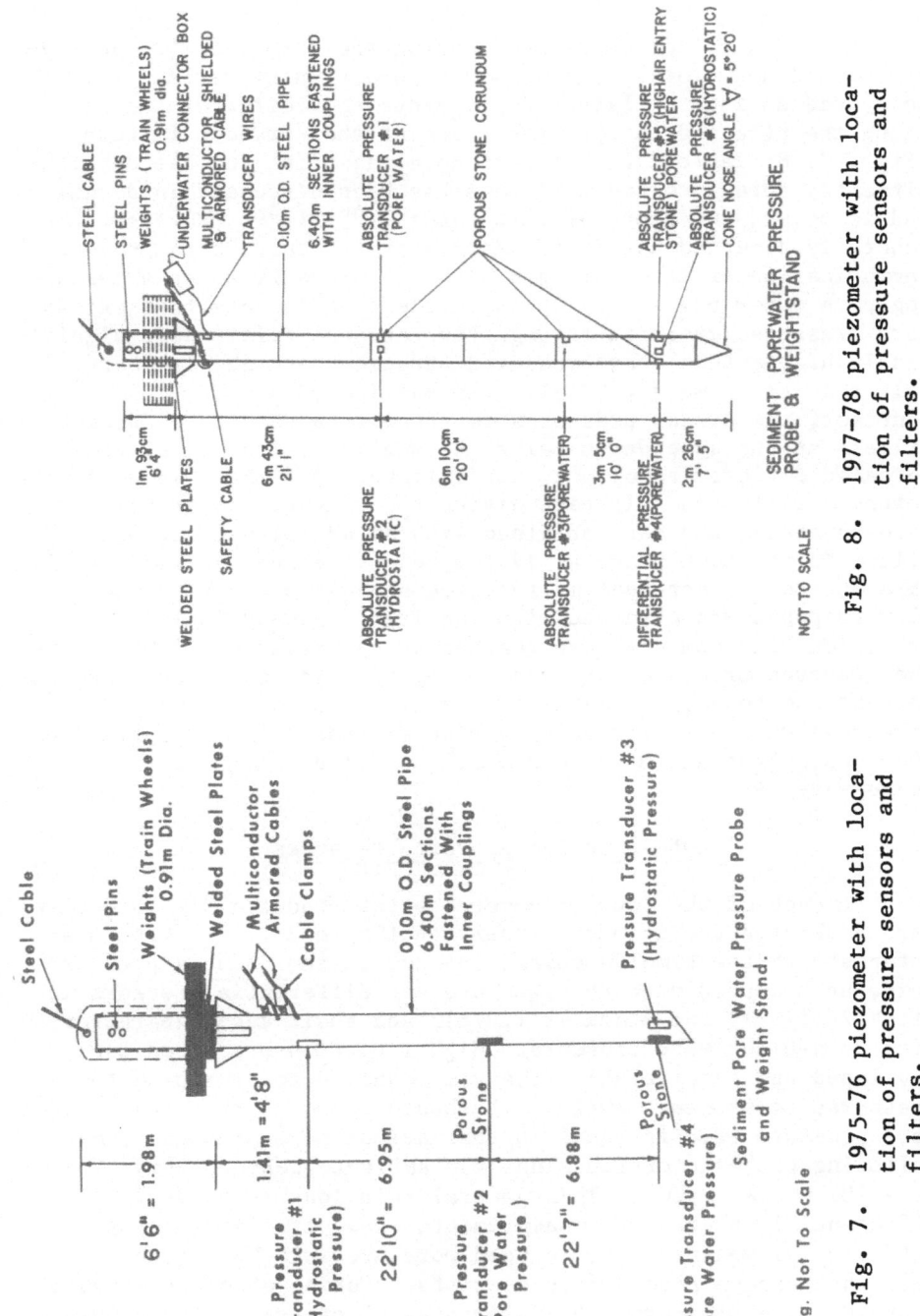

Fig. 8. 1977-78 piezometer with location of pressure sensors and filters.

Fig. 7. 1975-76 piezometer with location of pressure sensors and filters.

1830 m northeast of the 1975 site and of similar sediment type
(Table 1 and Table 2).

Corundum filters were used during the 1975 and 1976 deployments, and additional sensors and high-air-entry ceramic filters
were used during the latter deployments at similar distances
along the pipe and on opposite sides of the corundum filters
(Figs. 7, 8; Table 2). The rationale regarding the use of high-
air-entry versus corundum filters has been discussed in detail
and is beyond the scope of this paper[43,46,47,48]. However, the
generally accepted use of high-air-entry filters is to prevent
communication of free pore gas with the pressure sensors measuring pore water pressure. It is interesting to note that excess
pore pressures measured through the corundum filter were significantly higher than those measured through the high-air-entry
filter at the same depth below the mudline (Block 28, 1977). In
contrast, the excess pressures measured with the two different
filters at the same depth below the mudline (Block 73, 1978)
revealed differences of only approximately 3.4 kPa (0.5 psi), the
corundum filter in this case giving the slightly lower value.
Unfortunately, the data obtained with an additional ceramic type
filter in the same probe at 12.6 m below the mudline were unreliable due to the apparent malfunction of this pressure sensor.
Slow response was observed with the 12.6 m sensor following probe
insertion (maximum pressure reached at approximately 0.05 h) and
the observed pressures decayed to negative values which appeared
to continue to drop steadily (Table 2). Some anomalous pore
pressures were observed using a high-air-entry filter during the
1977 deployment and have been discussed in detail elsewhere[43,45,48].

Observed Excess Pore Pressures

At each of the four piezometer sites, induced pore pressures
were generated due to probe insertion (Table 2). Of particular
interest are the time-dependent changes in the induced pore pressure, as measured with the absolute and differential sensors in
Block 28 (1977) and Block 73 (1978), and their decay characteristics to ambient pore pressure. High insertion pressures were
developed and considerable time was required to dissipate these
pressures to ambient levels. It should be noted that the induced
pore pressure (U_i) is equal to the maximum pore pressure (Umax)
following probe insertion minus the ambient pressure (Uw), therefore (U_i = Umax - Uw). This general relationship holds for
differential or absolute measurements; however, in cases of
differential measurements induced pore pressure U_i = (Umax -
Ue), when excess pore pressure exists. During absolute measurements of pore pressure, the "hydrostatic" or free water column
pressure, must be subtracted from the absolute pore pressure
measurements in order to obtain ambient excess pressure (Ue).

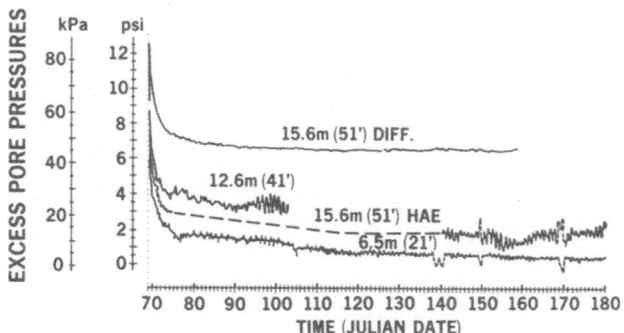

Fig. 9. Excess pore pressures measured in Block 28 (1977) in
13.5 m of water. Dashed lines represent time of anomalous
date (see Bennett and Faris, 1979). East Bay area.

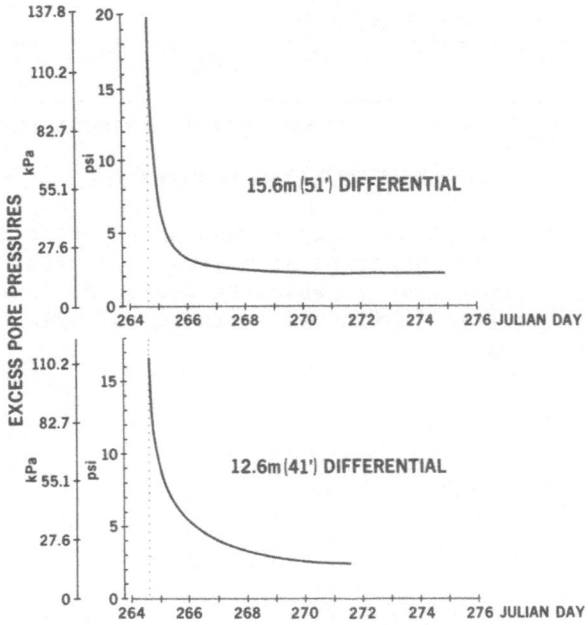

Fig. 10. Dissipation of excess pore pressures due to probe
insertion. Block 73 (1978), Main Pass area.

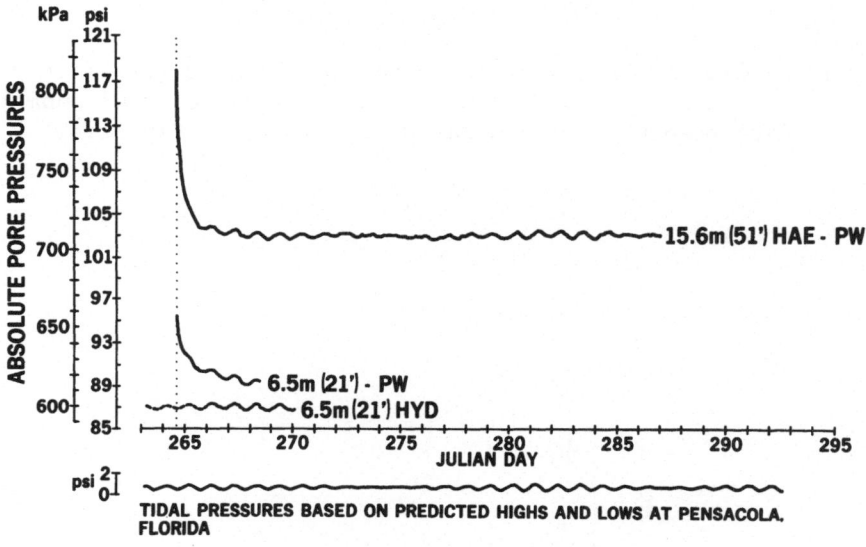

Fig. 11.   Dissipation of pressures induced by probe insertion and
ambient pore pressure as measured by absolute sensors.
Note correlation of absolute pressures (pore water and
"hydrostatic") with tidal pressures.   Block 73, Main
Pass area.

On the other hand, differential pressure sensors measure total
excess pore pressure at probe insertion and the ambient excess
directly, following the decay of insertion pressures. Obviously,
the decay characteristics of the insertion pressures measured
with differential sensors give smooth, "clean" curves compared
with those calculated from absolute sensor measurements (compare
the 15.5 m [51 ft.] differential measurements with the 6.4, 12.5,
and 15.5 m absolute sensors, Fig. 9). Although differential
sensors give direct measurements of excess pressures, absolute
sensors are required to monitor the effects of surface wave
activity on pore pressures because differential sensors measure
simultaneously the combined wave pressure through the free water
column and through the sediment. Thus, it is difficult to delin-
eate surface wave pressures in the sediment column with differential sensors. This will be shown later in this paper.

Excess pore pressures are relatively high at the 1977 probe
site with maximum values reaching 45.6 kPa (6.6 psi) at the
15.6 m depth (Table 2; Fig. 9). Excess pore pressures of 19.3
and 4.8 kPa were observed for the 12.6 and 6.5 m depths, respectively. Figure 2 represents the actual plots of pore pressure
data with depth at the 1977 site, and it is readily apparent that
the effective stress is substantially reduced by the excess pore
pressure.

Excess pore pressures measured in the East Bay area of the
Delta are considerably higher than were found in the Main Pass
area. In Block 73, the excess pore pressures were found to be
14.5 - 17.9 kPa at 15.6 m depth and 15.8 kPa at 12.6 meters below
the mudline (Figs. 10, 11; Table 2). An estimate based on graphical extrapolation of the decay curve for the 6.5 m sensor gives
an excess pore pressure of approximately 9.0 kPa. In contrast,
however, probe insertion pressures (U_i) are substantially
higher in the Main Pass sediments than those in East Bay
(Table 2). This is due to the differences in sediment shear
strength, textures, and related geotechnical properties as
observed for these two areas (Table 1). If the predicted relationships hold, insertion pressures would be expected to be highest in the sediments having the greatest shear strengths[42] as
observed for the Main Pass sediments.

In order to test the predicted relationship between maximum
induced pore pressure and undrained shear strength where U_i =
6 x (S_u) for lean inorganic clays[42], data from Block 73 were used
in addition to an earlier analysis[43] made for Block 28. Data are
summarized in Table 3. Estimates of undrained shear strength
derived from insertion pressures are in reasonable agreement with
the measured strengths for both study areas. With additional
testing using insertion data, piezometers may provide a valuable

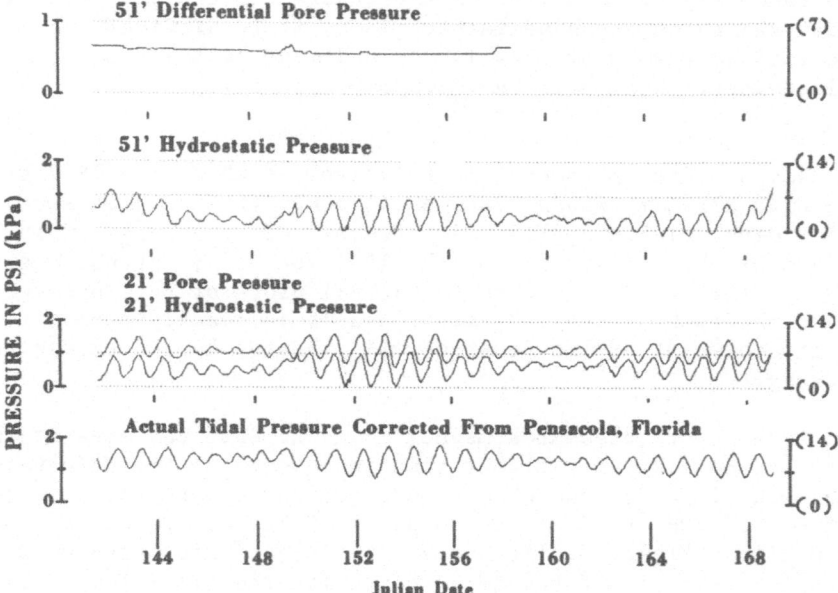

Fig. 12.   Comparison of absolute and differential pressures in
            response to tidal activity in Block 28, East Bay area.
            Note "flat" response of differential sensor due to
            pressure response through sediment and water column.
            Pressure fluctuation induced by tides in 13.5 m of
            water.

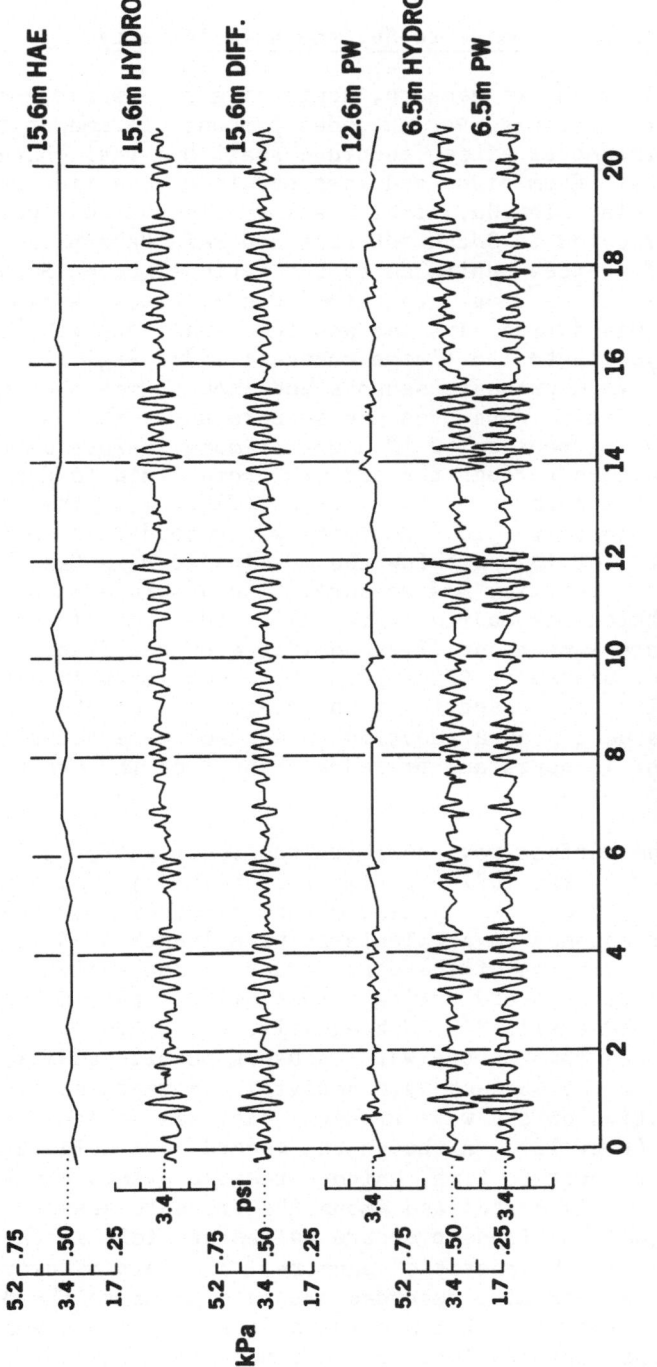

Fig. 13. Response of absolute and differential sensors due to short-period surface wave activity (~6.5 sec) in East Bay in 13.5 m of water. Note synchronous events and attenuation of pressure with sediment depth.

means for obtaining in situ undrained shear strengths of sediments.

Piezometer Response to Surface Wave Activity

The absolute pressure sensors, monitoring pore water and
free water column pressure over extended periods of time at the
1977 and 1978 piezometer sites, recorded small but real changes
in tidal activity. From highs and lows obtained at a tide station near Pensacola, Florida, tidal pressure fluctuations were
computed and corrected to Southwest Pass and Main Pass based on
the Table of Differences published by the National Ocean Survey.
Small ($\sim$1.4 kPa) as well as larger ($\sim$7.0 kPa) pressure fluctuations were recorded during days 146 and 160 (small changes; 1977,
Fig. 12) and days 151 to 156 (large changes; 1977, Fig. 12). The
response in the 6.5 m pressure sensors and lack of response in
the 15.6 m differential pore pressure sensors shows that sea surface activity having periods of 12 hours or more is felt as real
pressure fluctuations through the sediment pore fluid to depths
of at least 15-16 meters below the mudline (Fig. 12). Similar
pore pressure response to tidal activity was recorded in 43.6 m
of water to depths of 15.6 m below the mudline at Main Pass
(1978). The tidal activity was measured with absolute sensors at
6.5 and 15.6 m below the mudline (Fig. 11). Additionally, differential pressure sensors at 12.6 and 15.6 m showed flat
response to tidal pressures (Fig. 10). Data sets at both piezometer sites confirm that essentially no change in theoretical
effective stress occurs as a function of surface wave activity
having periods of 12 hours and pressure changes on the order of
7 kPa or less.

Short-period surface wave activity was recorded for a 30-
minute period during the 1977 experiment in East Bay. Auto-cor-
relation of the "hydrostatic" pressure records gives a period of
6.4 s, which corresponds to a calculated wave length of 53.3 m in
13.7 m of water. Details of the data reduction and analysis of
four of the pressure sensors during a three-minute period of time
have been discussed earlier[43]. Subseqently, reduction of data
from the 12.6 m and 15.6 m pore water sensors permits additional
comments regarding the surface wave activity. A representative
20-21 minute portion of the wave pressure data was selected for
this discussion (Fig. 13). Although the record is considerably
compressed with respect to time, unique pressure events can be
discerned and visually correlated among the pressure sensors.
Approximately equal amplitude pressure pulses are felt in the
6.5 m pore water and "hydrostatic" sensors. The 15.6 m "hydrostatic" pressure sensor also recorded events similar to the 6.5 m
sensors. In contrast, the 12.6 m pore water sensor shows small,
irregular pressure perturbations that are not generally correlated with the 6.5 m sensors. Likewise, the 15.6 m (HAE)

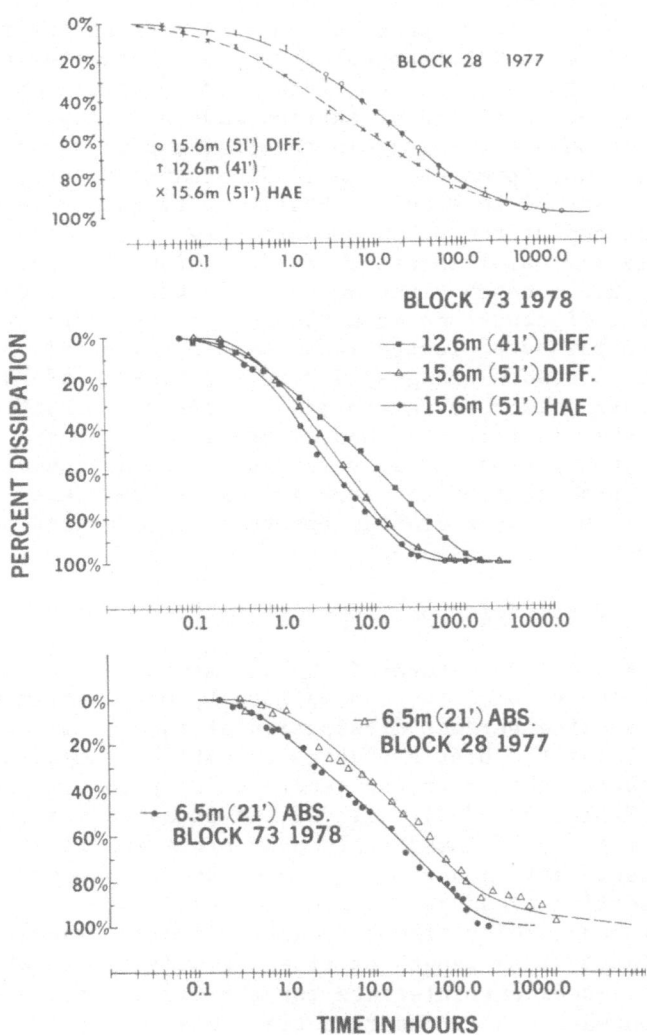

Fig. 14.   Comparison of times to dissipate pressures induced by
           probe insertion normalized in terms of percent of pres-
           sure induced for different sensor depths below the mid-
           line. Note considerably greater times required to
           reach ambient pressure for Block 28 versus 73
           sediments.

absolute sensors show small, irregular pressure undulations with
virtually "flat" response. These two absolute sensors (12.6 and
15.6 m) indicate a significant pressure attenuation with depth
below mudline due to the observed short-period surface wave
activity (Fig. 13). In addition, the 15.6 m differential sensor
shows a negative cross-correlation (-0.68) indicating the excess
pressure is dominated by the fluctuations in the "hydrostatic"
(free water column) instantaneous pressure port on the reference
side of the differential transducer[43]. This indicates essentially no short-term change in the total pore pressure at 15.6 m as
a function of the short-period surface wave activity. It is
interesting to note that although the 12.6 m pore water sensor
record shows minor (pressure change) fluctuations, the greatest
activity is observed to occur generally at times during maximum
activity recorded at the 6.5 m sensors (Fig. 13). This suggests
that possibly some small pressure perturbations were actually
felt at the 12.6 m depth. During the 1975 pore pressure experiment, pressure fluctuations were recorded by the pore pressure
sensors (pore pressure fluctuations were approximately 50% of the
"hydrostatic" pressure changes) during the passage of Hurricane
Eloise; however, discrete waveform data were not possible at that
time due to the inability to record data at a high data rate
during the storm period[44]. Improvements in data acquisition systems and the placement of pressure sensors in future experiments
hopefully will provide a greater opportunity to assess potential
wave effects on pore pressure.

Dissipation of Induced Pore Pressures

     Comparison of the nature of the sediments in East Bay and
Main Pass at the probe sites can be made by examination of decay
curves representing percent dissipation of induced insertion pore
pressures. Using the best estimates of ambient excess and total
induced pressure, the pressures generated by probe insertion were
determined (Tables 2 and 3). All of the pressure sensors did not
reach maximum pressure immediately upon probe insertion and the
maximum pressure attained shortly thereafter was used to calculate the insertion pressure for each sensor. The 6.5 m sensors
showed the longest delay time to reach maximum pressure. The
apparent slightly long length of time for sediment remolding at
the probe pipe-sediment interface for the 6.5 m sensor depth may
have contributed to their observed time delay.

     Insertion pressures were normalized to produce a plot of
percent dissipation versus the logarithm of time (Fig. 14). The
time required for the insertion pressures to dissipate to ambient
pressure is, to a first approximation, a function of the probe
pipe radius and the sediment coefficient of consolidation, which
is highly dependent upon the permeability of the particular
material[49,50].

TABLE III CALCULATED SHEAR STRENGTHS VERSUS MEASURED STRENGTHS

	Pressure Sensor Depth Below Mudline	U_i (psi)	$U_i/6$ (psi)	S_u kPa (psi)	Depth Below M.L.	Test Type
Block 73	6.5 m	7.4	1.23	9.4 (1.36)	(5.9-7 m)	lab. vane on six cores
	12,6 m	14.4	2.40	20.0 (2.9)	(12.4 m)	in situ vane
	15.6 m	15.6	2.60	24.1 (3.5)	(15.5 m)	in situ vane
	15.6 m	17.7	2.95	24.1 (3.5)	(15.5 m)	in situ vane
Block 28	6.5 m	5.1	0.85	5.8 (0.84)	(~5-7 m)	in situ vane (ave. of 3)
	12.6 m	8.4	1.40	9.0-11.7 (1.3-1.7)*	(~12.6 m)	in situ vane
	15.6 m	8.0	1.33	9.0-11.7 (1.3-1.7)*	(~15.6 m)	in situ vane
	15.6 m	10.8	1.80	9.0-11.7 (1.3-1.7)	(~15.6 m)	in situ vane
				7.9-9.5 (1.0-1.4)*		lab. vane

S_u = undrained shear strength

$U_i/6$ = calculated undrained shear strength

TABLE IV

t_{100} VALUES DERIVED USING LOG-FITTING TECHNIQUE
FROM CONSILIDATION THEORY

	Depth	t_{100}
Block 73	6.5 m	140 hrs.
	12.6 m	98 hrs.
	15.6 m	23 hrs.
	15.6 m	19 hrs.
Block 28	6.5 m	190 hrs.
	12.6 m	~200 hrs.
	15.6 m	~200 hrs.
	15.6 m	~200 hrs.

TABLE V

PERCENT REDUCTION (%Δ) IN VERTICAL STRESS
DUE TO EXCESS PORE PRESSURE

| | Depth Meters | Total Effective Vertical Geostatic Stress $\overline{\sigma}vg$ | | Effective Vertical Stress $\overline{\sigma}v_e$ | | % Δ |
		kPa	PSI	kPa	PSI	
1977 Site Block 28	6.5	24.66	3.58	19.80	2.87	20
	12.6	48.91	7.10	29.61	4.29	40
	15.6	61.32	8.90	15.82	2.29	76
1978 Site Block 23	6.5	23.77	3.45	14.77	2.14	38
	12.6	54.91	7.94	39.11	5.67	29
	15.6	71.90	10.44	54.00	7.83	25

Plots showing the time-related dissipation of induced pressures indicate significant differences in decay times for the two
probe sites. Using the log-fitting technique from consolidation
theory[51], t_{100} values were determined (Table 4). Main Pass sediments at the 15.6 m depth indicate t_{100} values approximatley 10
times faster than for East Bay sediments (Fig. 14). The 12.6 m
depth shows t_{100} to be about 2 times faster for the Main Pass
sediment. On the other hand, the 6.5 m dissipation curves are
only about 25-26% faster than at East Bay. These differences
are undoubtedly a function of the sediment types and related
properties characteristic of each site. Obviously, the overall
coarser textures of the sediments at Main Pass (Table 1), and the
presence of sand layers observed in these cores, result in high
permeabilities and local drainage paths for pore water dissipation to occur rather rapidly. Boring data obtained near the
probe site (Block 73) indicated sediments are predominantly clay
to depths of 12.6 m with silt pockets and seams. Below 13.0 m,
there is a marked increase in sand content with sandy silt
layers. Some shell fragments were observed below 13.0 meters.
Also, these factors probably account for the occurrence of considerably lower ambient excess pore pressures at Main Pass as
compared to East Bay.

Effective Stress Estimates and Stability Considerations

The vertical effective geostatic stress ($\overline{\sigma}$vg) from eq. (3)
was calculated from the available sediment core and borehole data
for the two probe sites (1977 and 1978). Using the best estimates of the ambient excess pore pressure (Table 2) at the
respective depths below the mudline for each site, the vertical
effective stress ($\overline{\sigma}v_e$) from eq. (4) was determined as well as the
percent reduction in effective geostatic stress (Table 5). The
reduction in effective stress for the 1977 site at 6.5, 12.6 and
15.6 m is 20%, 40% and 76%, respectively, which agree with
earlier estimates[43] of 19%, 39% and 74% for the same site. These
data show an increase in the reduction in effective stress with
depth whereas the 1978 site data show a decrease in effective
stress reduction with depth below the mudline (Table 5). The
reduction in effective stress at the 1978 site for the same
equivalent depths are 38%, 29% and 25%. These data agree with
what would be expected based on the characteristics of the sediment and coarsening of texture with depth. In each case, the
reduction in effective stress is significant.

In order to evaluate the possible influence of excess pore
pressures on the calculations of seafloor stability, a few preliminary analyses were made using the infinite slope[52] and
Bishop[53] slip circle techniques for computing factors of safety
(F.S.) for potential slope failures. Data from the Block 73 site
were used in the analysis which was performed with an in-house

modified version of a computer program described earlier[54].
Input of the excess pore pressures in one case and recomputing
F.S. values based on no excess pore pressures gave significant
differences in results. Computing factors of safety to depths of
20 m below the mudline showed reductions in F.S. values of up to
17.5% for a 0.2° slope when excess pore pressures were considered
with no surface wave activity. Factors of safety for a 20° slope
were found to be reduced by up to 20% when excess pore pressures
were considered. Although these calculations were first approxi-
mations based on simplified methods of analysis with no wave
effects, the exercise demonstrated the importance in knowing the
state of stress and the importance in obtaining pore pressure
measurements when assessing the stability of certain submarine
sediments. A knowledge of pore pressures and state of stress
could be particularly important in sedimentary environments where
factors of safety are low based on assumptions of zero excess
pore pressure (calculations based on $\bar{\sigma}$vg, eq. [3]) and where
storm waves frequently affect the bottom. An earlier analysis[32]
using infinite slope techniques to assess stability in the Mississippi Delta suggested that the reduction in effective stress
may be sufficient to cause seafloor failure by gravitational
stresses alone. The piezometer data clearly demonstrate the
feasibility of making pore pressure measurements and the criticality of excess pore pressures in Mississippi Delta sediments.

CONCLUSIONS

    The feasibility of measuring absolute and excess pore pres-
sures in submarine sediments on continental shelves has been
demonstrated using multisensor piezometer probes. Relatively
high excess pore pressures were found for the sediments in East
Bay and somewhat lower values were found to be characteristic of
the Main Pass sediment in the Mississippi Delta. The excess
pressures reduced the calculated effective stress significantly
at both sites. Coarser sediment textures and higher shear
strengths were found for the Main Pass area compared to East Bay.
These selected geotechnical properties were correlated with the
pore pressure data including observed insertion pressures, decay
times for the induced pore pressure, and the ambient excess pressures.

    Decay time for the induced pressures to reach ambient pres-
sure appears to be highly dependent upon the sediment type and
texture for similar probe pipe diameters. The induced pore pressures were found to be directly related to the undrained shear
strength for the two sites evaluated. With additional testing,
the piezometer probe may prove to be a valuable tool in evaluating the in situ undrained shear strength of submarine sediments.
Derived values for shear strength from induced pore pressure data
appear, to a first approximation, to agree with in situ remote

vane data, as well as or better than is commonly observed between laboratory and in situ tests on Mississippi Delta sediments. Undoubtedly, smaller diameter probes than were used during these experiments would reduce the induced pore pressure decay time considerably if several short-term tests were desired in a local study area. This would be particularly important in studies designed to map out the pore pressure fields as was undertaken during the 1978 NOAA experiment in cooperation with Lehigh University. Data from these tests using the Lehigh mini-piezometer are currently being analyzed at Lehigh University.

Preliminary stability analyses comparing factors of safety computed with and without excess pore pressure were found to show significant differences. Up to 20% reduction in calculated factors of safety resulted when excess pore pressures were included in the stability calculations.

Limited dynamic surface wave activity due to tides and short-period waves were observed to have an effect on the pore pressures. Tidal pressures were found to be felt to depths of at least 15-16 m below the mudline having little or no effect on the theoretical effective stress. Short-period (approximately 6.5 s) waves were felt to depths of between 6.5 and 12.6 m below the mudline with severe pressure attenuation at depths of 12.6 m and greater. Considerably more analysis is necessary in order to assess the potential time-dependent changes in stress with depth below the mudline for the short-period wave activity.

ACKNOWLEDGEMENTS

The authors appreciate the assistance of a number of colleagues during various phases of this project. Messrs. Douglas Lambert and William Sawyer of AOML contributed greatly to the success of the field activities. Special thanks are due to Ms. Frances Nastav and Mr. George Merrill of AOML for their tireless and expert assistance during data reduction and drafting of the illustrations. The authors appreciated the support of several individuals at Texas A&M University, Louisiana State University, Shell Oil Co., Mobil Oil Corp., and the U.S. Geological Survey during various phases of these studies. The piezometer investigations and funding were supported primarily by NOAA with assistance from various other organizations mentioned above. The authors appreciate critical reviews of the manuscript by Drs. George H. Keller, David B. Prior, R. Chaney, and Kjell Karlsrud.

REFERENCES

1. Bea, R. G. and P. Arnold, Movements and forces developed by
    wave-induced slides in soft clays, Fifth Annual Offshore
    Technology Conference, p. 731-742 (1973).
2. Bea, R. G., H. A. Bernard, P. Arnold, and E. H. Doyle, Soil
    movements and forces developed by wave-induced slides in
    the Mississippi Delta, Journal of Petrol. Technol.,
    p. 500-514 (1975).
3. Jacobi, R., Sediment slides on the northwestern continental
    margin of Africa, Mar. Geol. 22, p. 157-173 (1976).
4. Coleman, J. M. and D. B. Prior, Submarine landslides in the
    Mississippi River Delta, 10th Annual Offshore Technology
    Conference, p. 1067-1071 (1978).
5. McGregor, B. A. and R. H. Bennett, Continental slope sedi-
    ment instability northeast of Wilmington Canyon, Amer.
    Assoc. of Petrol. Geol. Bull., 61, p. 918-928 (1977).
6. Bennett, R. H., D. N. Lambert, B. A. McGregor, E. B. Forde,
    and G. F. Merrill, Slope map: A major submarine slide on
    the U.S. Atlantic continental slope east of Cape May,
    A-5787, U.S. COMM-NOAA-DC (1978).
7. Coleman, J. M. and L. E. Garrison, Geological aspects of
    marine slope stability, northwestern Gulf of Mexico, Mar.
    Geotech., 2, p. 9-44 (1977).
8. Embley, R. W. and R. D. Jacobi, Distribution and morphology
    of large submarine sediment slides and slumps on the
    Atlantic continental margins, Mar. Geotech., 2, p. 205-
    228 (1977).
9. Hampton, M. A. and A. H. Bouma, Slope instability near the
    shelf break, western Gulf of Alaska, Mar. Geotech., 2,
    p. 309-331 (1977).
10. Almagor, G. and G. Wiseman, Analysis of submarine slumping
    in the continental slope off the southern coast of
    Israel, Mar. Geotech., 2, p. 349-388 (1977).
11. Prior, D. B., J. M. Coleman, J. N. Suhayda, and L. E. Gar-
    rison, Subaqueous landslides as they affect bottom struc-
    tures, Port and Ocean Eng. Under Arctic Conds. (POAC 79),
    p. 921-933 (1979).
12. Prior, D. B. and J. M. Coleman, Submarine landslides on the
    Mississippi River Delta-front slope, Geoscience and Man
    19, p. 41-53 (1978).
13. Prior, D. B., J. M. Coleman, and L. E. Garrison, Digitally
    acquired undistorted side-scan sonar images of submarine
    landslides, Mississippi River Delta, Geology 7, p. 423-
    425 (1979).
14. Knebel, H. T. and B. Carson, Small-scale slump deposits,
    Middle Atlantic continental slope, off eastern United
    States, Mar. Geol. 29, p. 221-236 (1979).

15. Keller, G. H., D. N. Lambert, and R. H. Bennett, Geotechnical properties of continental slope deposits--Cape Hatteras to Hydrographer Canyon, Soc. Econ. Paleon. & Min. Special Pub. No. 27, p. 131-151 (1979).

16. McGregor, B. A. and R. H. Bennett, Mass movement of sediment on the continental slope and rise seaward of the Baltimore Canyon Trough, Mar. Geol. 33, p. 163-174 (1979).

17. McGregor, B. A., R. H. Bennett, and D. N. Lambert, Bottom processes, morphology, and geotechnical properties of the continental slope south of Baltimore Canyon, Appl. Ocn. Res., 1, p. 177-187 (1979).

18. Summerhayes, C. P., B. D. Bornhold, and R. W. Embley, Surficial slides and slumps on the continental slope and rise of southwest Africa: A reconnaissance study, Mar. Geol. 31, p. 265-277 (1979).

19. Almagor, G., Halokinetic deep-seated slumping on the Mediterranean slope of northern Sinai and southern Israel, Mar. Geotech., 4, p. 83-105 (1980).

20. Bunn, A. R. and B. A. McGregor, Morphology of the North Carolina continental slope, western North Atlantic, shaped by deltaic sedimentation and slumping, Mar. Geol. 37, p. 253-266 (1980).

21. Bennett, R. H., D. N. Lambert, and M. H. Hulbert, Geotechnical properties of a submarine slide area on the U.S. continental slope northeast of Wilmington Canyon, Mar. Geotech., 2, p. 245-261 (1977).

22. Prior, D. B. and J. M. Coleman, Disintegrating retrogressive landslides on very-low-angle subaqueous slopes, Mississippi Delta, Mar. Geol., 3, p. 37-60 (1977).

23. Stanley, D. J. and N. Silverberg, Recent slumping on the continental slope of Sable Island Bank, southeast Canada, Earth Plant. Sci. Lett. 6, p. 123-133 (1969).

24. Bohlke, B. M. and R. H. Bennett, Mississippi Prodelta crusts: A clay fabric and geotechnical analysis, Mar. Geotech., 4, p. 55-82 (1980).

25. Bea, R. G., S. G. Wright, P. Sircar, and A. W. Niedoroda, Wave-induced slides in South Pass Block 70, Mississippi Delta, Preprint 80-506, Am. Soc. Civ. Eng. Convention & Exposition, 23 pp., 1980.

26. Malahoff, A., R. W. Embley, R. B. Perry, and C. Fefe, Submarine mass-wasting of sediments on the continental slope and upper rise south of Baltimore Canyon, Earth Planet. Sci. Lett. 49, p. 1-7 (1980).

27. Wright, S. G. and R. S. Dunham, Bottom stability under wave-induced loading, 4th Annual Offshore Technology Conference, 1, p. 853-862 (1972).

28. Wright, S. G., Analyses for wave induced sea-floor movements, Proc. 8th Annual Offshore Technology Conference, 1, p. 41-50 (1976).

29. Yamamoto, T., Wave induced instability in seabeds, Coastal
     Sediments '77, Am. Soc. Civ. Eng., p. 898-913 (1977).
30. Yamamoto, T., H. L. Koning, H. Sellmeijer, and E. van Hijum,
     On the response of a poro-elastic bed to water waves, J.
     Fluid Mech., 87, p. 193-206 (1978).
31. Schapery, R. A. and W. A. Dunlap, Prediction of storm-
     induced sea bottom movement and platform forces, Proc.
     10th Annual Offshore Technology Conference, p. 1789-1793
     (1978).
32. Prior, D. B. and J. N. Suhayda, Application of infinite
     slope analysis to subaqueous sediment instability, Mis-
     sissippi Delta, Engr. Geol. 14, p. 1-10 (1979).
33. Prior, D. B. and J. N. Suhayda, Submarine mudslide morphol-
     ogy and development mechanics, Mississippi Delta, Proc.
     11th Annual Offshore Technology Conference, p. 1055-1058
     (1979).
34. Yamamoto, T., Sea bed instability from waves, Proc. 10th
     Annual Offshore Technology Conference, p. 1819-1824
     (1978).
35. Richards, A. F. and H. Matlock, eds., Present and recom-
     mended U.S. Government research in seafloor engineering,
     NOAA, NOAA-S/T 78-413, U.S. Gov't Print. Off., 93 pp.
     (1978).
36. Esrig, M. I., R. S. Ladd, and R. G. Bea, Material properties
     of submarine Mississippi Delta sediments under simulated
     wave loadings, Proc. 7th Annual Offshore Technology Con-
     ference, p. 399-404 (1975).
37. Henkel, D. J., The role of waves in causing submarine land-
     slides, Geotech., 20, p. 75-80 (1970).
38. Bennett, R. H., W. R. Bryant, W. A. Dunlap, and G. H.
     Keller, Initial results and progress of the Mississippi
     Delta sediment pore water pressure experiment, Mar.
     Geotech., 1, p. 327-335 (1975).
39. Suhayda, J. N., T. Whelan III, J. M. Coleman, J. S. Booth,
     and L. E. Garrison, Marine sediment instability: Inter-
     action of hydrodynamic forces and bottom sediments, Proc.
     8th Annual Offshore Technology Conference, p. 30-33
     (1976).
40. Yamamoto, T., H. Sellmeijer, and E. van Hijum, Wave induced
     pressure, stress and strain in sand beds, 7th Int'l
     HavenKongres, K.V.I.V., p. 1.03/1-1.03/11 (1978).
41. Yamamoto, T., Wave induced stress instabilities in inhomo-
     geneous seabeds--stability analysis of seafloor founda-
     tions, 17th Annual Conf. on Coastal Engineering,
     I.C.E.E., Australia, 2 pp. (1980).
42. Esrig, M. I., R. C. Kirby, and R. G. Bea, Initial develop-
     ment of a general effective stress method for the predic-
     tion of axial capacity for driven piles in clay, Proc.
     9th Annual Offshore Technology Conference, p. 495-501
     (1977).

43. Bennett, R. H. and J. R. Faris, Ambient and dynamic pore
    pressures in fine-grained submarine sediments: Missis-
    sippi Delta, Appl. Ocn. Res., 1, p. 115-123 (1979).
44. Bennett, R. H., Pore-water pressure measurements: Missis-
    sippi Delta submarine sediments, Mar. Geotech., 2,
    p. 177-189 (1977).
45. Dunlap, W. A., W. R. Bryant, R. H. Bennett, and A. F.
    Richards, Pore pressure measurements in underconsolidated
    sediments, Proc. 10th Annual Offshore Technology Confer-
    ence, 2, p. 1049-1056 (1978).
46. Vaughn, P. R., The measurement of pore pressures with
    piezometers, in: "Field Instrumentation in Geotechnical
    Engineering," John Wiley, New York, p. 411-422 (1973).
47. Bishop, A. W., Pore pressure measurements in the field and
    in the laboratory, Proc. 7th Int'l Conf. on Soil Mech.
    and Foundation Eng., 3, p. 427 (1969).
48. Hulbert, M. H. and R. H. Bennett, Anomalous pore pressures
    in Mississippi Delta sediments: Gas and electrochemical
    effects, submitted to Mar. Geotech.
49. Soderberg, L. O., Consolidation theory applied to foundation
    pile time effects, Geotech. 12, p. 217-225 (1962).
50. Verruijt, A., A simple formula for the estimation of pore
    pressures and their dissipation, Appl. Ocn. Res. 2,
    p. 57-62 (1980).
51. Lamb, T. W. and R. V. Whitman, "Soil Mechanics," John Wiley,
    New York, 553 pp. (1969).
52. Taylor, D. W., "Fundamentals of Soil Mechanics," Wiley, New
    York, 322 pp. (1948).
53. Bishop, A. W., The use of the slip circle in the stability
    of earth slopes, Geotech., 5, p. 7-17 (1955).
54. Haffner, J. C., Influence of ocean surface waves on the
    instability of marine sediments, In-house report, Marine
    Geotechnical Laboratory, Lehigh University, 85 pp.
    (1977).

MARINE SLOPE STABILITY - A GEOLOGICAL APPROACH

Adrian F. Richards and Ronald C. Chaney

Marine Geotechnical Laboratory
Lehigh University
Bethlehem, PA 18015

INTRODUCTION

On one hand, the literature of marine slope instability and
marine slides is voluminous. Extensive bibliographies are given in
the many papers comprising this volume. On the other hand, basic
information available to geoscientists on how marine slides or slope
instability can be investigated is widely scattered. This paper
attempts to fill the gap by briefly discussing mthods for preliminary surveys, sampling, and analysis. Emphasis is placed on
providing elementary information for geoscientists who might be
starting geotechnical studies. More elegant or detailed treatments
on this subject are available elsewhere in this volume and in the
science and engineering literature.

The proposed general methods of marine slide investigation in
this paper were developed by Kjell Karlsrud and Adrian Richards at
the end of the NATO Workshop on Marine Slides. Members of the Workshop reviewed this information and provided a number of useful
additions. We acknowledge their assistance as well as the initial
co-authorship by Mr. Karlsrud. To complete the general methods,
Ronald Chaney has added a section on elementary methods of slope
stability determination and analysis. For convenience, each step
will be described separately.

SURVEY AND MAPPING PROGRAM

Geophysical surveys and bathymetric mapping are essential first
steps for any investigation of seafloor sediment instability.
Methods and relevant literature were previously cited by Richards
et al. (1976). Table 1 summarizes geophysical survey methods and

tools yielding data that eventually will result in bathymetric and
geologic maps adequate for the intended investigation. However,
before any map can be considered valid a ground truth must be determined. This is conventionally done by means of bottom observations from submersibles, or by sampling, or by both methods.

Once a suitable bathymetric map and, preferably, a geologic map
have been prepared and verified for ground truth, the next step is
to sample for geotechnical purposes and/or to undertake in situ
testing. Usually both methods are conducted concurrently if in situ
testing is to be performed. A decision analysis, including a cost/
benefit analysis, is customary to decide what types of sampling
and/or in situ testing methods are most appropriate and can or will
be deployed for the investigation.

Table 1. Geophyscial Survey Methods and Equipment

I. Side Scan Sonar Mapping Methods

Range	Examples	Maximum Water Depth, km	Maximum Slant Range, km
Long	GLORIA	7.5	70
Intermediate	Sea MARC	6	6
Short	E.G.&G.-type	1-6	0.5

II. Bathymetric Mapping Methods
Wide-beam (conventional) echo sounders
Narrow-beam echo sounders

III. Subbottom Profiling Methods
Low frequency for penetration: multichannel devices
High frequency for stratigraphic detail: "sparker"; 3.5 kHz;
multichannel devices; Deep Tow or similar type of remote
device

SOIL SAMPLING TECHNIQUES

Soil sampling is defined by a seismic survey of the site based
on project requirements. A minimum sampling program for a typical
site would consist of at least one sample from each soil layer
within the project area.

The four basic soil sampling methods applicable for marine
geotechnical investigations are summarized in Table 2. Piston and
gravity cores are normally employed for sampling soils at shallow
depths (<20m) while a combination of rotary drilling with wireline
sampling is used at deeper depths (>20m). When piston coring, the
piston should be fixed so that it does not move vertically during
the drive stroke. Reasons why this is essential for high quality
sampling have been given by Richards and Parker (1968).

Table 2. Sampling Methods

Typical Penetration Depths, m	Tools
<5	Gravity Type "Drop" Corer
<20	Piston Corer
<50 to 100	Giant Corer[a]
>10 to 20	Rotary Drilling and Wireline Sampling

[a]Device is currently under development.

IN-SITU TESTING TECHNIQUES

For the solution of most geotechnical problems in slope instability investigations, the key parameters required to be measured are shear strength, bulk density, and excess pore pressure. These parameters can be used for either total stress or effective stress methods of analysis, which will be described later. Table 3 summarizes relevant information on tools and methods.

Dynamic methods of analysis at present usually include shear moduli determined from laboratory tests. In situ equipment, listed in Table 3, is under development and may be soon ready for field deployment.

The pressuremeter, which provides a measure of stress-strain information, sometimes is used, but not commonly in slope instability investigations; nevertheless, it also is listed in Table 3.

Most in situ equipment has been used in water depths of less than a few hundred meters. At these depths, the CPT and vane have been used at penetration depths generally less than about 50 m from remote devices and to greater penetration depths through a drill-string. Richards and Chaney (in press) summarized equipment availability for in situ testing in water depths of thousands of meters. They observed that the vane, static CPT, and nuclear transmission densitometer have been used only for very shallow penetration (less than 5 m) from remote vehicles. The dynamic CPT has a greater penetration depth capability, but the resulting data are presently less reliable than comparable data from the static CPT. The Deep Sea Drilling Project has used conventional oil company logging tools in comparable water depths. These tools have the capability of deep penetration through a drilled open hole in the seafloor.

Table 3. In Situ Tools and Applications

Geotechnical Property to be Measured	Tool or Method	Direct or Indirect Measurement	Continuous or Discontinuous or Measurement	Operational Experimental	Common Deployment Method	
					Drillship	Remote/ Tethered
Shear Strength	Vane	Direct	Discontinuous	Operational	X	X
Shear Strength	CPT*-Static CPT-Dynamic	Indirect Indirect	Continuous Continuous	Operational Operational/ Experimental	X	X X
Excess Pore Pressure	Piezometer	Direct	Usually Discontinuous	Operational	X	X
Bulk Density	Electrical Resistivity	Indirect	Continuous	Operational	X	X
Bulk Density	Nuclear Backscatter	Direct	Continuous	Operational	X	X
Bulk Density	Nuclear Transmission	Direct	Continuous	Experimental		X
Compressional Wave	Wave Propagation	Indirect	Continuous	Operational	X	X
Shear Wave	Wave Propagation	Indirect	Discontinuous	Experimental	X	X
Stress Strain	Menard Pressuremeter	Direct	Discontinuous	Operational/ Experimental	X	

*CPT = Cone Penetration Test

RECOMMENDED SAMPLING AND IN SITU TESTING METHODS

Recommendations for sampling and in situ testing for most slide or stability investigations using state-of-the art equipment (1981) are presented in Table 4. This listing does not necessarily represent the most desirable method, instead it reflects present capabilities. In the next few years, it is likely that deep water in situ testing tools will be developed and used for remote testing with penetrations of 6-10 m and for use from a drillship with substantially greater penetration.

Table 4. Recommended Methods for Sampling and In Situ Testing

I. Sampling

Water Depth, m	Penetration Depth, m	Tool
Any	<5	Gravity corer
Any	1-20	Fixed piston corer
10-6,000	5-200+	Drillship rotary corer

II. In Situ Testing

Water Depth, m	Penetration Depth, m	Tool	Deployment Ship/ Remote Device	Deployment Drillship Tool
<200	<50	"Static" CPT	X	X
<200	<50	Differential Pressure Piezometer	X	X
>200	<15	Total Pressure Piezometer	X	X

PREDICTIVE MODELING AND TESTING

The estimation of the stability of a slope against movement (i.e. failure) consists of four interrelated elements. These elements are (1) the characterization of the seabed, (2) the characterization of external environmental loading, (3) the analysis methodology, and (4) the determination of the safety against failure. These various elements will be discussed in detail in the following sections. The interrelationships between the four elements are summarized in Fig. 1.

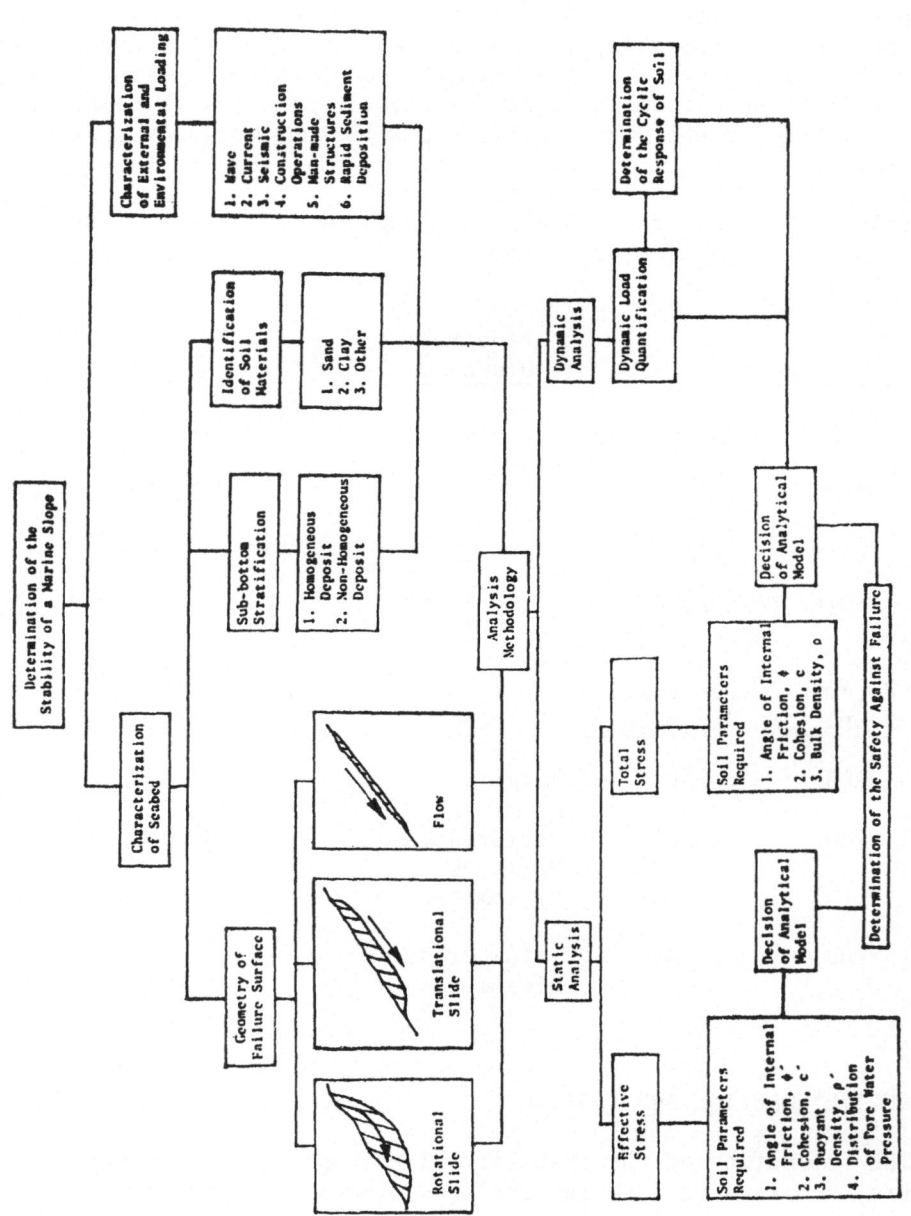

Figure 1. Determination of the Stability of a Marine Slope

Characterization of the Seabed

The characterization of the seabed for use in a slope stability analysis requires that the topography and stratigraphy of the slope under study be defined. This knowledge is necessary to describe both the geometry of a potential failure surface (i.e. rotational, translational, or flow) and its material composition. This information is required for the selection of the appropriate mathematical model to describe the behavior of the slope.

Characterization of External and Environmental Loading

Movement of a marine slope is associated with the application of external and environmental loading applied to it (i.e. driving forces), Fig. 2. External and environmental loading consists of loads resulting from (1) waves, (2) currents, (3) seismic, (4) construction operations, (5) man-made structures, (6) rapid sediment deposition, and (7) gravity loads. The effect of these loadings on the behavior of a slope will depend upon their rate of application (static or dynamic) and duration.

Analysis Methodology

The analysis methodology to be utilized to determine the stability of a marine slope will depend upon (1) the type of external and environmental loading (i.e. static and/or dynamic), (2) the need to determine long or short term stability, (3) geometry of failure surface, (4) sub-bottom stratification, and (5) type of soil material. The ultimate purpose of the analysis methodology is to compare the resisting forces (F_R) preventing the movement of the slope to the driving forces (F_D) attempting to cause movement. Many analytical techniques are available to analyze slope stability problems. A recent compilation is given by Schuster (1978) describing how these various methods are utilized. This information will not be repeated here.

Marine slopes are normally evaluated using simpler analytical techniques than employed on terrestrial slopes. The primary reasons for this approach is (1) a usual lack of detailed information about marine slides and (2) a predominance of translational and flow slides as opposed to rotational slides. As a result of the flat geometry of most marine slides, several varieties of infinite slope analysis have been developed (Morgenstern, 1967, and Finn, 1966). For slides involving the more complex rotational movement, procedures using the method of slices are normally utilized (Bishop, 1955). A typical rotational slide divided into a series of slices is presented schematically in Fig. 3a. The various analytical techniques available to handle rotational slides differ predominantly in the manner in which they handle the normal side forces (E) and the shear forces (T) acting along the sides of the various slices, Fig. 3b.

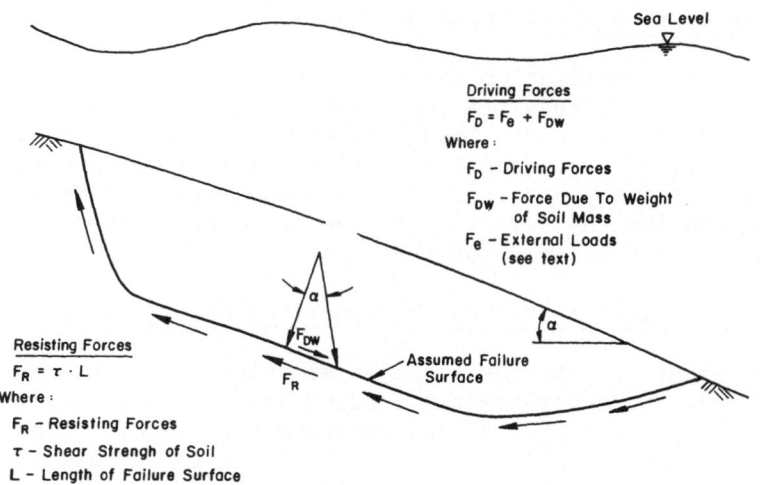

Fig. 2.   Resisting and Driving Forces on a Marine Slope

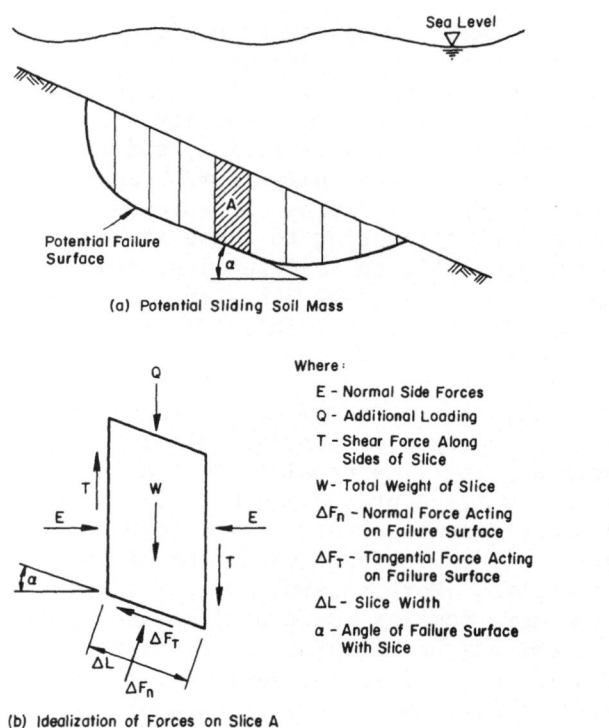

(a) Potential Sliding Soil Mass

Where:
   E – Normal Side Forces
   Q – Additional Loading
   T – Shear Force Along
       Sides of Slice
   W – Total Weight of Slice
   ΔF_n – Normal Force Acting
       on Failure Surface
   ΔF_T – Tangential Force Acting
       on Failure Surface
   ΔL – Slice Width
   α – Angle of Failure Surface
       With Slice

(b) Idealization of Forces on Slice A

Fig. 3.   Slope Stability Analysis by
          Method of Slices

The development of soil parameters for use in these various
analytical techniques depend upon (1) whether the soil condition
being modelled are drained or undrained, (2) whether the loading is
static or dynamic, and (3) if the pore water pressure (u) is measured.
Recalling that the relationship between total and effective stress
is defined by the following,

$$\sigma = \sigma' + u \tag{1}$$

where

σ = total stress

σ' = effective stress

u = pore water pressure,

then depending on the combination of the above elements an effective
or total stress analysis will be employed. A simplified summary of
the various possible combinations is presented in Table 5.

Table 5. Typical Types of Analysis

Loading Condition	Pore Water Pressure (u)	Type of Analysis
Static	Drained	Effective Stress
Static	Undrained (u not measured)	Total Stress
Static	Undrained (u measured or predicted)	Effective Stress
Dynamic	Undrained (u not measured)	Total Stress
Dynamic	Undrained (u measured or predicted)	Effective Stress

Determination of the Safety Against Failure

An acceptable level of performance of a marine slope can be described either by a factor of safety or by its resultant deformation. A factor of safety is defined as the ratio of the resisting forces (F_R) to driving forces (F_D). A factor of safety less than 1.0 indicates a failure or un-safe condition while a factor of safety greater than 1.0 indicates a non-failure or safe condition. In contrast, the safety of the slope can be evaluated also by estimating the gross deformations at various key locations, such as a point on the slope surface. Failure or non-failure would then depend on the effect of the deformation on the safe performance of the slope.

SUMMARY

In summary, this paper has tried to present in a general form the overall program for the conduction of a slope stability analysis in the marine environment. This sequential program consists of (1) survey and mapping, (2) soil sampling, (3) in-situ testing, and (4) analysis.

REFERENCES

Bishop, A. W., 1955, The analysis of stability of slopes, Géotech. 5:7.

Finn, W. D. L., 1966, Earthquake stability of cohesive slopes, Am. Soc. Civil Eng., Jour. Soil Mech. and Found. Div., 92, SM1:1.

Morgenstern, N. R., 1967, Submarine slumping and the initiation of turbidity currents, in: "Marine Geotechnique," A. F. Richards, ed., Univ. Illinois Press, Urbana.

Richards, A. F. and Chaney, R. C., (in press) Present and future geotechnical research needs in deep ocean mining, Marine Mining, 2.

Richards, A. F., Ling, S. C., and Gerwick, B. C., 1976, Site selection for offshore facilities, Ocean Eng., 3:189.

Richards, A. F. and Parker, H. W., 1968, Surface coring for shear strength measurements, in: "Civil Engineering In the Oceans," Am. Soc. Civil Eng., New York.

Schuster, R. L., ed., 1978, "Landslides: analysis and control," National Academy of Sciences/Engineering, Transportation Research Board.

GEOPHYSICS

When Alfred Wegener published his first paper on continental drift (1915) the ocean floor was at that time virtually unknown.

In the subsequent half century many new techniques have been developed and employed in the geophysical exploration of oceanic areas. One major advantage was the echo sounder which permitted the mapping of deep-sea physiography (Heezen et al., 1959). This visual presentation of the ocean floors has been greatly appreciated by a generation of students.

Another milestone was the applicability of magnetic surveys, airborne as well as marine, which related to the magnetic timetable of reversals of the Earth's magnetic field over the past 4.5 million years. This development has given information relative to the composition of the oceanic crust as well as important data of horizontal displacement along sea-floor fault systems.

The study of earthquake surface waves and epicentres has contributed to our knowledge of the structure of the ocean floor and the upper mantle, and to the probable contrast between the oceanic and the continental mantle as well as to the delineation of the seismic zones of recent tectonic activity.

Measurements of the heat flow through the sea floor and marine gravity measurements have improved our knowledge of structure and composition. Combined with age determinations, rock dredging, and core sampling we are completing the geological picture of the ocean floor. However, without seismic investigations, seismic-refraction and seismic-reflection studies as well, we would not have been at the present stage of development. The very fast technical development in electronics during the last 20 years has not only improved the accuracy of measurements and the speed of surveying but also facilitated the processing and interpretation of data assuring conclusions of at least one order higher.

In his contribution William Ryan describes the wide-swath sonar system in imaging of submarine land-slides, the swath width being 5 km. The system has been applied to explore the continental slope of New England and the mid-Atlantic margin of the United States.

Robert Embley in his paper on the anatomy of some Atlantic margin sediment slides and some comments on age and mechanisms describes two areas, one off Northwest Africa and another off eastern North America. He concludes that off Northwest Africa the slide zone represents an area of long-term instability, while the slides off Maryland occurred around the flanks of a series of old submarine channels.

In a short note Auffret et al. discuss evidences of late quaternary mass movements in the Bay of Biscay.

Ref.: Heezen, B.C., M. Tharp, and M. Ewing, 1959,
       The Floors of the Oceans. I: The North Atlantic.
       Geological Society of America, New York.

       Wegener, A., 1915, Die Entstehung der Kontinente
       und Ozeane, Vieveg & Sohn, Brunswich.

IMAGING OF SUBMARINE LANDSLIDES WITH

WIDE-SWATH SONAR

William B. F. Ryan

Lamont-Doherty Geological Observatory
of Columbia University
Palisades, New York

INTRODUCTION

Avalanche scars, landslide scree and debris flow
tongues have been imaged in plan view on the continental slope of the western margin of the North
Atlantic. Their detection and mapping has been
accomplished with a tethered deep-ocean side-scan
sonar having a swath width of 5 km. The sonar was
constructed to search for and locate the wreck of the
S. S. TITANIC (Wilford, 1980). We have subsequently
used it in cooperation with the U.S. Geological Survey
to explore the continental slope of New England and
the mid-Atlantic margin of the United States (Robb et
al., 1981, a, b; McGreggor et al., 1981).

DESCRIPTION OF THE SONAR SYSTEM

The sonar system is contained in a neutrally
buoyant vehicle that is tethered 100 m behind a 1,000
kg depressor (Figure 1). The depressor is guided within 200-400 m of the seafloor behind an oceanographic
survey vessel with a 7,500 m armored co-axial cable.
The sonar vehicle is towed at speeds of 1-2.5 knots,
in sea states of up to Force 6 and winds up to 40
knots, although in such conditions the survey vessel
needs a bow thruster or active stern rudders for
maneuverability.
The sonar vehicle is equipped with left and right
looking transducers that operate with a horizontal

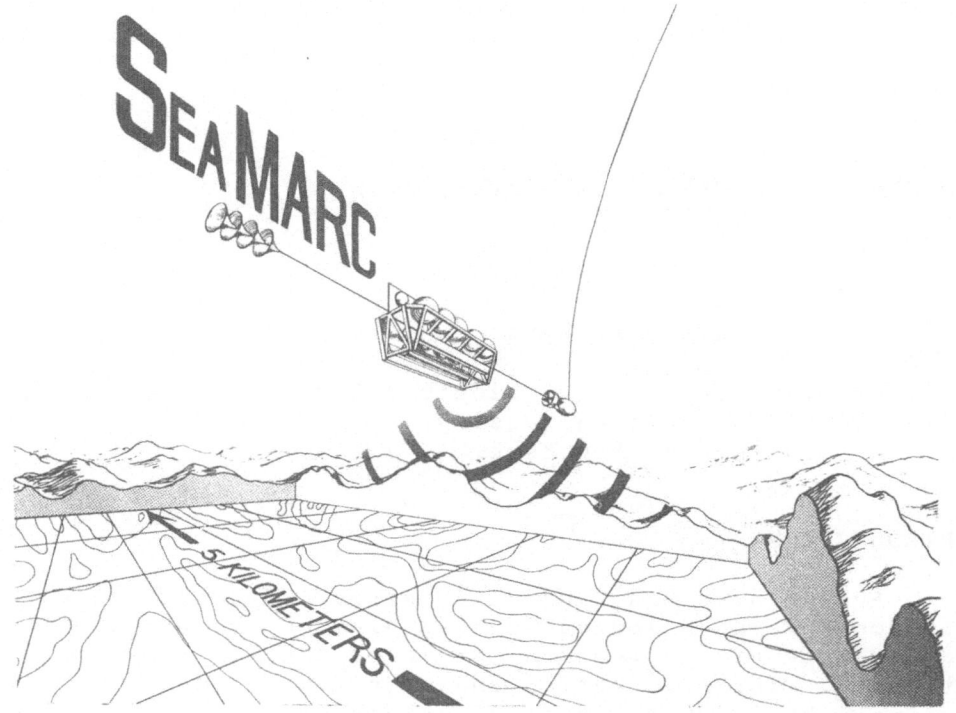

Fig. 1.   The tethered survey vehicle with depressor.
Its tradename is SEA MARC which stands for Sea Mapping
and ·Remote Characterization. (Courtesy of International
Submarine Technology).

beam width of 1.7° and a vertical beam of 50°. The
side-looking transducers transmit at 30 and 27 kHz
with variable ping lengths. Acoustic backscattering
in the slant range time domain is converted by real
time digital processing to horizontal range and is
displayed on electro-sensitive graphic recorders. The
depth of the vehicle is measured with a semi-conductor
strain gauge and altitude is detected with a 4.5 kHz
sub-bottom profiler. The vehicle has instrumentation
to determine its heading, its speed through the water
and its acoustic slant range from the surface ship
and bottom-moored acoustic transponders.
        Approximately 400 square km of sea floor can be
mapped per operation day. Using careful navigation
and control, such as that provided by radio positioning
systems, mosaics of the study area can be constructed
which show the geomorphological features without
geometrical distortion (Lowenstein et al., 1980). The

visual display can be enhanced by post-cruise signal
processing to recover additional information from the
large dynamic range of the sonar equipment.

The sonar system was designed and fabricated by
International Submarine Technology of Redmond, Washington. The purpose was to build instrumentation that
satisfies the compromise between seeing a large enough
area to image entire geological features (i.e., solve
the "elephant and hide" problem) and at the same time
hold a resolution capable of detecting geological
bedforms or individual rock outcrops. The side-scan
signals have a wavelength of 5 cm, which is the
dimension of ripple marks, animal mounds, or the
average bedding thickness in pelagic or terrigenous
sediments. The second objective was to achieve a
high degree of reliability, ease of maintenance and
component miniturization so that the fully equipped
vehicle could be relatively small and safe to handle
from academic-size research vessels in harsh weather.

The side-looking sonar has a growth potential to
allow for future quantitative bathymetric measurements
perpendicular to the track of the vehicle and synthetic
aperture processing.

FIELD EXPERIENCE

The wide-swath sonar has been deployed for 3 months
in the field and more than 1200 linear miles of survey
track has been obtained (Figure 2). The most significant feature of the new equipment is the passively
stable tow vehicle which minimizes yaw and pitch to
less than $0.3°$ over an interval of 30 seconds. We have
learned that we can tow the sonar at relatively high
heights above the seabed (i.e., up to 10% of the swath
width) and detect reflectivity variations that correspond to substrate textural differences in addition to
mapping acoustic shadows to determine topographic
trends.

The system resolves 256 levels of signal intensity
and displays the signals in pixels that represent
approximately 15 m^2 each. Intensity variations
indicate that the sonar is resolving bottom roughness
variations of the size of the acoustic wave length
and individual morphological features less than 10 m
across. Because of the excellent horizontal stability,
image resolution is not significantly degraded at far
range.

Before being displayed the intensity of the acoustic
signal is corrected for beam pattern variations,

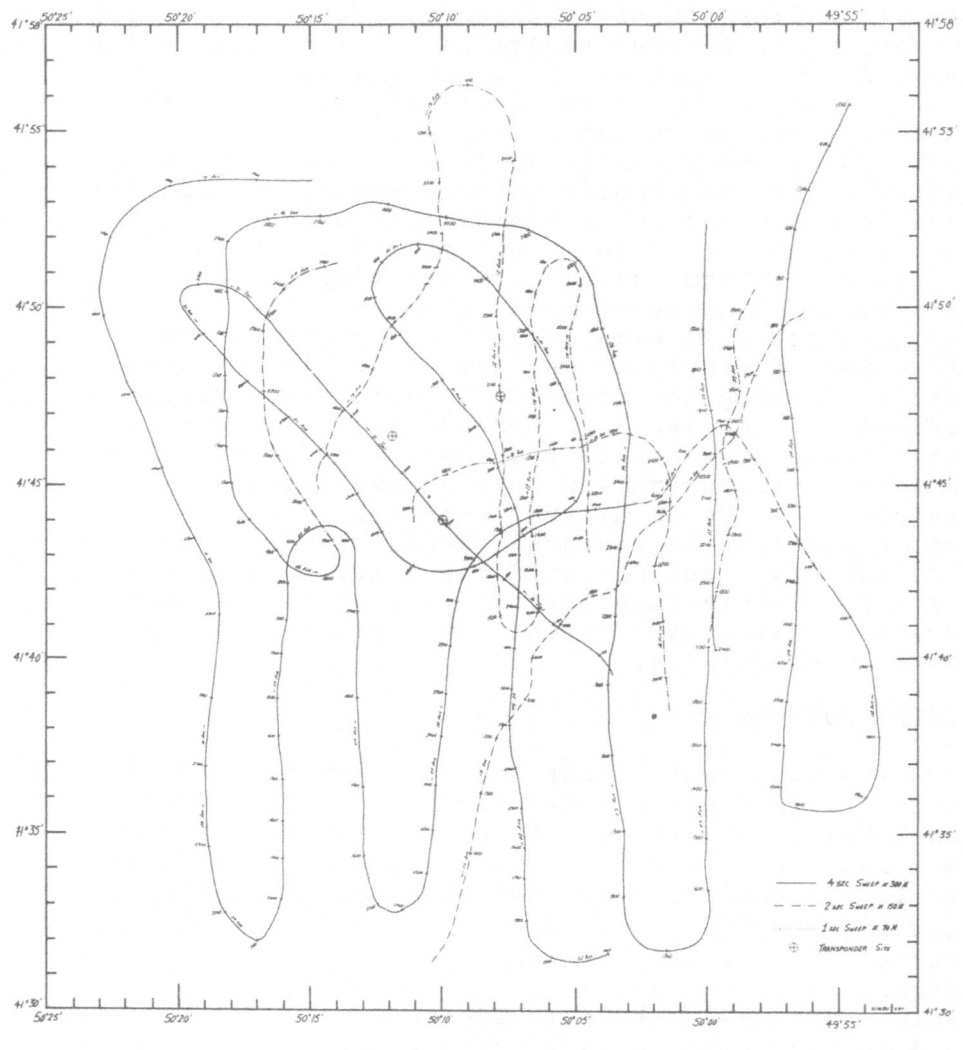

TITANIC 1980 'FISH' TRACKS

Fig. 2. Near-bottom "fish" tracks in the Titanic survey area southeast of Newfoundland. Approximately 350 line miles of profiling was accomplished in 2 weeks on site.

spherical spreading loss, water path attenuation and incident grazing angle.

RESULTS

The continental slope is incised by almost wall to wall canyons (Uchupi and Emery, 1967). The upper

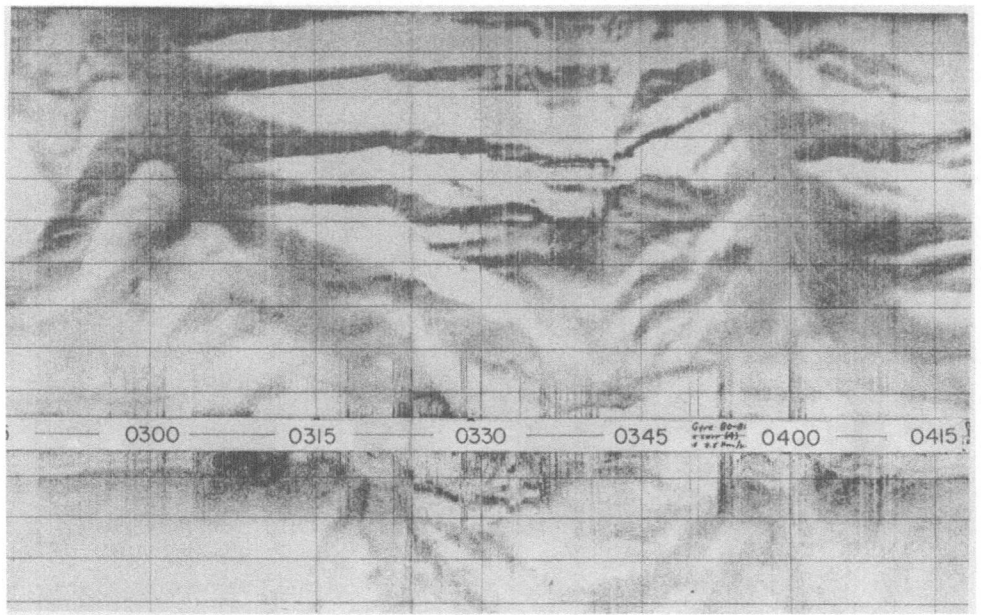

0300 ——— 0315 ——— 0330 ——— 0345 —— 0400 ——— 0415

Fig. 3. Trellis-like drainage pattern of side tributaries on the upper continental slope south of Wilmington
Canyon. Scale lines are spaced at 250 meters.

reaches of the canyons in depths generally shallower
than 1000 m display second and third order tributaries
which form a trellis-like drainage network (Figure 3).
The second order tributaries occur approximately every
300 m. along the main canyon axis, and the third order
tributaries occur every 100 m along the second order
tributaries. The major canyons that indent the continental shelf are remarkable because their thalwegs
display tight meanders (Figure 4). Where the slope
canyon passes to a channel on the upper continental
rise elevated terraces are seen which indicate that the
channel has migrated laterally across the subaqueous
flood plain. The channel walls are scalloped in
appearance (Figure 5). The scallops resemble crescent-
shaped slump scars. Rotational slump blocks exist at
the base of the channel walls and small oblate debris
flow tongues radiate across the channel floor from the
fresher looking slump blocks.

Canyon and channel meanders have a mean radius of
about 400-500 m. Some "oxbow lakes" are seen. The
channel walls are steepest on the outside of the
meanders. The channel floor shows a greater acoustic
roughness than the external leveas. The higher levels
of back-scattering probably correlate with the presence

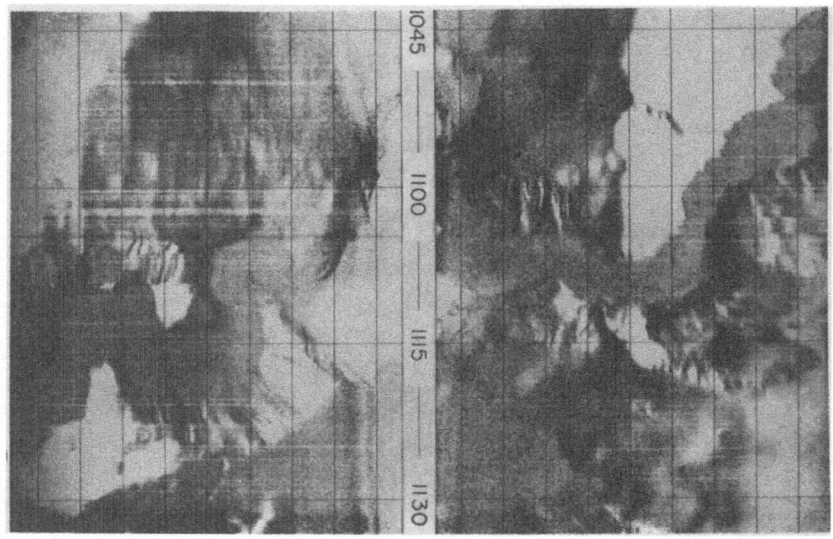

Fig. 4. Meanders in the thalweg of Wilmington Canyon
at mid slope depths.

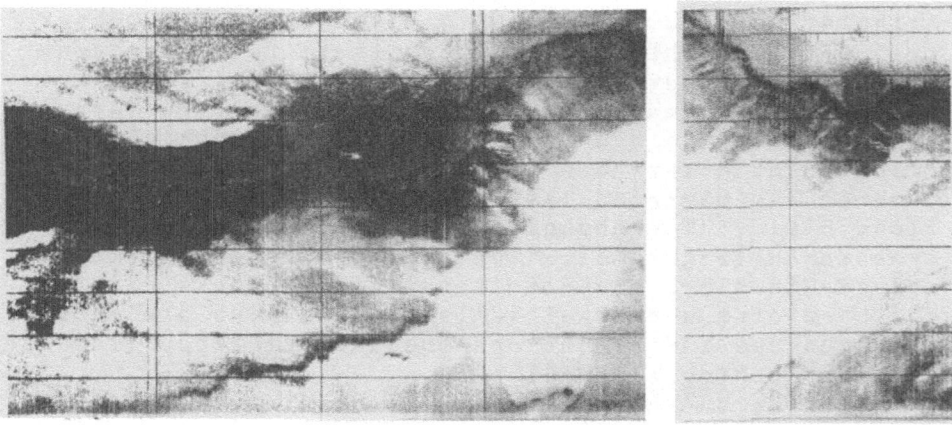

Fig.5. Scallops in the wall of channel on the upper
continental rise. Notice the local debris flow tongue
on the thalweg floor and the former and shallower level
of incision.

of rippled sand which has been photographed from
submersibles and bottom-towed photographic sleds.
     The canyons which do not indent the slope are
separated from each other by knife-edged promontories.
The heads of second order tributaries of adjacent
canyons meet at the crests of the promontory, indicating
that the sediment which moves down the side tributaries
is derived from the slope and not from the continental
shelf. The canyons which do not indent the slope
generally do not display meanders. The main axial
thalwegs widen from a width of less than 50 m in the
canyon heads to a width greater than 500 m near the base
of the slope. These thalwegs display longitudinal
ridges which appear to emanate from the side tributaries
analogous to lateral moraines in a glacial valley
(Figure 6). The ridges have a wavelength of approximately
50 m and elevations of less than 5 meters.

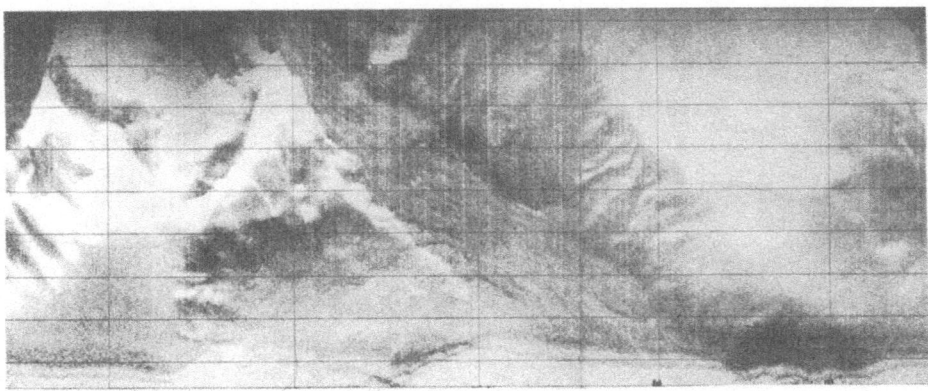

Fig. 6. Longitudinal ridges on the thalweg floor of a
slope canyon. Note the numerous side gullies.

     The regional slope between the major canyons is
sculptured by large lenticular avalanche scars. The
headlands of the scars are amphitheater-shaped (Figure
7). In many cases the seafloor immediately upslope of
the amphitheater is cracked with orthogonal fault
patterns (Figure 8). The downthrow sides are invariably
on the downslope side of the faults, giving them the
resemblance of "growth faults" or "rotational faults".
     Some horizontal transport of up to 50 m of intact
sheets of sediment cover is observed in the vicinity of
the slump source areas (Figure 9). Avalanche scars
reach 1.5 km in width and they can be traced along their
length for distances greater than 7 km. They generally

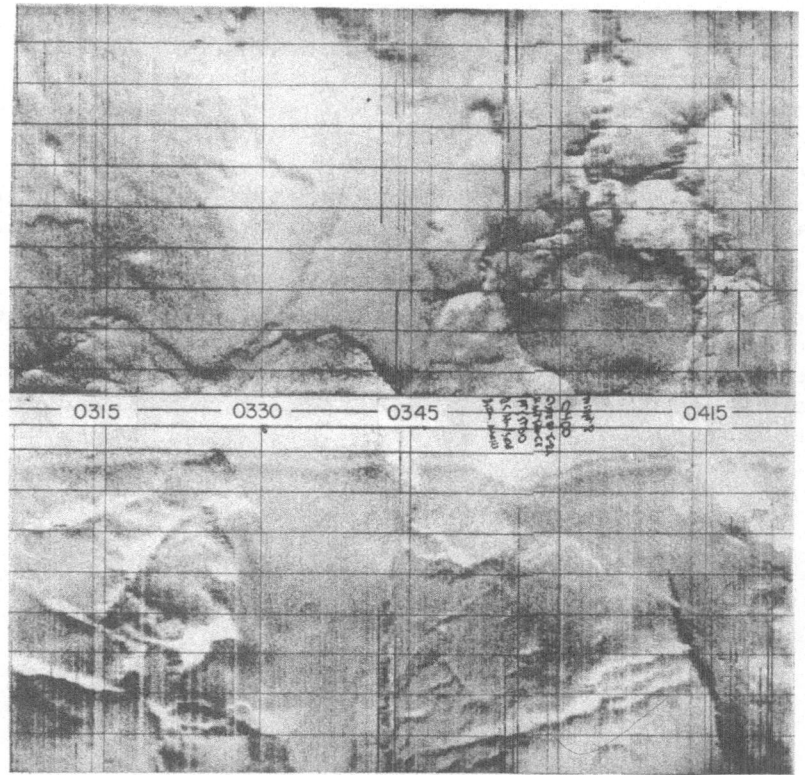

Fig. 7.   The amphitheater or "bowl" shape of an
avalanche source region.   Note the abruptness of the
headland face of the slump scar.

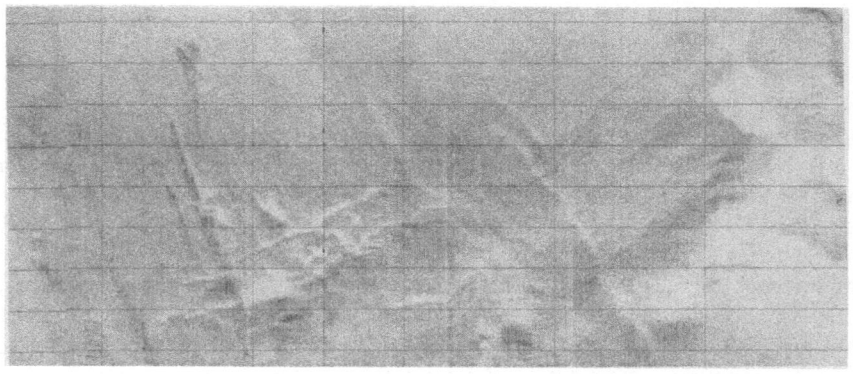

Fig. 8.   Orthogonal faults near the head of an
avalanche scar.

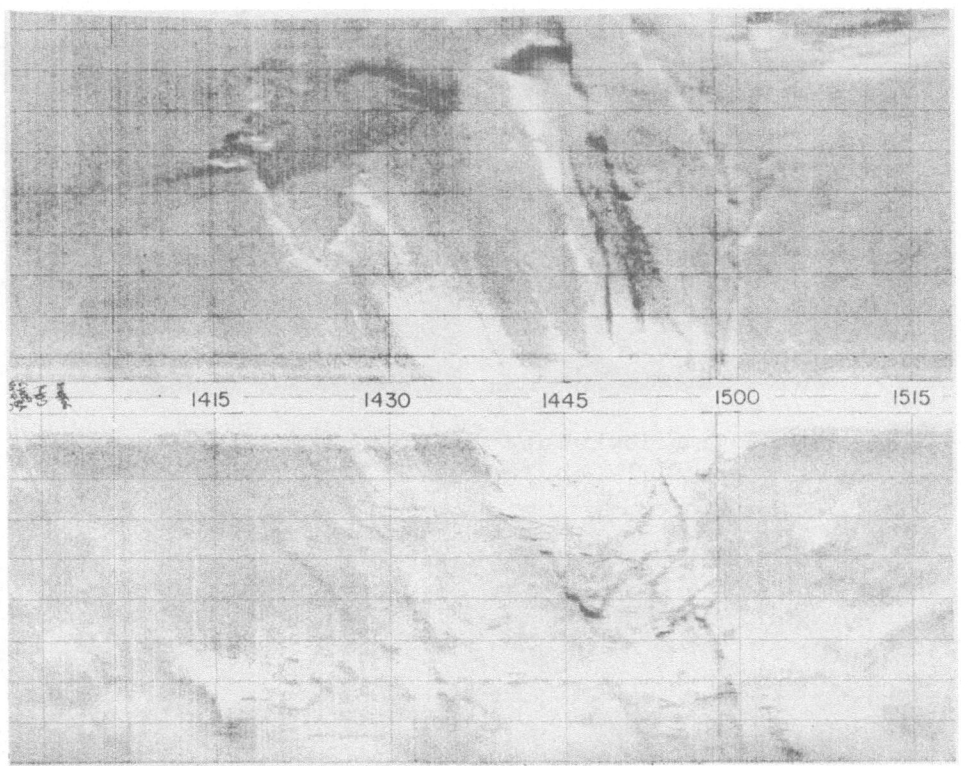

Fig. 9. Example of detached sheets of the sediment
cover in the head of an avalanche scar.

run downslope perpendicular to the regional contours.
The middle part of the slide chute possesses downslope
trending "rays" that might be gouge marks or very thin
debris flow streams (Figure 10). The rays are
remarkably parallel and evenly spaced. At the base of
the slope the rays transition into a hummocky terrain
that contains large blocks more than 10 m in height
and length. Many lumps and pits are observed here on
a feature that looks like the debris flow deposit
(Figure 11).
     There is a spectrum of freshness in the avalanche
scars. The most conspicuous property is the outcrop
pattern which indicates that the sediment has been
stripped away parallel to bedding surfaces. The slump
disturbance appears to propagate upslope analagous to
the landslides observed in the shallow prodelta
settings of the Gulf of Mexico (Prior and Coleman,
1977). Second order and even third order funnels are
recognized. The vertical relief between the top of the

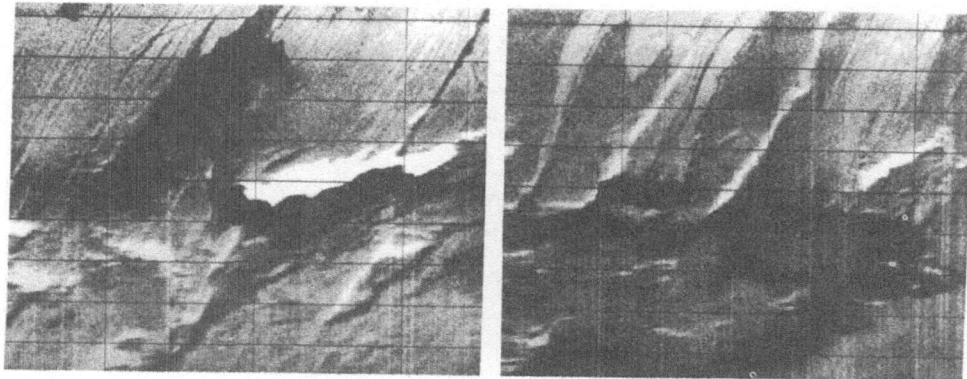

Fig. 10. The downslope trending streaks or rays
observed commonly on the lower continental slope.

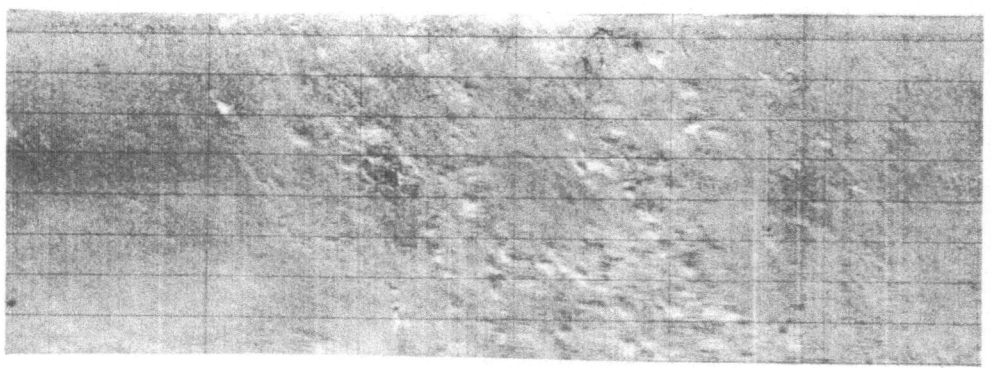

Fig. 11. Allochthonous blocks and pits on the surface
of the debris flow deposit at the base of the continental slope.

headland scar face and the floor of the amphitheater
ranges from 20 to 100 meters. Slumps occur within
slumps but slump scars rarely cross other slump scars.
      One has the impression that sediments fail progres-
sively layer by layer (Figure 12). In the areas surveyed
the freshest scars are near the base of the slope and
the older and partly buried features are in the mid
and upper slope. Large allochthonous blocks with
dimensions greater than 200 m have not yet been
confirmed. On the mid-Atlantic continental slope
slump scars comprise over 30% of the area between
1000 and 2000 m water depth.

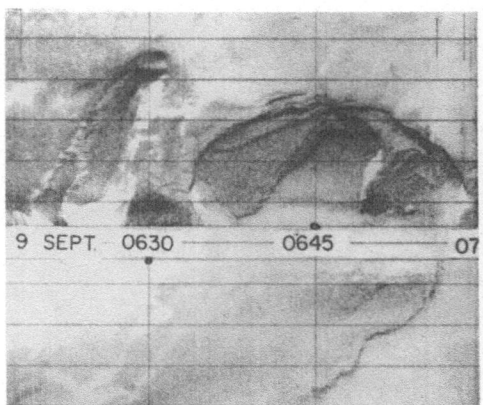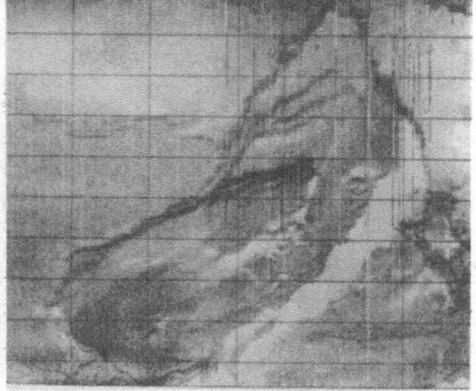

Fig. 12. The stripping away of sediment along bedding
planes within the avalanche scar.

INTERPRETATION

     The avalanches terrain is distinguished on the
sonar records from the submarine canyons because of
the absence of drainage tributaries, the presence
of cross-cutting faults, the stepwise excavation
that follows bedding plains, and the talus chutes.
The lumpy terrain at the foot of the chutes displays
the same type of transparent sub-bottom reflectivity
as noted by Embley, 1976 and Summerhayes et al.,
1979. However, small displaced blocks are noted on
canyon walls.
     The higher order canyon tributaries are sympto-
matic of a mass wasting process inherent in the slope
province itself. There is no suggestion that the
canyon tributarary excavation is made by agents that
use the gullies as bypass routes from the continental
shelf. Explanation is needed to satisfactorily
explain the trellis pattern of the second order
tributaries. It is clear, however, that the observed
drainage pattern resembles terrestrial counterparts
where the erosion proceeds in the headward direction.
In the upper slope more than 75% of the seafloor forms
part of the canyon drainage system. There are very
few sites spared from mass wasting.
     South of Newfoundland where the continental slope
extends below 3.5 km water depth the erosion is
expressed less by gullies and more by regional
bevelling across a gently dipping peneplain. A large
box-shaped submarine canyon bisects the Titanic survey
area (Figure 13). In this region the Western Boundary

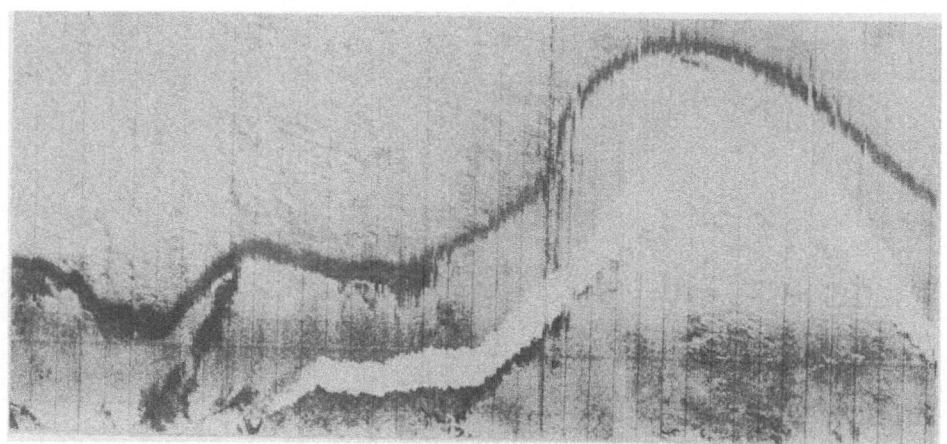

Fig. 13.   Dissection of an erosional peneplain
by the Titanic Canyon.

Sub-bottom profiling reveals a low angle truncation
of previously deposited strata. The dominant regional
reflectivity pattern on the side-scan records is
interpreted to be low linear ridges of resistant
outcrop. Crescent-shaped slump scars are present,
but they are less common here on the New England or mid-
Atlantic margins. The south Newfoundland region is
essentially a modern day active unconformity analagous
to widespread seismic Horizon A^u that bevels Cenozoic
and Mesozoic strata in the subsurface of the Western
Atlantic continental margins (Tucholke and Mountain,
1979). Although low lying rock ledges are ubiquitous,
loose talus is conspicuously sparse or too finely
broken up to be resolved. Erosional grooves and furrows
trend more or less parallel to the regional contours.

ACKNOWLEDGMENTS

James Kosalos directed the overall design and
fabrication of the sonar equipment. Considerable help
was provided by Dale Chayes, Larry Robinson, Brent
Miller, Robert Bookbinder and Jack Lee. Fred N. Spiess
was chief scientist during the project to search for
the Titanic. Vince O'Leary, James Robb and Bonnie
McGreggor were co-chief scientists of the surveys off
Georges Bank and the Mid-Atlantic. Discussions with
the leaders of these cruises and with Steve Eittreim
and Ellen Herron have been very informative.
Logistics were managed and expedited by Michael
Rawson and Thomas Aldrich. Financial support has been

provided in part by the U.S.G.S. through NSF-OCE-77-
23274B and the Office of Naval Research through
contract N00014-80-C-0098.
    Special thanks is given to Jack Grimm who
sponsored the Titanic project and who donated the survey
equipment to Columbia University.

REFERENCES

Embley, R.W., 1976. New Evidence for the occurrence of
        debris flow deposits in the deep sea. Geology,
        v4, p. 371-374.
Lowenstein, C. D., Kastens, K. A. and Spiess, F. N.,
        1980. Display processing for side-scan sonar
        images. Proceedings of the 12 Annual Offshore
        Technology Conference, Houston Texas, May 5-8,
        1980.
McGreggor, B. A., Twichell, D. C. and Ryan, W.B.F.,
        1981. Side-scan sonar images of the continental
        slope around Wilmington Canyon off the eastern
        United States. Abstract. Amer. Assoc. Petrol.
        Geologists.
Prior, D. B. and Coleman, J. M., 1977. Disintegrating
        retrogressive landslides on very-low-angle sub-
        aqueous slopes, Mississippi Delta. Marine
        Geotechnology, v. 3, p. 36-60.
Robb, J. M., Booth, J. S., Ryan, W. B. F. and Hampson,
        J. C., 1981. History and processes of the•
        continental slope off New Jersey: results of
        geophysical and sedimentological surveys.
        Abstract. Geol. Soc. America.
Robb, J. M., Hampson, J. C,, Ryan, W. B. F. and Kirby,
        J. R., 1981. Geological mapping of the conti-
        nental slope off New Jersey: Structure, history,
        processes, and geologic hazards. Abstract.
        Amer. Assoc. Petrol. Geologists.
Summerhayes, C., Barnhold, B., and Embley, R. W., 1979.
        Surficial slides and slumps on the continental
        slope and rise off Southwest Africa. Marine
        Geology, v. 31, p. 265-277.
Tucholke, B. E. and Mountain, G. S., 1979. Seismic
        stratigraphy, lithostratigraphy and paleo-
        sedimentation patterns in the North American
        Basin. Maurice Ewing Series 3, Amer. Geoph.
        Union, v. 3, p. 58-86.
Uchupi, E. and Emery, K. O.,1967. Structure of
        continental margin off Atlantic. Bull. Amer.
        Assoc. Petrol. Geologists, v. 55, p. 687-704.

Wilford, J. N., 1980. Scientists begin a quest for
    the Titanic. <u>New York Times</u>, June 24, 1980.

ANATOMY OF SOME ATLANTIC MARGIN SEDIMENT SLIDES AND SOME COMMENTS

ON AGES AND MECHANISMS

Robert W. Embley

National Ocean Survey/NOAA
Rockville, Maryland 20852

ABSTRACT

Two areas of submarine slides are described; one off Northwest Africa and another off of eastern North America. The Northwest African slide consists of a 18,000 km^2 slide scar with a 30,000 km^2 area of slide deposits. The total displaced volume is of the order of 1100 km^3. The scar is bounded on its eastern and southern sides by 10 to 100 m scarps. The failure occurred over a slope of about 1° to 1.5° and the mass flows moved over slopes as gentle as 0.1°. The available evidence suggests that the slide zone represents an area of long term instability and that individual failures are related to tectonics associated with a fracture zone. In contrast, the slides off Maryland occurred around the flanks of a series of old submarine channels. Distinct glide planes are correlated with subbottom reflectors. Possible triggering mechanisms include earthquakes and undercutting by turbidity currents.

Available age dates for submarine slides on the North Atlantic margin suggest that they have occurred randomly during the past 25,000 years.

INTRODUCTION

Recognition of the importance and extent of marine slides and their deposits has evolved over a long period of time. As recently as the 1960's such catastrophic events were thought to be rare in the ocean basins except on very steep slopes, in deltaic environments or in active seismic areas (Moore, 1961).

The first evidence for large scale failures on the seafloor came primarily from either anomalous topography (e.g., Shepart, 1955)

or displaced material (e.g. Archanguelsky, 1930). Development
of seismic, side scan and sampling techniques since the 1950's
coupled with increasing interest in the continental margin as a
resource area has resulted in the accumulation of a large amount of
data pertinent to the evaluation of this phenomenon on passive
continental margins. Most recently, the development of multibeam
sonar systems and long range side scan has advanced the state of
the art greatly.

It is now clear that slides are very common on the tectonically
passive, Atlantic-type margins and even on mid-ocean rises (Embley
and Jacobi, 1977; Moore, 1977; Embley, 1980).

This paper will discuss and compare the anatomy of sediment
slides off of Northwest Africa and Northeast North America. Also
included is an attempt to summarize the available age data on more
recent features, and a discussion about some of the various
mechanisms for triggering slides.

SLIDE ZONE OFF NORTHWEST AFRICA

General Setting

More than 45,000 km^2 of the continental margin off Northwest
Africa and the Canary Island archipelago has failed or been over-
ridden by a giant submarine sediment slide or series of slides
(Figs. 1 and 2). The disturbed zone is comprised of a slide scar
where there has been net sediment removal and a depositional zone
which has been covered by a zone of distinctive mass-flow deposits.
Although the slide strikes roughly parallel with the continental
margin, its boundaries maintain a perpendicular strike to the
regional bathymetric contours. The head of the slide (southern
part) begins on the eastern part of a broad northwest-trending
bulge in the continental rise which is most apparent on the
2200-3400 meter bathymetric contours (Figs. 2 and 3).

At present the area from about 18°N to 28°N (northern Mauritania
and former Spanish Sahara) is desert. The Holocene (Ericson et al.,
1961, zone Z) has been a time of minimal terrigenous-sediment
supply to the ocean along this margin. The primary sediment input
has been planktonic tests supplied from overlying high-productivity
zones and a windblown terrigenous component. The sedimentation rate
averaged 2.5 to 6.0 cm/1000 yrs. (Seibold et al., 1976). In
contrast, the Wisconsin (Ericson et al., 1961, zone Y) was characterized by more humid conditions on land and, coupled with lowered
sea level, this resulted in a somewhat increased rate of terrigenous
sediment input to the oceans. Increased upwelling also supplied
some additional organic and shell material to the sea floor.
Sedimentation rates averaged between 5.0 and 20.0 cm/1000 yrs.
(Seibold et al., 1976).

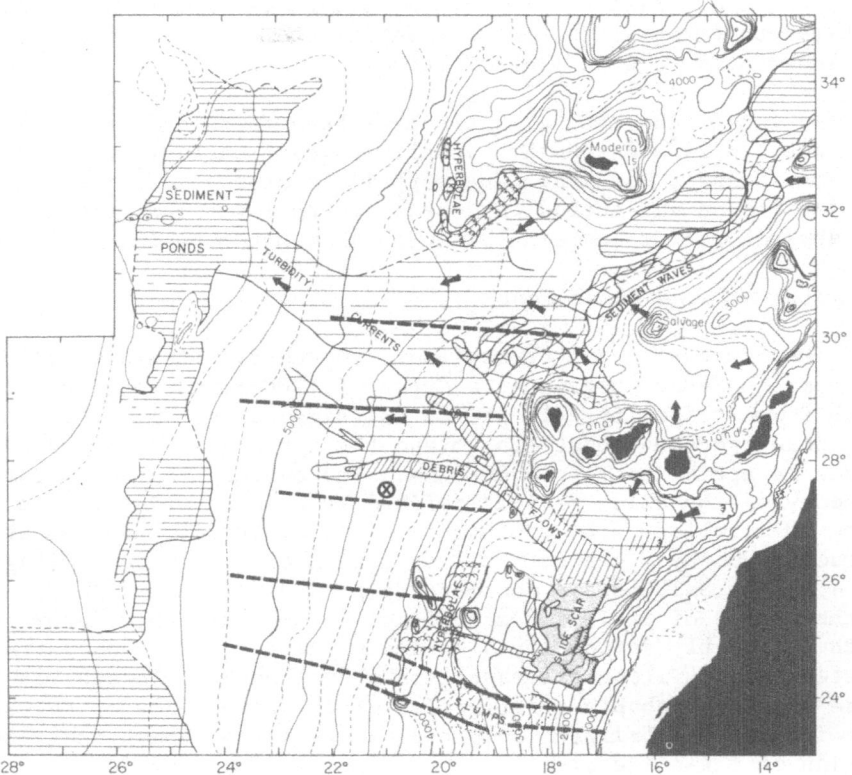

Figure 1.   Summary of some sedimentary features and turbidite
            dispersal paths (arrows) of Canary Basin. Closely spaced
            horizontal lines show flat sediment ponds with greater
            than 50 m of continuous subbottom reflectors. Wide-spaced
            horizontal lines indicate poor acoustic penetration with
            less than 20 m of continuous subbottom reflectors. Thick
            dashed lines show possible extensions of oceanic fracture
            zones in region (after Hayes and Rabinowitz, 1975; Rona
            and Fleming, 1973). Circle with cross in center locates
            1959 earthquake of magnitude 6.1 (Sykes, 1978).

        According to Seibold et al. (1976), the general effect of a
glacial climate in this region was to move the various climatic
belts about 5°-7° south and to increase the intensity of upwelling.
These authors also conclude that the area from 23°N to 26°N was
subjected to more humid conditions during this period.

        The low level of chemical weathering on the adjacent continent
and, consequently, the reduced supply of sediment available to the
deep ocean have resulted in low sedimentation rates throughout the

northern Canary Basin. The abyssal plains are relatively small
compared to the western North Atlantic, and Pleistocene sediment
thicknesses average only about 100 m.

No large submarine channels have been found on the continental
rise between 23°N and 28°N. Locally, however, active volcanic centers
(Canary Islands, Madeira) have supplied coarse volcanogenic sands to
the deep basin and turbidites cored on the lower continental rise
and abyssal plains contain volcanogenic mineral suites.

Morphology of the Slide Scar

The depression in the seafloor created by the slope failures(s)
off Spanish Sahara begins at a depth of 1700 m and continues to the
north (down-slope) for almost 200 km (Figs. 2 and 3). The width of
this zone averages about 125 km and the total area affected by the
slides is approximately 18,000 km^2. The eastern and southeastern
boundary of the slide scar is defined by a scarp which varies from
20 to 80 m relief. The southwestern and western boundaries are
characterized by a gentle swell or series of small scarps (Figs. 3
and 4). The bathymetric contours (Fig. 3) and the profiles in the
southern part of the slide scar (Fig. 5) show a deeper inner trough
which is roughly defined by a series of scarps. The concentration
of steep scarps along the eastern and southeastern boundaries of the
slide scar is probably a reflection of the kinematics of the slide.
That is, initial failures occurred along the southeast and eastern
margins of the slide area.

The floor of the slide scar has variable local relief. The
southern, upslope zone is relatively smooth. There are no large
blocks (hyperbolic reflections) present on the seafloor in this
region and the echo return is characterized by a "fuzzy" overcast,
presumably from small-scale relief not resolved on the depth recorder.
Continuous subbottom reflectors can be traced through this region
onto the undisturbed continental rise (Fig. 5 profile A). This
relatively smooth seafloor is also present between the inner and
outer scarps on the eastern side of the scar (Fig. 6 profiles 5 and
8). The floor of the remainder of the slide scar is characterized by
irregular undulations, small scarps and hyperbolic reflections
(Fig. 6). There is also a "fuzzy" return superimposed over this
region, but no persistent subbottom reflections are recorded.

Most of the 3.5 kHz profiles across the slide scar north of
about 24°45'N cross numerous flat reflective patches (e.g. Fig. 6,
profile 6). They occur in the trough of the irregular topography
and are probably either local ponds of sediments or channels created
by small turbidity currents generated by the slides. The track
spacing is too wide and the "channels" are too small to be traced as
continuous features. North of 26°N, the slide zone is characterized
by acoustically unstratified mass-flow deposits which onlap the

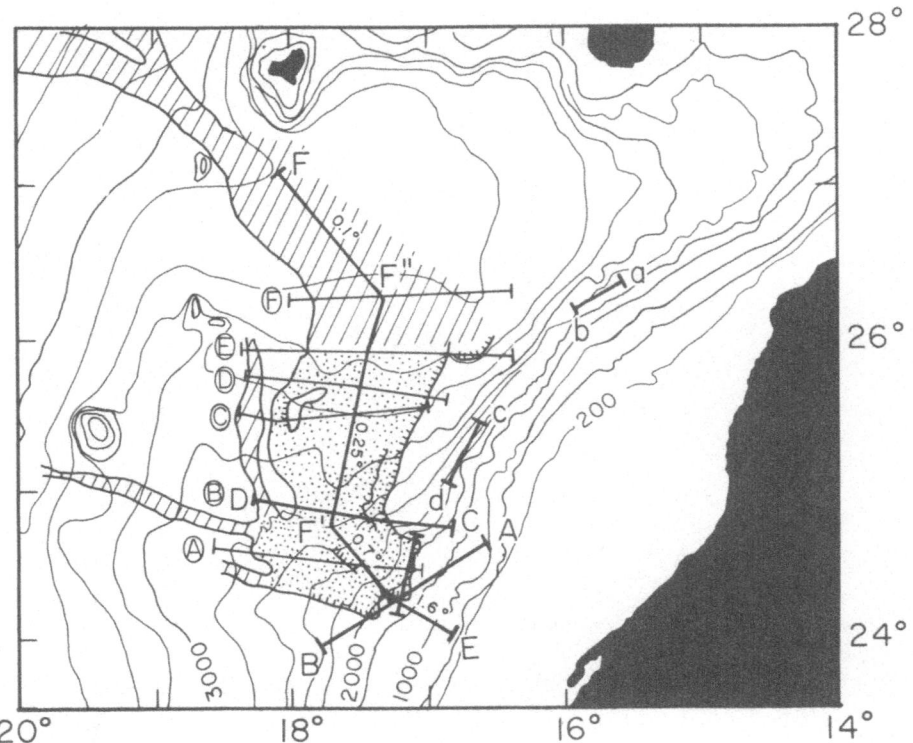

Figure 2.   Location map for topographic and seismic profiles shown
            in figures 4 and 5.   Regional slopes shown along profile
            E-F.   Locations of METEOR profiles (Seibold and Hinz,
            1974) a-b, c-d, e-f are also shown.

stratified continental rise hemipelagic sediments.   Several (mass-
flow) channels also emanate from the upper head region of the
slide scar.   These deposits are discussed in a later section.

     The most direct evidence of recent dislocations within the slide
scar comes from a series of bottom-camera stations taken in 1975
(Fig. 7).   On most of the more than 250 bottom photographs from ten
stations on the floor of the slide scar, irregular mound topography
and scarps are clearly visible on the photographs.   In most cases
the steep scarps found in the bottom photographs coincide with the
presence of small scarps which are barely resolvable on the 3.5 kHz
records.   In contrast, all of the bottom photographs taken on the
undisturbed continental rise and slope show a flat and tranquil
seafloor (Embley, 1975).

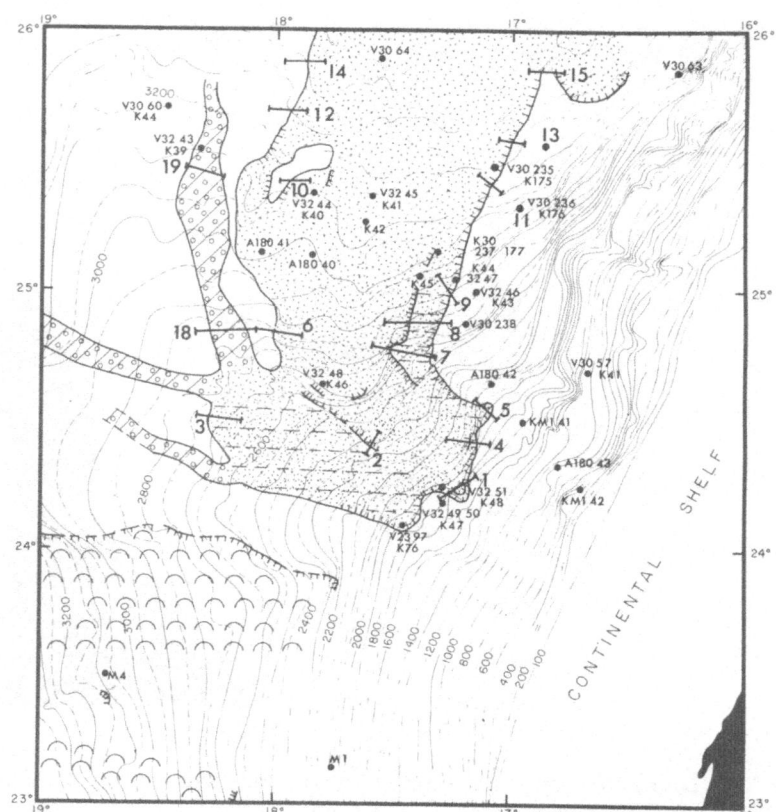

Figure 3.   Close-up view of slide scar shown in Figures 2 and 3.
Contours in corrected meters (after Embley, 1975).
Hachures show location of scarps having relief of 20-
100 m.   Dashed lines indicate that some subbottom
reflectors are recorded and diagonal lines show location
of overflow channels containing 10-20 m-thick wedges of
material characterized by a prolonged echo.   The seafloor
across the remainder of the zone outlined by a solid
line is characterized by irregular hyperbolae, a prolonged
echo and no visible subbottom reflectors.   Dark lines
with numbers refer to 3.5 kHz profiles.   Also shown are
locations of core and camera stations (K) area.

      Examination of piston cores from the continental slope and upper
continental rise off Spanish Sahara suggests that the slide scar is
a zone of relatively recent sediment displacement.   Cores taken in
the zone of sediment removal contain sharp angular contacts and
hiatuses where as those taken on the undisturbed seafloor contain
normal sequences of pelagic or hemipelagic sediments (Embley, 1975).

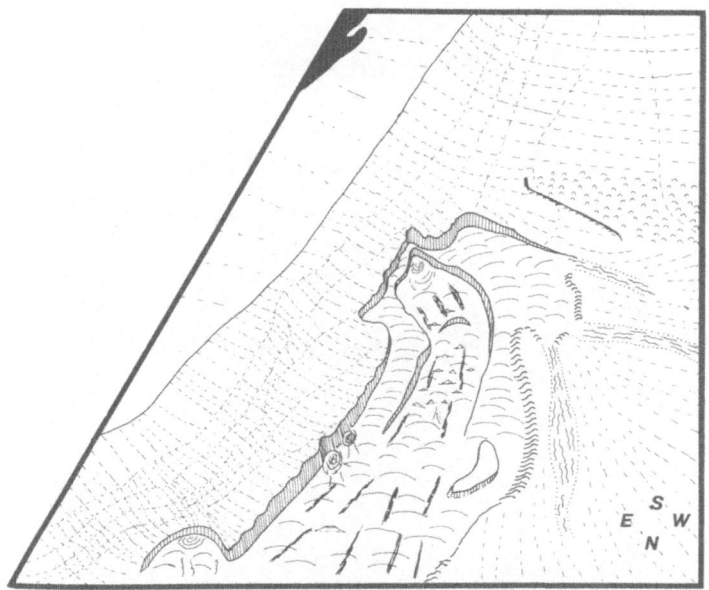

Figure 4. Perspective sketch of slide scar on Figure 3. Relief is
exaggerated. Note inner and outer scarps, secondary
slumps along main scarps, and channels (wavy lines).

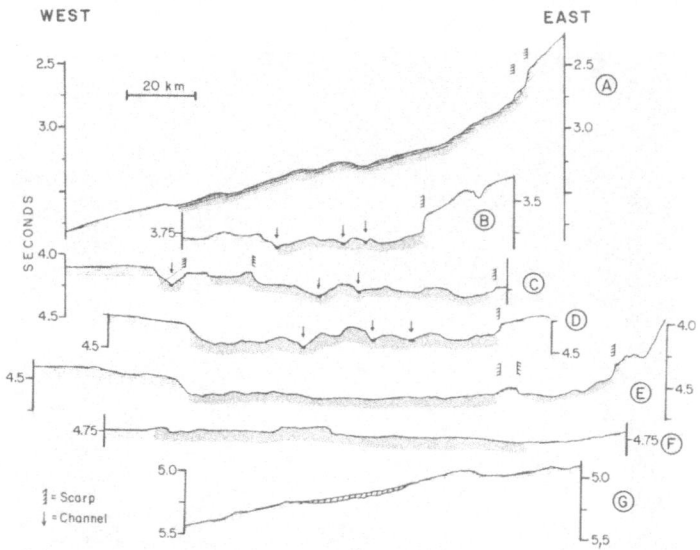

Figure 5. Bathymetric profiles across slide (see Fig. 2). Hachures
locate depression with flat reflective floors, horizontal
lines indicate presence of subbottom reflectors and
stipples indicate their absence.

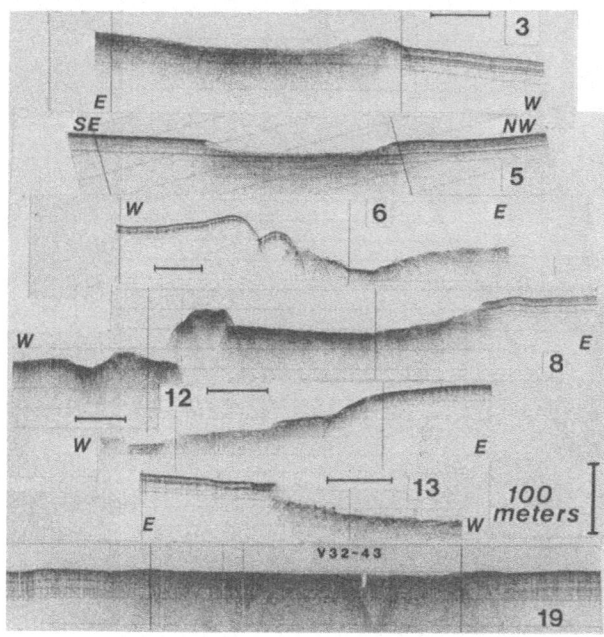

Figure 6.   Photographs of selected original 3.5 kHz echograms across
boundaries of slide scars.   Location on Fig. 3. Profile 19
is across channel where V32-43 was taken.

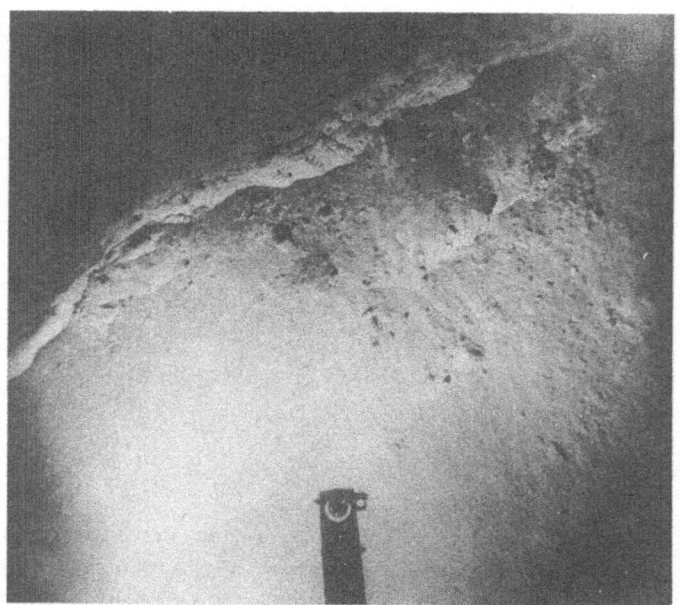

Figure 7.   Bottom photograph of seafloor within slide scar.
(V32-K40).   Position of this station shown on Figure 3.

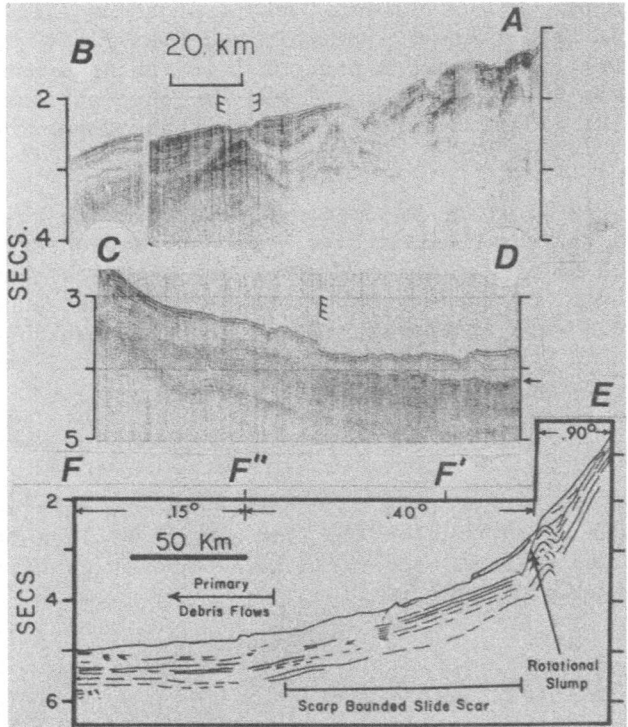

Figure 8.   Seismic (airgun) profiles recorded across sediment slide
            zone off Northwest Africa. Hachures show position of
            scarp on profile. Arrow on right indicated Lower
            Miocene reflector.

Seismic Profiles

     Low-frequency seismic profiles (20-300 Hz) were recorded on the
VEMA cruise in the area concurrently with the 3.5 kHz PDR and these
reveal some underlying structural features which may be related to
the slide. The east-west profiles (Fig. 8, Profiles A-B and C-D)
across the shallower part of the slide scar show the "gap" in the
sediments created by the slide. The slide appears to have involved
only sediments from the upper acoustically transparent interval. The
top of the reflective sequence below the section is approximately mid-
Miocene (Hayes, Pimm, et al., 1972) so the slide has primarily
involved Neogene sediments.

     A striking anticlinal structure occurs directly below the
shallowest part of the slide scar (profiles A-B and E-F, Fig. 8).

Our data does not define the exact shape of the anticlinal feature
and its origin is not known. Seibold and Hinz (1974) suggested that
these structures are buried canyons related to an extensive Neogene
drainage system in this region. They also suggest that the bulge
in the contours (Figs. 2 and 3) represents the depositional lobe of
this system.

Profile E-F (Fig. 8) was made perpendicular to the regional slope
and along the center of the slide. The association of the head scarp
with the flank of the anticlinal structure suggests that initial
seafloor movement began as a rotational slump. Subsequently, large
sediment slides were triggered. The northwest flank of the buried
feature shown in profile E-F (Fig. 8) may have acted as a plane of
décollement for the slump.

Slide Deposits

The bulk of the deposits resulting from the massive sediment
failures on the upper continental rise reside as lens-shaped,
acoustically-unstratified sediments on the lower continental slope
(Embley, 1976; Figs. 9 and 10).

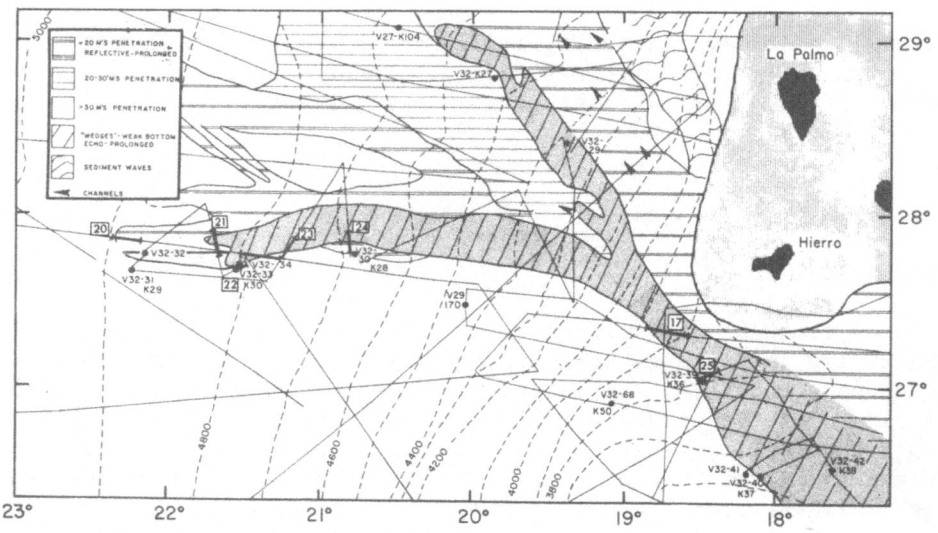

Figure 9.   Acoustic zones south and west of the Canary Islands show-
            ing debris flows (diagonal lines with stippling) and
            turbidites (horizontal lines). Also shown are locations
            of 3.5 kHz profiles shown in Figure 10 and core and
            camera (K) stations.

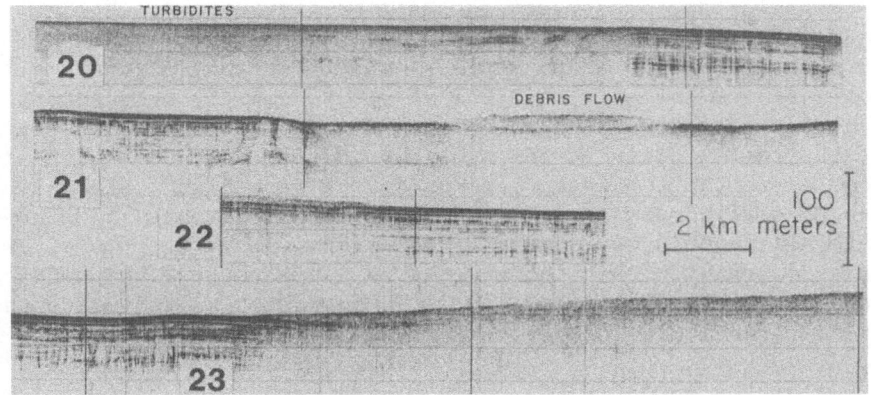

Figure 10.   Echograms (3.5 kHz) across boundaries of mass-flow
             deposit shown in Figure 9.

     All of the 3.5 kHz profiles across the edges of the slide zone
north of about 26°N show a characteristic lens of acoustically
unstratified sediment onlapping the stratified, hemipelagic
continental rise sediments (Fig. 10). The top of the hemipelagic
section can be followed beneath the slide deposits until the latter
reaches a thickness of approximately 20 m, at which point the subbottom reflector is obscured. Subbottom reflectors are occasionally
seen in the interior of the slide zone but cannot be followed
continuously. The slide sheet is not clearly defined by airgun
records (Fig. 8, Profile E-F), probably because its thickness is
at or below the effective resolution of the system ($\sim$ 50 m).

     Farther northwest, the slide deposits divide into two long
"fingerlike" lobes which terminate the continental rise at about
4600-4800 m. Apparently, the slight levelling of the seafloor
gradient seaward of the 4800 m contour where the gradient changes
from $\geq 0.1°$ to $< 0.1°$ was sufficient to stop the mass-flow from
advancing farther downslope.

     The acoustic province-bathymetric map of this area (Fig. 9)
shows that the mass-flows followed pre-existing lows in the
continental rise. In this region the mass-flow lenses lie directly
over volcanogenic turbidites.

     The piston cores (V32-32,-39,-40, and -42) taken in the acoustic
zone which had been acoustically identified as mass-flows (Fig. 11)
all contain distinct primary sedimentary structures which do not
appear in the cores taken in the other acoustic zones (Figs. 9 and
11 and Fig. 4 of Embley, 1980). These include (1) angular bedding
contacts, (2) exotic pebbles and pyrite worm tubes, (3) irregularly-
shaped exotic-sediment inclusions, and (4) small folds.

Overall, the cores taken in the mass-flows convey the im-
pression of a complex interfingering of sediments of various
lithologies (gray and brown marls, oozes). Because of the small
diameter of the coring tube, (2 1/2" = 6.25 cm) the sedimentary
structures which were penetrated are, in most cases, impossible
to reconstruct. However, the numerous angular contacts which are
obvious at contacts between various colors of sediment probably
represent small cross-sections of convoluted sediments. The absence
of laterally continuous internal reflectors within the mass-flow
deposits agree with this interpretation. Apparently, the bedding
planes within the (mass flows) are too contorted to reflect sound
with any lateral consistency.

Two of the cores (V32-40 and V32-43) penetrated through the
mass flows and into a pelagic section beneath. These two cores
were purposefully taken in locations where the 3.5 kHz record showed
the top of the stratified continental rise sequence buried at a depth
of 10 m or less, beneath the seafloor. V32-40 was taken at the edge
of the main lobe of the mass flow at 3680 m and V32-43 was taken
in one of the channelized deposits west of this slide scar (Profile
18, Fig. 6) at a depth of 3111 m. The boundary between the convoluted structures and sharp contacts of the various units of the
mass-flows and the rather structureless units and gradational
burrowed contacts in the pelagic sediments is well defined. The
contact in V32-43 is particularly striking because it is marked by a
sharp change from olive marls (slide deposits) to brown marls (hemipelagic sediments). A nearby core, taken on the stratified sequence
(V30-60) also consists of brown marls and white chalk oozes.

Bottom photographs were taken in several places on the mass-
flow deposits as well as on the "normal" continental rise. Photos
taken on the "normal" continental rise show the usual tranquil seafloor with a subdued organically produced micro-relief. Photographs
taken on the surface of the mass flows, however, reveal striking
mounds and ridges (See Embley, 1975, 1976, 1980 for examples of these
photographs). These are most likely deformation features produced
when the sediment was still in a mobile state. The prolonged echo
return of these deposits can be explained by acoustic scattering
from these features.

It is obvious from the 3.5 kHz records, bottom photographs, and
core data that the slide is fairly recent in origin. The bottom
photographs, both within the slide scar and on the mass-flows,
show primary features associated with the slide (eg. scarps,
undulations). The cores show only a thin cap of hemipelagic sediment
above the slide deposits. The slide scar contains multiple scarps
so it is likely that following the main slide, there were secondary

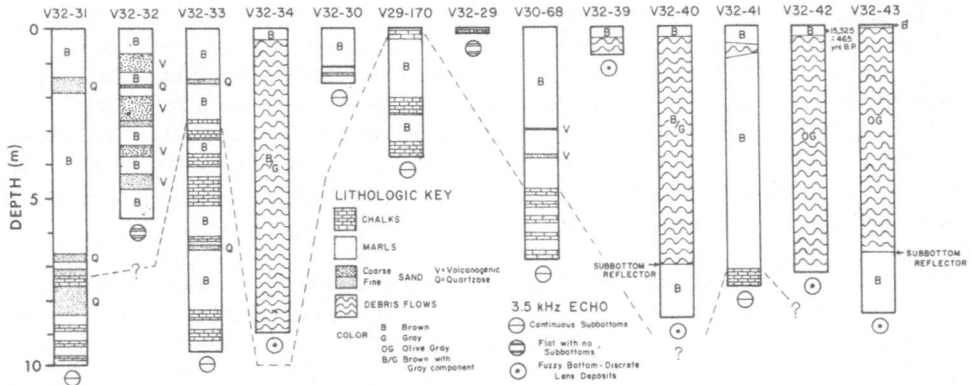

Figure 11.   Lithologic logs of cores from area of Figure 8 and for
             V32-43 which was taken in a mass-flow channel near the
             primary slide scar (See Fig. 3 for location).

events occurring over an unknown period to time.   Cores from near
the western scarps (eg. V30-235, 237) contain mass-flow structures
as do several other cores from the floor of the slide scar (eg. V32-
45).   Some of the bottom photographs also show scarps and ridges.

      In order to date the main slide, all the cores from the slide
deposits were examined to determine the upper boundary between the
slide deposits and the hemipelagic cap.   In practice this boundary
could be chosen with confidence in only two cores, V32-42 and
V32-43.   The section chosen for radiocarbon dating in V32-42 was
from 15 cm to 19.5 cm.   From 20 cm to 25 cm a foram iniferal ooze
layer is present and below 25 cm mass-flow structures are obvious
in the core.   The foram ooze layer was not dated because it was
thought to be a turbidite, possibly generated by the slide.   The
radiocarbon date for the 15-19 cm interval is 15,325 ± 475 yrs. B.P.
Assuming a constant accumulation rate and carbonate percentage for
the sampled interval, the sediment just above the mass-flow is
probably slightly older than average for the date for the total

4.5 cm interval.   Assuming a Holocene sedimentation rate of 2 cm/1000
years this would make the age of the debris flow between 16,000
and 17,000 yrs.B.P.

There is only about 4 cm of pelagic sediment overlying the
debris flows in V32-43. The trigger-weight core also shows about
this much brown sediment on top of the olive gray sediment below.
It is obvious that the channelized slide deposits on the upper rise
represent significantly younger flows than those in the main deposits
downslope. Assuming a sedimentation rate of 2-4 cm/1000 years
(Diester-Haas et al., 1973), this would date the last debris flow
in the channel at 1000-2000 yrs.B.P. Cores taken within the slide
scar also commonly show very thin caps ($\leq$5 cm) of pelagic sediment
above the displaced material.

The distal position of V32-42 with respect to the probable
source suggests that the primary slide occurred during the Wisconsin
lowered sea level. The existence of the younger, channelized deposits (V32-43) indicated continued instability, at least up until
recently. The morphology of the slide scar (Fig. 4), which suggests a multiphased event, agrees with this "growth through time"
hypothesis.

MASS WASTING ON THE CONTINENTAL SLOPE AND
RISE SOUTH OF BALTIMORE CANYON

On the continental slope and upper rise between Washington and
Baltimore Canyons a distinct 50 m to 100 m layer of acoustically
stratified sediments has been draped over the tops of the ridges
and the flanks of the valleys (Figs. 12 and 13). Subsequent failure
of large areas of this stratified sequence has resulted in the
creation of scarps dipping 15°-25° on the flanks of the ridges and
the emplacement of large volumes of slumped sediment in the valleys
(Malahoff et al., 1980, Malahoff et al., 1981).

The valley walls have very steep 10 m to 50 m scarps where
the strata are abruptly truncated (Fig. 13). In some places, the
scarps can be traced continuously around the flanks of the hills
over distances of 10 km or more; in other places, they appear to
be less than 2 km in length. In general, the scarps cut obliquely
across isobaths.

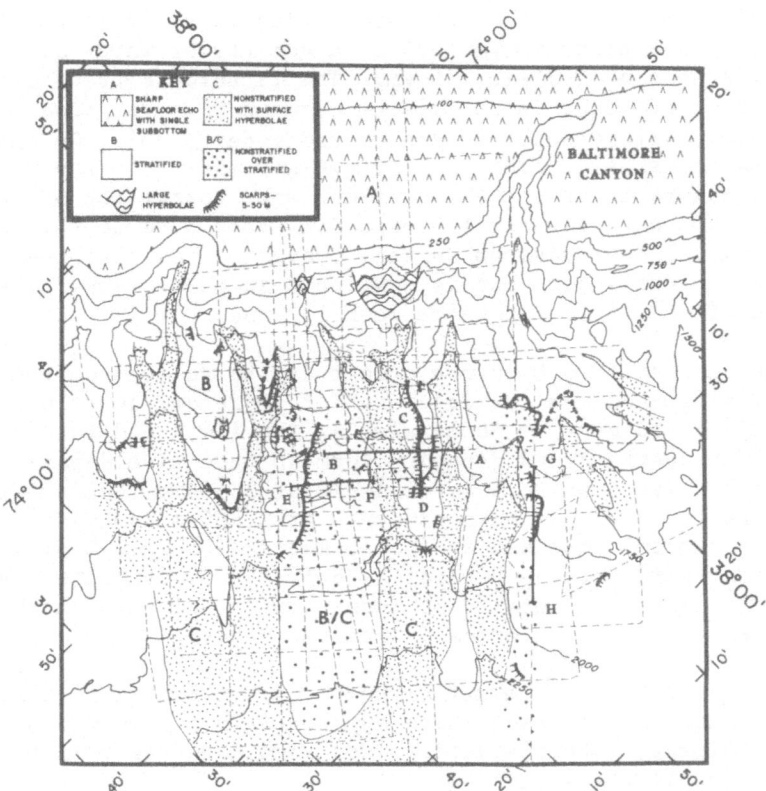

Figure 12.   Acoustic facies map of continental margin south of
            Baltimore Canyon.

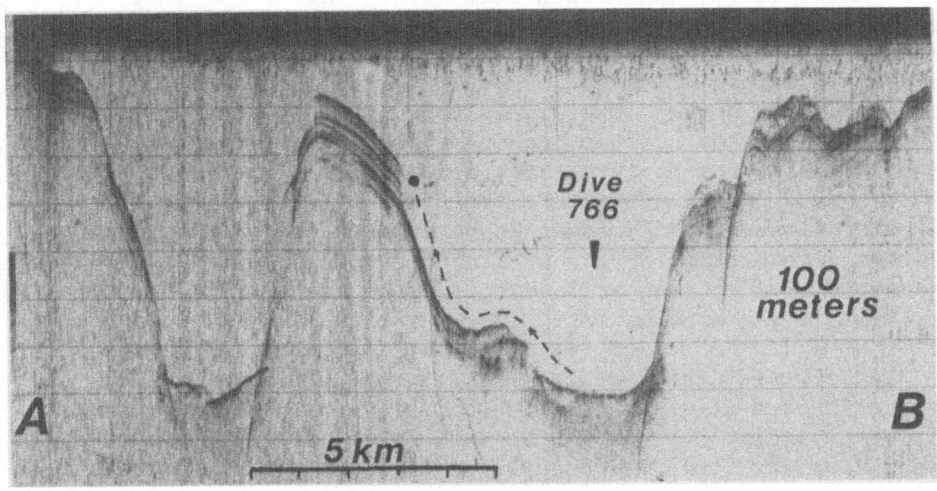

Figure 13.   3.5 kHz profile parallel to shelf edge south of
            Baltimore Canyon (See Fig. 12 for location).

Failure of sections of the stratified layer has occurred along the same "glide plane reflector" at a variety of water depths and over a distance of at least 15 km. The lithologic transition at this glide plane reflector is at present unknown. The regional slopes along which failure occurred are of the order 1°-3°. On ALVIN dive 766 (Malahoff et al., 1981) a traverse was made from the valley shown in Figure 13 up the flank of the hill to the top of the scarp. The scarp was subsequently followed for about 5 km to the southeast and the visual observations on this traverse show that the trace of the scarp is not smooth but is scalloped in plane view. The local gradient of the scarp ranges between 15° and 20°.

The unstratified sequence occurs in the valleys on the continental slope and, as the local relief decreases downslope, the unstratified sediment is found also on the flank and tops of the hills. The unstratified deposits form an extensive deposit on the continental rise down to a water depth of 3,100 m (Embley, 1980). A box core (RC19-49) in a water depth of 2,800 m contains large clasts ($\sim$ 2-5 cm) imbedded in a clay matrix (Embley, 1980). This sedimentary structure is typical of debris flow deposits. Embley (1980) discusses the details of mass-flows along the continental rise of Eastern North America in greater detail.

To summarize, in the region south of the Baltimore Canyon, numerous small-scale slides have apparently occurred which transported sediments into the valleys and then down the valleys onto the upper continental rise.

SOME COMMENTS ON TRIGGERING MECHANISMS

Moore (1961) and Morgenstern (1967) have shown that, as long as no excess pore pressures are present within the sedimentary column, slope failure of submarine sediments does not occur on slopes less than about 20°. However, slumping does occur on much smaller slope angles when excess pore pressures develop suddenly in the sedimentary column and the effective shear strength is reduced below the failure limit for a given slope angle. The most common causes of failure are probably from earthquake accelerations, rapid changes in pore water pressures from the sudden introduction of gas or fluids, rapid changes in the slope geometry from erosion or loading from material which has come from upslope, and diapirism.

The classic Grand Banks study (Heezen and Ewing, 1952) established a definite relationship between the generation of a large turbidity current and an earthquake, but no slump or slide scar was ever found which could be definitely shown to have been contemporaneously formed with the turbidity current. Large slump masses were, however, later found in the eipcentral region by

seismic profiling. It appears likely, particularly in view of
Hampton's work (1972) in the transition between debris flows and
turbidity currents, that a large slump was triggered which generated
a debris flow that was subsequently transformed into a turbidity
current. Relationships between several historical earthquakes and
the generation of turbidity currents are well documented through
the use of submarine cable break records (Heezen and Ewing, 1952;
Heezen and Hollister; 1971, p. 299; Heezen, 1957; Ryan and Heezen
1965). Good cases for earthquake-induced submarine failures have
also been reported by Houtz (1962) and Houtz and Wellman (1962)
for the 1953 earthquake in Suva, Fiji, by Shepard (1933), and
Menard (1964) for the 1923 Kwanto earthquake, and by Coulter and
Migliaccio (1966) for the 1964 Alaskan earthquake. In all of these
cases the epicenters were within 100 km of the probably failure
zone, and all involved earthquakes of magnitude 6 or more. All
but the Grand Banks earthquakes occurred in areas near the edges of
plate boundaries where earthquakes are common. The 1929 Grand
Banks event demonstrates that significant seismic events do occur
periodically even on "passive" continental margins. Because of the
low frequency of such events the underlying causes and geographic
patterns of seismic events in intraplate areas is poorly understood
although recent work has been increasing in this field (Sbar and
Sykes, 1973; Sykes, 1978).

     There is some evidence that moderate size earthquakes (magnitude
6) preferentially occur along old lines of weakness within a plate
(Sykes, 1978). Examples of such lines of weakness include old
sections of transform faults and the old boundary faults associated
with the original rifting. Stein et al (1979) have suggested that
many of the large historical earthquakes along the northern Atlantic
margins (eg., the 1929 Grand Banks and the 1933 Buffin Bay events)
resulted from reactivation of some of these ancient rifts from the
stress drop associated with glacial rebound. Stein also suggests
(pers. commun.) that sediment loading could induce enough stress
along sections of the margin where glaciation has not affected the
contiguous continental area. To date, all of these causal mechanisms
for passive margin earthquakes are rather circumstantial, primarily
because the long recurrence time of intraplate earthquakes makes
it difficult to see a "true" seismicity pattern. It is clear,
however, that, with the present state of understanding of intraplate
seismicity, no area can be judged aseismic over a time span of
thousands of years.

     The fracture-zone pattern off northwest Africa (Fig. 1), as
deduced from magnetic anomalies (Rona and Fleming, 1973; Hayes and
Rabinowitz, 1975), suggests a possible relationship between the
position of a major transform fault striking southeast-northwest
at 24° -26°N (Fig. 1) and the position of both the head of the
large sediment slide described here and the position of a series

of *en echelon* scarps which apparently bound a slump block (Rona, 1970; Embley and Jacobi, 1977). Rona and Fleming (1973) have mapped this area as a graben and used this feature to define the strike of the fracture zone nearest the continent (Figs. 1,3 and 4). There are no historical teleseismic earthquakes located along this trend although a magnitude 6.2 earthquake occurred in 1959 about 400 km to the northwest (Fig. 1) and is located near a possible small transform fault extension of about 20-30 km offset (Hayes and Rabinowitz, 1975). An earthquake (magnitude 5.9) also occurred near Agidir, Morocco in 1960. This is located about 1000 km to the northeast of the slide area. It may also be significant that Watkins and Hoppe (1979) have recognized several large slumps of Miocene and Oligocene age in this region, suggesting a long-range record of slope instability.

Thus, the fact is that there have been a number of direct or indirect observations which correlate large scale submarine slope failure to earthquakes. Predicting the magnitude, distance from epicenter etc., that will cause a particular submarine slope to fail is a rather complex problem. Morgenstern (1967) Almagor (1977) and Marks (1980) have attempted to use real observations to relate slope failure and earthquakes in a quantitative way. However, as Spudich and Orcutt (this volume) point out, estimation of earthquake induced seabed motion is difficult for a number of reasons and that research on the problem is at an infancy.

Shifting in the level of the gas hydrate (lathrate) free gas boundary has also been suggested as a mechnaism to induce slope instability. McIver (1977) and Summerhayes et al., (1979) suggested that pressure and temperature changes induced by sea level variations could change the depth at which clathrate forms and induce instabilities, possibly by the generation of excess free gas in the sediments above.

Summerhayes et al., (1979) have mapped several slides similar in morphology and depth to the Saharan slide along the continental margin of Southwest Africa. The southwest African margin is also contiguous with a desert area (Namib), and is also an area of intense upwelling. Summerhayes et al., (1979) have suggested that pressure-temperature changes induced by alternating sea level stands and fluctuating bottom water may have shifted the gas-hydrate boundary and altered the stability of the sediment column. Although associations of gas saturated sediments and slope instabilities have been commonly noted in studies of the shallow regions of the Mississippi delta area (eg. Roberts et al., 1976), it is unclear whether instabilities resulting from migration of the gas hydrate boundary in deep water are a viable triggering mechanism for submarine slides. Recently, however, Carpenter (in press) has shown a direct correlation between a clathrate horizon and a sediment

slide, so, although the exact mechanism is not understood there may
indeed be a relationship between clathrate and submarine slides in
some areas.

Undercutting of the lower part of a slope and subsequent
failures from progressive slumping is a common cause of subaerial
slope failure and this mechanism has also been suggested as a
probable cause of failure of some submarine slopes (Berger and
Johnson, 1976; Arthur et al., 1979). A similar mechanism is one
possibility to explain the multiphase slumping south of Baltimore
canyon. Such a scenario is illustrated in Figure 14. In stage
I an initial phase of valley cutting from turbidity currents has
been completed. The truncated reflectors are inferred from
sparker records and from the study of McGregor et al., (1979),
who infer a complex history of erosion and deposition for this
area. In Phase II a period of hemipelagic sedimentation has
resulted in a draping of sediment over the erosional surface.
During phase III reinitiation of turbidity currents through the
channels erodes the hemipelagic sequence, which results in undercutting and oversteepening of the draped sequences. This results
in mass transport of the hemipelagic draped sequence into and down
through the valleys in phase IV. In this scenario, different stages
in this geomorphic evolution may be observed off the eastern U.S.
For example, the deep-water tributary valleys described by Ryan
(this volume) may in fact be narrow sediment.

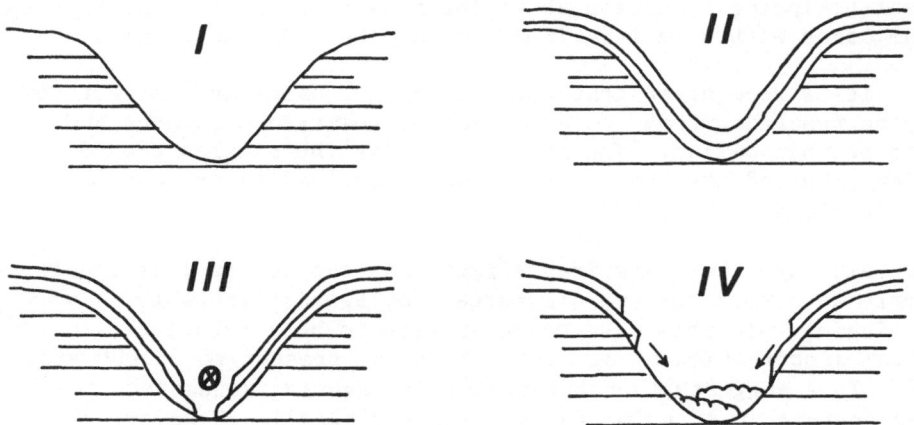

Figure 14.   A possible scenario for successive stages in geomorphic
             evolution of continental slope and rise valleys south
             of Baltimore canyon.

slides such as described by Prior and Coleman (1980). Lateral
growth of these features could result in a stripping off of the
material from the hillsides and development of a fan of deposits
downslope such as is observed South of Baltimore Canyon.

Other possibilities for triggering mechanisms include:
(1) migration of fresh water through aquifers outcropping on
the continental slope (Veatch and Smith, 1939; Manheim, 1967)
(2) Loading of the seafloor by slumps from shallower water or oversteepening by rapid sedimentation (Prior and Coleman, 1978),
(3) cyclic wave loading in shallow water (Sangrey et al., 1969)
and (4) diapirism (Shih et al., 1977).

In a given area, slope failure can be due to a combination
of these factors. For example, in an active region such as the
Mississippi delta region, subaqueous slope failure can be due
to a combination of factors including wave loading, sediment loading
and gas (Prior and Coleman, 1978).

AGE OF SLIDES

One of the major questions of interest regarding submarine
slides is their age distribution. The age distribution of the
initial failures of submarine slides is important for a better
understanding of the triggering mechanism. Many authors have
either assumed or concluded on the basis of rather scanty data
that most of these slumps and slides which occur on continental
slopes are of Late Wisconsin age. A summary of available age
data for Atlantic slides is shown in Table 1. The age data are
derived either from extrapolation of sedimentation rates or from
radiocarbon dates. The objective was to carefully select the
basal hemipelagic section above the mass flow unit for age dating
from cores with a well-defined boundary between these units.

It is recognized that the radiocarbon dates are subject to
errors resulting from contamination of reworked carbonate and
from benthic mixing. The former can bias the samples toward
older ages and the latter effect can also lead to systematic
biasing (Peng et al, 1977).

Although these combined effects can subject an individual
sample to error, the overall pattern of ages is probably
meaningful. In this case there appears to be a relatively
random distribution of age dates from the present to 24,000 yrs.
B.P. This suggests that the primary triggering mechanisms may
also operate in a random fashion (earthquakes?). However, the
small number of samples precludes any definitive conclusion.

TABLE I   AGES FOR SOME NORTH ATLANTIC SLIDES

Location	Sample No. Name, etc.	Latitude Longitude	Depth (m)	Age (Yrs. B.P.)
Upper Continental Rise between Baltimore and Norfolk Canyons	MM 78-11	32 49.5'N 73 42.4'W	1867	10,000 ± 450 [1]
	MM 78-14	37 5.4'N 74 23.0'W	1740	6,680 ± 280
Middle Continental Rise between Baltimore and Norfolk Canyons	RC 19-49	33 33.7'N 72 59.9'W	2810	5,280 ± 150 [1]
	KZ 78-03	37 47.5'N 73 35.7'W	2060	7,285 ± 290
Lower Continental Rise Blake Slide Zone	VIF-208	32 35.0'N 74 23.4'W	4671	20,830 ± 690 [1]
	RC 1-10	32 30.0'N 74 1.0'W	4892	12,125 ± 290
Upper Continental Rise South of New England	V23-1	39 44.0'N 70 21.0'W	1997	23,440 ± 2770 [2] ± 2060
Northeast Atlantic E. Flank Rockall Bank Feni Ridge	KNS1-17GPC	56 16.5'N 12 30.5'W	2565	15,200 [3]
	KNS1-19GPC	50 17.7'N 12 31.9'W	2560	15,600
Northwest African Margin	V30-42	26 31.1'N 17 36.2'W	3655	15,000 - 16,000
	V30-43	25 32.9'N	3105	2,000 - 5,000

[1]Embley (1980)
[2]Embley (unpublished data)
[3]Flood et al. (1979)

ACKNOWLEDGEMENTS

    Support for most of the work off Northwest Africa was through
grants from the Office of Naval Research to Lamont-Doherty Geological
Observatory (Columbia University) and the National Science
Foundation. The NOAA allowed me the opportunity to attend the NATO
conference and to prepare the manuscript. I thank Dennis Hayes,
Robert Jacobi, Phillip Rabinowitz, Lawrence Sullivan, Richard Bogart,
Ed thorndike and the officers and crew of VEMA for assistance in
various stages of data collection and analysis. I am grateful to
John Dormuth, William Ruddimore, Steve Hammond and John Ewing for
valuable review and discussions and to Paula McConville for typing
the manuscript.

REFERENCES

Almagor, G., and Wiseman, G., 1977. Analysis of submarine slumping
        in the continental slope off the southern coast of Israel,
        Marine Geotechnology, 2:349-389.
Archanguelsky, A.D., 1930. Slides of Sediments on the Black Sea
        bottom and of the importnace of the phenomenon for geology,
        Bull. Soc. Nat. Moscow (Sci. Geol.), 38:38-80.
Arthur, M., VonRad, V., Conford, C., McCoy, F., and Sarnthein, M.,
        1979. Evolution and sedimentary history of the Cape Bojador
        continental margin, Northwestern Africa, in: VonRad, V.
        and Ryan, W., et al, 1979, Initial Reports of the Deep
        Sea Drilling Project, V. 47, pt. 1, p. 773-816.
Berger, W., and Johnson, T.C., 1976. Deep-sea carbonates: dissolution
        and mass-wasting on the Ontong Java Plateau, Science,
        192: 785-787.
Carpenter, G., 1981, Coincident Sediment Slump-Clatrate Complexes on
        the U.S. Atlantic Continental Slope Geo-Marine Letts. in
        press.
Coulter, H.W., and Migliaccio, R.R., 1966. Effects of the earthquake
        of March 27, 1964 at Valdez, Alaska, U.S. Geol. Survey
        Prof. Paper 542-C.
Diester-Hass, L., Schrader, H.J., and Thiede, J., 1973. Sedimento-
        logical and paleoclimatological investigations of two
        pelagic ooze cores off Cape Barbas, North-west Africa,
        "Meteor" Frosch.-Ergebn., C, 16: 19-66.
Embley, R.W., 1975. Studies of deep-sea sedimentation processes
        using high frequency seismic data: Ph.D. Thesis, Columbia
        University, N.Y. 334 pp.
Embley, R.W., 1976. New evidence for the occurrence of debris flow
        deposits in deep sea, Geology, 4: 371-374.
Embley, R.W., 1980. The role of mass transport in the distribution
        and character of deep-ocean sediments in the Atlantic,
        Mar. Geol., 38: 23-50.
Embley, R.W., and Jacobi, R., 1977. Distribution and morphology of
        large submarine sediment slides and slumps on Atlantic
        continental margins, Marine Geotechnology, 2: 205-228.

Ericson, D., Ewing, M., Wollin, G., and Heezen, B.C., 1961. Atlantic
    deep-sea sediment cores, Geol. Soc. Amer. Bull., 72: 193-286.
Flood, R.D., Hollister, C.D., and Lonsdale, P., 1979. Disruption of
    the Feni sediment drift by debris flows from Rockall Bank,
    Mar. Geol. 32:311-334.
Hampton, M., 1972. The role of subaqueous debris flows in generating
    turbidity currents, J. Sediment. Petrol., 42, 775-793.
Hayes, D.E., and Pimm, A., et. al., 1972. Initial Reports, Deep-Sea
    Drilling Project, XIV.
Hayes, D.E., and Rabinowitz, P., 1975. Mesozoic magnetic lineations
    and the magnetic quiet zone off northwest Africa, Earth Plan.
    Sci. Lett., 28:105-115.
Heezen, B.C., 1957. Messina earthquake, tsunami and turbidity current
    (abstract), Bull. Geol. Soc. Amer., 68:1743.
Heezen, B.C., and Ewing, M., 1952. Turbidity currents, submarine
    slumps and the 1929 Grand Banks turbidity current, Amer.
    Jour. Science, 250:849-873.
Heezen, B.C., and Hollister, C.D., 1971. The Face of the Deep.
    Oxford Univ. Press, New York 659 pp.
Houtz, R., 1952. The 1953 Suva earthquake and tsunami. Bull. Seism.
    Soc. Amer., 52:1-12
Houtz, R., and Wellman, H.W., 1962. Turbidity current at Kadavu
    Passage, Geol. Mag., 99:52-62.
Malahoff, A., Embley, R.W., and Fornari, D., 1981 in press. Geomorphology
    of Norfolk and Washington Canyons and the surrounding
    continental slope and upper rise as observed from DSRV ALVIN
    in: Scrutton, R., (ed.), Bruce Heezen Memorial Volume
    (in press).
Malahoff, A., Embley, R.W., Perry, R., and Fefe, C., 1980. Submarine
    mass-wasting of sediments on the continental slope and upper
    rise south of Baltimore Canyon, Earth Plan. Sci. Letts.,
    49:1-7.
Manheim, F.T., 1967. Evidence for discharge of water on the Atlantic
    continental slope of the southern United States and sugges-
    tions for further search. New York Acad. Sci. Trans,
    (Ser. 22), 29:839-853.
Marks, D., 1980, Slope stability analysis of the Baltimore Canyon
    region, North Atlantic ocean, MSc. thesis, Cornell Univ.,
    99 pp.
McGregor, B.A., Bennett, R.H., and Lambert, D.N., 1979. Bottom
    processes, morphology and geotechnical properties of the
    continental slope south of Baltimore canyon, Appl. ocean
    Res, 1:177-187.
McIver, R.D., 1977. Hydrates of natural gas - important agents in
    geologic processes, Geol. Soc. Amer., Abstracts with
    Programs, 9:1089.
Menard, H.W., 1964. Marine Geology of the Pacific, McGraw-Hill,
    New York, 271 pp.

Moore, D.G., 1961. Submarine slumps, J.Sed. Petrol., 31:343-357.

Moore, D.G., 1977. The geology of large submarine slides, in:
    Voigt (Ed.), Mechanics of Landslides and Avalanches, Geol.
    Soc. Am., Spec. Paper.

Morgenstern, N.R., 1967. Submarine slumping and the initiation of
    turbidity currents, in: Richards, A., (ed.), Marine
    Geotechnique. Univ. Illinois Press, Urbana: 189-220.

Peng, T.H., Broecker, W.S., Kipphut, G., and Shackleton, N., 1977.
    Benthic mixing in deep-sea cores as determined by C14 dating
    and its implications regarding climate stratigraphy and the
    fate of fossil fuel CO_2, in: N.R. Andersen and A. Malahoff
    (editors), the Fate of Fossil Fuel CO_2 in the Oceans. Plenum
    Pub. Co., N.Y., N.Y., p. 355-373.

Prior, D., and Coleman, J., 1980. Sonograph mosaics of submarine
    slope instabilities, Mar. Geol., 36:227-239.

Prior, D., and Coleman, D., 1978. Submarine landslides on the
    Mississippi River Delta-front slope, Geoscience and Man,
    XIX: 41-53.

Roberts, H., Cratsley, D., Whelan, T., III, and Coleman, J., 1976.
    Stability of Mississippi delta sediments as evaluated by
    analysis of structural fractures in sediment borings, Eighth
    annual Offshore Technology Conference Proceedings, 1:9-28.

Rona, P., 1970. Comparisons of continental of eastern North America
    at Cape Hatteras and northwestern Africa at Cape Blanc, Am.
    Assoc. Petr. Geol. Bull., 54:129-258.

Rona, P., and Fleming, H.S., 1973. Mesozoic plate motions in the
    eastern central North Atlantic, Mar. Geol., 14:239-252.

Ryan, W.B.F., and Heezen, B.C., 1965. Ionian Sea submarine canyons
    and the 1908 Messina turbidity current, Geol. Soc. Amer. Bull.,
    76:915-932.

Ryan, W.B.F., Imaging of submarine landslides with wide swath sonar
    this volume.

Sangrey, D.A., Henriel, D.J., and Esrig, M.I., 1969. The effective
    stress response of a saturated clay soil to repeated loading,
    Canadian Geotech. Jour., 6:241-252.

Sbar, M.L., and Sykes, L.R., 1973. Contemporary compressive stress
    and seismicity in eastern North America: an example of intra-
    plate tectonics, Bull. Geol. Soc. Amer., 84:1861-1882.

Siebold, E., Diester-Haas, L., Futterer, D., Hartmann, M., Kogler,
    F.C., Lange, H., Muller, P.J., Pflaumann, U., Schrader, H.J.,
    and Suess, E., 1976. Late Quaternary sedimentation off the
    Western Sahara, in: Annais N' a Academia Braileira de Ciencias,
    Supplemento, 48:287-296.

Siebold, E., and Hinz, K., 1974. Continental slope construction and
    destruction, in: Geology of Continental Margins, C.A. Burk
    and C.L. Drake, (eds.), Springer - Verlag, p. 179-196.

Shepard, F., 1933. Depth changes in Sagami Bay during the great
    Japanese earthquake, Jour. Geology, 1:527-536.

Shepard, F.P., 1955. Delta Front Valleys bordering the Mississippi
    distributaries, Geol. Soc. Amer. Bull., 66:1489-1498.
Shih, T., and Worzel, J.L., and Watkins, J., 1977. Northeastern
    extension of Sigsbee Scarp, Gulf of Mexico, Am. Assoc. Petr.
    Geol. Bull., 61:1962-1978.
Stein, S., Sleep, N.H., Geller, R.J., Wang, S., and Kroeger, G.C.,
    1979. Earthquakes along the passive margin of eastern Canada,
    Geophys. Res. Letts., 6:537-540.
Spudich, P., and Occutt, Estimation of earthquake ground motions
    relevant to the triggering of marine mass movements, this
    volume.
Summerhayes, C., Bornhold, B., and Embley, R.W., 1978. Surficial
    slides and slumps on the continental slope and rise of
    southwest Africa, Mar. Geol., in press.
Sykes, L.R., 1978. Intra-plate seismicity, reactivation of pre-
    existing zones of weakness, alkaline magmatism, and other
    tectonism postdating continental fragmentation, Rev. Geophys.
    and Space Phys., 4:621-688.
Veatch, A.C., and Smith, P.A., 1939. Atlantic Submarine Valleys
    of the United States and the Congo Submarine Valley, Geol.
    Soc. Amer. Spec. Paper 7:101 pp.
Watkins, J.S., and Hoppe, K.W., 1979. Seismic reflection re-
    connaissance of the Atlantic Margin of Morocco, in: M. Talwani,
    W. Hay and W.B.F. Ryan (Editors), Deep Drilling Results in
    the Atlantic Ocean: Continental Margins and Paleoenvironment.
    Amer. Geophys. Union Maurice Ewing Series 3, p. 205-217.

EVIDENCES OF LATE QUATERNARY MASS MOVEMENTS IN THE BAY OF BISCAY

G.A. Auffret, J.P. Foucher, L. Pastouret, V. Renard

Centre Océanologique de Bretagne

B.P. 337, 29273 Brest cedex, France

In the Bay of Biscay that opened during Upper Cretaceous and belong to an area tectonically inactive since middle Miocene, the northern and southern margins offer a great contrast. The broad Celtic and Armorican shelf decreases in wide toward the south-east, while the Iberian margin is characterized by a very narrow shelf and important canyons heading near the coast.

Indirect evidences of slope instability are provided by a seismic profile from the Celtic margin : the upper slope there is apparently affected, by rotational slides of large amplitude. In situ observations of recent gravity induced movements affecting Cenozoic strata have also been made from the submersible Cyana.

Three 3.5 kHz profiles, provide also evidence of sliding and slumping on the lower continental slope of the Armorican margin. Cores from this zone also exhibit indications of mass movements. These events are probably of late Pleistocene age. In the Cap Ferret area high Holocene sedimentation rates and disturbed sea-floor indicate the activity of turbidity currents.

The distribution of carbonate contents in surface sediments from the Bay of Biscay suggests present transport from a high energy shelf (La Chapelle) toward the abyssal plain.

EARTHQUAKE SEISMOLOGY

An earthquake is known to human beings directly as a trembling or shaking of the ground, sometimes so violent as to crack or collapse strong buildings, break water and gas pipes, cause gaping cracks in the ground, and thus bring great loss of life and property.

The vast majority of important earthquakes are produced by faulting in the earth's crust or expressed otherwise by tectonic forces.

No one can doubt that there are earthquakes caused by volcanoes, but these earthquakes are usually very small in magnitude, and the opposite way round, earthquakes recorded have only given rise to volcanic eruptions in a few cases registered. There are, however, very special situations where self-increasing forces may develop a chain of incidents. An earthquake may cause a volcanic eruption which may cause a tsunami which may cause marine sliding or slumping; or an earthquake may cause strong oceanic waves which will influence the disequilibrium in the overbalanced and unstable marine sediment formations.

Therefore, besides enlarging our knowledge of the relationship earthquake - marine slides and slumping, it is also essential to obtain a much better knowledge of the prediction of earthquakes. For the time being we have to rely upon statistical information plus more field data.

We have three basic aspects to earthquake predictions:
1) location of the areas where large earthquakes are most likely to occur;
2) observations within these areas of measurable changes and determination of the area and time over which the earthquake will occur;
3) development of models of earthquake sources in order to reliably interpret the changes.

In prediction of marine slides and other mass movements we are at the beginner's stage.

Spudich and Orcutt discuss the earthquake ground
motions relevant to the triggering of marine mass movements. They emphasize the difference between ground-
motions on land and in the marine environment. Another
question is the effect of the water overburden on seafloor motions. Special attention should be given to the
propagation of seismic surface waves on the seafloor
because they may be important with regard to liquefaction
at relatively large distances from the source.

Maria Cita et al. postulate that an earthquake
followed by a volcanic eruption and a collapse of the
caldera followed by a tsunami has acted as the triggering mechanism of the distribution of homogenites in the
eastern Mediterranean. It concerns the volcano of Santorini in the Aegean Sea. A detailed analysis of eleven
piston cores gives the grounds for the statements.

ESTIMATION OF EARTHQUAKE GROUND MOTIONS RELEVANT TO

THE TRIGGERING OF MARINE MASS MOVEMENTS

Paul Spudich[*] and John Orcutt[†]

[*]U.S. Geological Survey
Menlo Park, CA 94025

[†]Geological Research Division
Scripps Institution of Oceanography
La Jolla, CA 92093

I. INTRODUCTION

Estimation of earthquake-induced motions in the marine environment is currently difficult for a variety of reasons. First, almost no seafloor ground-motion data exist for earthquake magnitudes greater than 3. Second, theoretical understanding of seismic wave propagation in the marine environment is incomplete. And third, it has not yet been clearly demonstrated how to extrapolate to a seafloor environment ground motions observed or predicted on land.

Estimation of earthquake ground-motions on land is somewhat easier owing to the relative abundance of empirical data and the more fully implemented theories of wave propagation appropriate to the land environment. However, in both the submarine and subaerial environments, earthquake ground motions are dependent upon so many factors, such as earthquake magnitude, epicentral range, hypocentral depth, and earth structure, that it is probably impossible to make generalizations about ground motions which apply in all cases.

Consequently, in this paper we will not describe earthquake ground motions in any specific way, but instead we will indicate appropriate sources of detailed information about ground motions in various circumstances. Because of the relative abundance of land ground-motion data compared to seafloor data, the methods of estimating ground motions presented in section III will all be appropriate to a land environment. In section IV we will discuss those aspects of knowledge about land ground motions which might be extrapolated to the marine environment, and in section II we will present the minimum amount of seismology necessary to understand the most important features of earthquake ground motions.

II. A BRIEF DESCRIPTION OF EARTHQUAKE GROUND MOTIONS

The seismic waves that emanate from earthquake sources can be approximately divided into two types of waves, body waves and surface waves, and it is the relative proportions of these two waves that determine the general character of earthquake ground motions. Body waves consist of compressional (P) and shear (S) waves, which radiate in all directions from earthquakes and whose propagation paths follow the laws of geometric optics at high frequencies. Because body waves radiate in all directions from earthquakes, the energy carried in a body wave decays as $1/r^2$, where r is distance along the propagation path, neglecting reflections and ray focusing and defocusing caused by variations in velocity structure along the

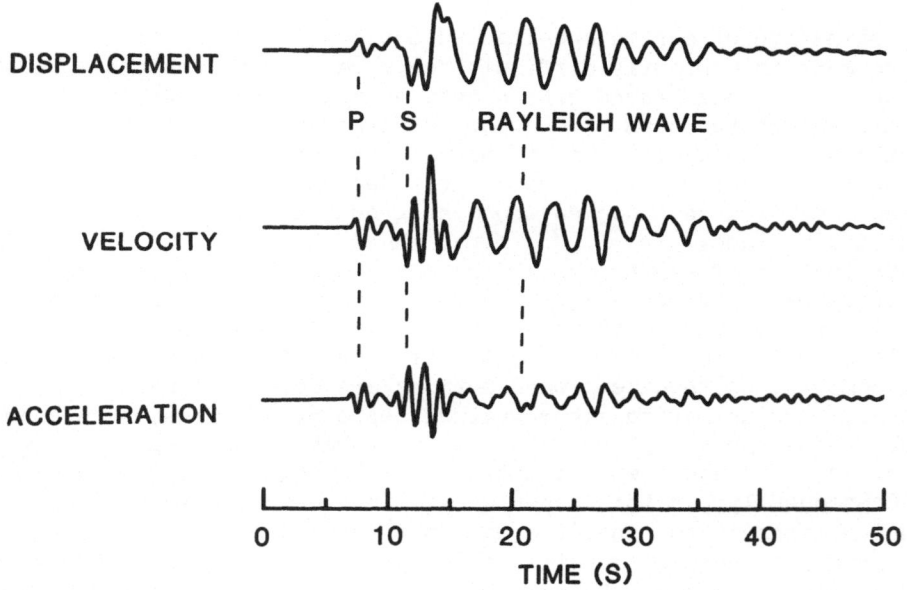

Figure 1. Vertical displacement, velocity, and acceleration calculated at 50 km distance from a thrust earthquake in a continental shelf model. The high-frequency P and S body waves dominate the acceleration record, whereas the low-frequency, oscillatory Rayleigh wave dominates the displacement trace.

path, and neglecting anelastic dissipation. Surface waves, on the other hand, are waves that are trapped on surfaces of major velocity discontinuity in the earth and cannot be observed far from the interfaces. On land the major discontinuity along which surface waves propagate is the earth's surface, although at sea surface waves can propagate along the sea floor. Because these waves spread along a surface, the energy they carry decays approximately as 1/r. Consequently, at large epicentral distance they may be the dominant wave observed. In general, surface waves start to become important at epicentral ranges which exceed three times the earthquake depth.

Surface waves differ significantly from body waves in frequency content and duration. Body waves tend to contain much higher frequencies than surface waves. In addition, whereas body waves are very weakly dispersive, surface waves are highly dispersive and travel with a group velocity less than that of shear waves. These attributes lead to the characteristic appearance of ground motions observed many source-depths away from an earthquake; high-frequency, impulsive body wave arrivals constitute the early part of the ground motions (such as the P and S waves in figure 1), and lower frequency sinusoidal oscillations of long duration (such as the Rayleigh wave train in figure 1) make up the later ground motions.

In addition, surface wave amplitude is strongly dependent on near-surface seismic velocity structure. Rapid increase of seismic velocity as a function of distance from the surface tends to trap seismic energy near the surface, thereby enhancing surface wave amplitudes. Hence, these waves tend to be more important in sedimentary basins, where shallow velocity gradients are often large, than at hard rock sites.

When reading ground motion literature, it is imperative to keep in mind the attributes of these two types of waves and their epicentral-range dependent proportions. For example, because of the higher frequency content of body waves, body waves will appear to dominate records of ground acceleration whereas surface waves may dominate records of ground displacement. This can lead to conflicting estimates of the duration of ground motion at a particular site, and a consequent misappraisal of hazard at the site. For example, shown in figure 1 are the vertical component of displacement, velocity, and acceleration recorded at a particular site at 50 km epicentral range. The relative amplitudes of the body waves (P and S) and surface waves (Rayleigh waves) vary considerably between the three traces. This is especially clear for the P wave, which is well separated from the Rayleigh waves. If one were to define the "duration" of the record to be the time interval during which maxima in the oscillations exceed half the peak value observed in the record, the displacement trace would be seen to have a duration of 17 s, but the acceleration trace would have a 3 s duration.

Before proceeding we will define a few terms that will be used in the subsequent discussion. Proximity to the earthquake source

introduces two complications into earthquake ground motion
estimation and analysis. The first of these is caused by the
physical extent of the source. When one is sufficiently far from an
earthquake source, the solid angle subtended by the source as seen
from the observation point can be very small. If at the same
location the travel-time difference between seismic waves emitted
from the nearest and farthest points on the source is smaller than
the shortest period ground motion of interest to the observer, then
we shall define the observation location to be in the "far-source"
region. Locations that are near enough to the source that either of
the above conditions is not true will be called "near-source." With
these terms we are defining a hypocentral distance range and a
frequency band in which an earthquake may be approximated by a point
source. The motivation for defining a near-source and far-source
region is that at a point in the near-source region (in a whole
space, for simplicity), seismic energy arrives from many different
directions simultaneously, making theoretical estimation of
near-source ground motions difficult. In the far-source region,
theoretical ground motion estimation is considerably easier and is
close to becoming a routine tool in hazard assessment.

In addition to the surface and body waves which decay as $1/r$ or
$1/r^2$, the so-called "near-field" waves decay even more rapidly
with distance, being important only with a few wavelengths of the
seismic source (Aki and Richards, 1980). They tend to be lower
frequency than the body or surface waves, and are not thought to be
as important in hazard estimation although they certainly appear in
data, as in Heaton and Helmberger (1979).

Ground motion literature unfortunately uses the term
"near-field" to mean both distances at which near-field waves are
observed and near-source distances. It is worthwhile to make a
distinction between the two concepts, because an observer may
simultaneously be in the near-field and far-source regions of an
earthquake.

III. GROUND MOTION ESTIMATION METHODS ON LAND

Although seismologists have been recording earthquake ground
motions since the early 1900's, only in the last two decades have
seismometers been systematically deployed that are capable of
remaining linear and yielding unclipped records when subjected to
hazardous levels of ground motion. Hence, the observational data
set of unclipped near-source ground motions is rather incomplete,
being especially poor in data recorded at ranges less than 10 km
from earthquakes with magnitudes greater than 7. Many of the basic
data have been assembled by Hudson (1976), and a partial compilation
of subsequent data acquisition is listed in Brady et al (1980).
These basic data consist of raw accelerograms, acceleration,
velocity, and displacement records corrected for instrument
response, response spectra, and various measures of duration.

Because of their significance in structural engineering, much attention has been focused on peak parameters such as peak acceleration, velocity, and displacement. Statistical analyses of these parameters as a function of earthquake magnitude and epicentral range have been prepared by numerous authors, such as Boore and Porcella (1980), Boore et al (1980), Boore et al (1978), Trifunac (1976), Seed et al (1976), Schnabel and Seed (1973), Page et al (1972), and others cited therein.

Owing to a lack of empirical data, one must turn to theoretical methods to obtain estimates of peak motions for certain combinations of earthquake magnitude and epicentral range. McGuire and Hanks (1980) have shown empirically that peak acceleration of the S wave pulse (which is generally the largest acceleration in the seismogram) is well correlated with a quantity they call a_{rms}, which is the root-mean-square value of acceleration of the S wave, and Hanks (1979) has shown that a_{rms} can be estimated from a simple, two-parameter model of the earthquake source. With additional calibrations, their system may provide a useful and simple method for estimating peak accelerations when empirical data are unavailable.

In addition to peak motions, the Fourier spectrum of ground motions may be useful for hazards assessment. Brune (1970, 1971) has developed a simple theory which predicts the shape of the far-field S body wave spectrum on the basis of a two-parameter model of the earthquake source. Numerous observational studies, such as those by Thatcher and Hanks (1973) and Fletcher (1980), have supported Brune's spectral theory and have provided a wealth of empirical observations of the two source parameters, seismic moment and stress drop, required by the theory. For observation locations that are sufficiently near the earthquake so that surface waves are small (i.e. epicentral ranges equal to a few source depths or less), the S body wave is often by far the largest contributor to the ground motions, and Brune's theory provides an acceptable approximation of the spectrum of the total ground motion at these locations.

For studies of liquefaction hazard, it may be useful to estimate anticipated duration of shaking caused by a hypothetical earthquake or, more specifically, to estimate the amplitude and number of cycles of loading that might be expected.

Numerous empirical compilations of the duration of shaking have been done by such authors as Vanmarcke and Lai (1980), McCann and Shah (1979), Dobry et al (1978), Trifunac and Brady (1975), and others. However, duration studies are plagued by several problems, such as the ambiguity in duration mentioned earlier in the discussion of surface wave and body wave characteristics, and the arbitrariness of the definition of duration, which, unlike a peak parameter, can be defined in a variety of ways. In fact, reviewing four different definitions of duration, McGuire and Barnhard (1979) claim that none of them is useful for specification of seismic shaking hazard when used in addition to peak motion parameters in studies in which only earthquake magnitude and epicentral range are used as independent variables. They state further that frequency-based definitions of duration incorporating geophysical characteristics such as seismic source properties and surface wave generation should be examined as the next step.

If one is interested not only in the duration of shaking but also in the number of cycles of motion that might be expected, little empirical data can be found in the literature to answer this question. Perez (1980) and Brady et al (1980) have attacked this question from a response spectrum point of view by determining the number of cycles of oscillation, above a specified amplitude, experienced by narrow-band, lightly damped harmonic oscillators excited by observed earthquake ground motions. Although this approach is relevant to structural response, its applicability to soil mechanics is unclear.

Great advances have been made in answering theoretically a question that is even more general than estimation of the number of cycles of ground motion that would result from a hypothetical earthquake. A number of different theoretical techniques for solution of the elastic wave equation are becoming available which enable direct calculation of the time history of ground displacement or stress at a specified epicentral range from a hypothesized earthquake in a specified crustal structure.

In general, these methods are currently limited in application to crustal structures in which elastic properties vary as a function of depth only, and in which linear elasticity is valid. Because of the restriction of linear elasticity, these methods could not be used directly to calculate motions within soils governed by nonlinear stress-strain relations. Rather, they could be used to calculate stresses or motions at a particular location in a crustal model, such as at the top of acoustic basement in a marine crustal model. This calculated motion of acoustic basement could then be used as an input to other methods for calculating the response of a stack of sediments having nonlinear mechanical properties, such as in Joyner (1975).

At present, the calculation of theoretical ground motions in the near-source region of earthquakes is computationally formidable and is hindered in its predictive powers by lack of detailed knowledge of how earthquake sources act. As examples of two rather different theoretical calculations being used to model observed ground motions in the near-source region, see Heaton and Helmberger (1979), and Archuleta and Day (1980).

In the far-source region, however, calculation of ground motion time histories is computationally easier and is on the verge of becoming a routine predictive tool for hazard evaluation. The potential effectiveness of this method in the far-source region can be seen in the work of Heaton and Helmberger (1977), Swanger and Boore (1978a), Priestly et al (1980), and others. The factors

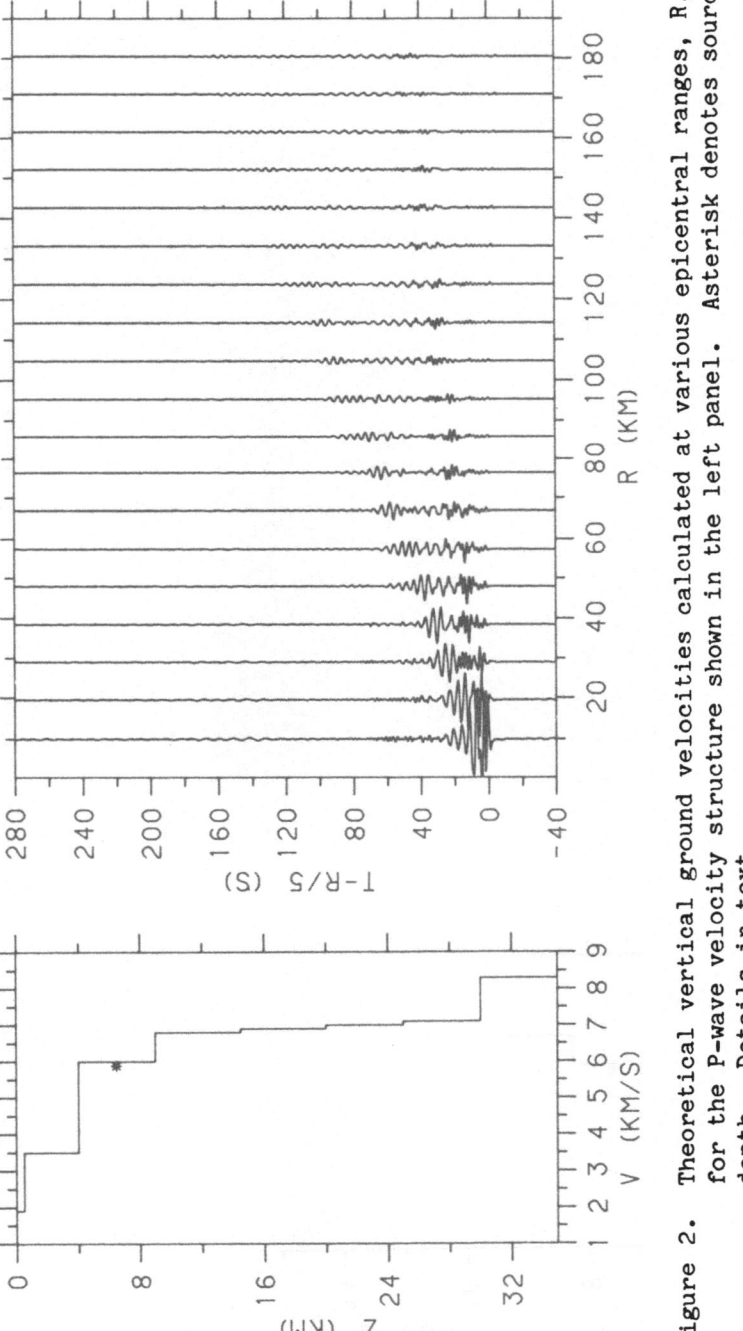

Figure 2. Theoretical vertical ground velocities calculated at various epicentral ranges, R, for the P-wave velocity structure shown in the left panel. Asterisk denotes source depth. Details in text.

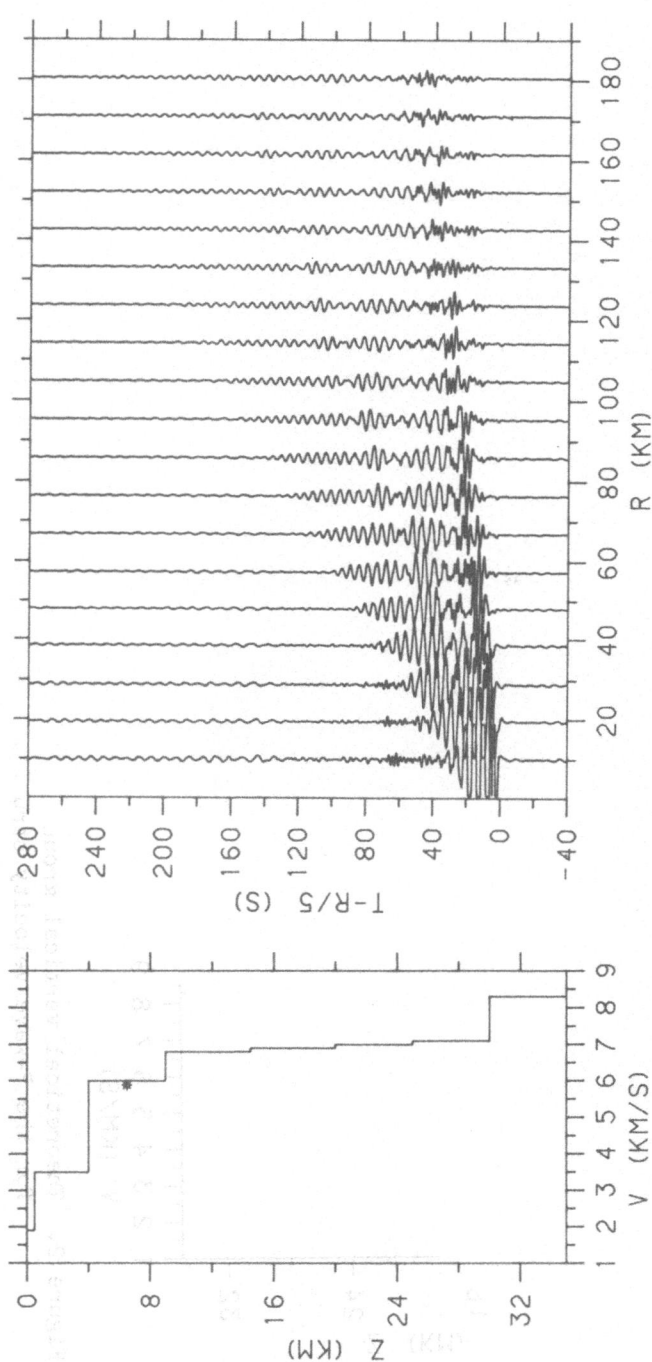

Figure 3. Theoretical radial ground velocities calculated for the velocity model shown in the left panel (same as figure 2). Details in text.

limiting accuracy in this type of calculation are lack of sufficient
knowledge of crustal velocity structure and the inability of current
methods to handle crustal structures that vary horizontally.

As an example of how such methods might be applied to a
realistic problem, we have used the method of Kind (1979) to
calculate the seismograms shown in figures 2 and 3. In the left
panel of each figure is shown the velocity structure used in the
calculation, and the star denotes the earthquake depth. Ground
velocities calculated for various values of epicentral range, R, are
plotted in the right panel of each figure. A time axis is shown
running vertically along the left side of the plot, and each
seismogram has been shifted so that seismic arrivals traveling
faster than 5 km/s appear to move closer to the bottom axis as
range, R, increases, and arrivals traveling slower than 5 km/s move
toward the upper axis with increasing R. This provides a quick
visual method for discriminating P waves, which usually travel
faster than 5 km/s, from S waves and surface waves, which travel
slower than 5 km/s.

A thrust earthquake with a left-lateral strike-slip component
was used for these calculations. The velocity model, based on the
work of Hersey et al (1959) and Hussong et al (1976), is meant to
approximate a continental shelf structure, except that there is no
water layer in this model. The thin top layer has the high
Poisson's ratio and low density appropriate to marine sediments.

Examining figures 2 and 3 we can see high-frequency body waves
in the early parts of the seismograms, followed by long, oscillatory
wave trains of conspicuously lower frequency surface waves. For
this particular velocity model and earthquake source, the horizontal
motions (radial component) are about three times larger than the
vertical motions. An intriguing aspect of the motion is that the
radial motions continue considerably longer than the vertical
motions, so that at 95 km epicentral range the last 60 s of motion
is almost entirely horizontal. This result cannot be generalized to
all continental shelf models, however, because we have calculated
models for which the surface waves are rather different, and Swanger
and Boore (1978b) have calculated surface waves in a continental
shelf model for which exactly the opposite result was obtained--the
vertical motions remained large long after the horizontal motions
had decayed to insignificance.

IV. EFFECT OF THE WATER OVERBURDEN ON SEAFLOOR MOTIONS

One question that has not yet been adequately investigated is
how ground motions experienced at sites with no water overburden
would be modified if the sites were covered by a layer of water of
given thickness and if all other factors remained the same. In lieu
of a systematic study of the problem, we can offer some speculations
based on physical arguments. These arguments apply to the
far-field, far-source region and to crustal structures that do not

VERTICAL DISPLACEMENT AT 120 KM

WITH OCEANIC OVERBURDEN

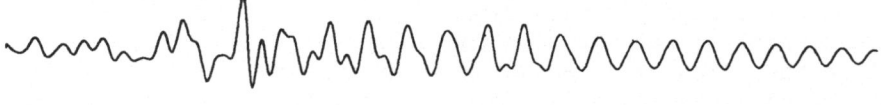

WITHOUT OCEANIC OVERBURDEN

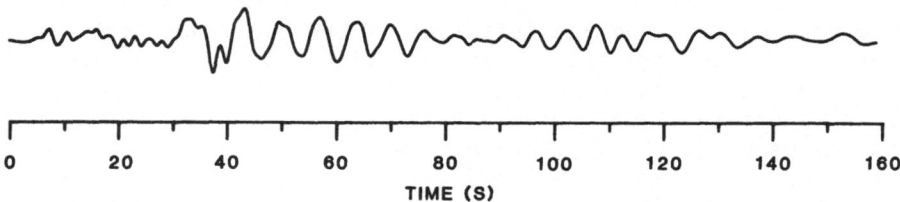

| 0 | 20 | 40 | 60 | 80 | 100 | 120 | 140 | 160 |

TIME (S)

Figure 4.   Theoretical vertical ground displacements in two velocity
             models differing only in the presence or absence of a
             1.5-km-thick ocean.

vary significantly in the horizontal direction.

From a mathematical point of view, the primary difference caused
by addition of a water layer to the problem of calculating ground
motions is that a boundary condition is being changed. Rather than
having a traction-free surface at which motions are not resisted
(the effects of the air being negligible), introduction of a water
layer causes vertical motions of the seafloor to be resisted while
horizontal motions are not. Hence, horizontal motions perpendicular
to a radial vector from the source (SH motions) will be unaffected
by the introduction of a water layer. Because vertical and radial
motions are coupled, however, radial motions will be affected by the
water layer, even though it does not directly resist them. Probably
vertical motions will be the most strongly affected on a
frequency-related basis. For high-frequency vertical motions, for
which the water's surface is many wavelengths away, the amplitude of
the motion may be reduced as much as 50%. The amplitude of
low-frequency vertical motions, for which the water's surface is
much less than a wavelength away, may be relatively unaffected.
General radial amplitudes may remain about the same, but the precise
motions in both the vertical and radial directions will change,
owing to P waves bouncing around in the water column and to altered
surface wave dispersion and excitation characteristics.

An example of this type of behavior for vertical motions is
shown in figure 4, where we show displacement seismograms calculated
at 120 km epicentral range from an earthquake source at 8 km depth

having the same mechanism as the source used in figures 2 and 3.
Both seismograms were calculated for the same crustal model, shown
in figures 2 and 3, but in one case a 1.5 km thick layer of
overlying water was included in the model. We used the method of
Adair and Orcutt (1980) to perform the calculation for the case with
the water layer. Two general differences can be seen between the
traces, both of which are consistent with the speculations offered
above. First, the upper trace (the oceanic case) is relatively
depleted in high-frequency energy compared with the lower trace;
this depletion is most evident in the body waves arriving in the
first 30 s of the record. Second, the surface waves are
substantially modified by the oceanic layer; they persist much
longer in the upper trace.

Whether these observations will hold true in general is
unknown. Although the theory exists for performing these types of
theoretical calculations of seafloor motions, it has not yet been
used as vigorously as it has for land seismology.

V. CONCLUSION

Clearly, very little is known of earthquake-induced seafloor
motions. Because it may be nearly impossible to acquire an
observational data set of seafloor motions that rivals in
completeness the admittedly incomplete data set of land ground
motions, it may be necessary to rely almost entirely on direct
theoretical calculations of seafloor motions, or it may suffice to
extrapolate from knowledge of ground motions on land, guided by a
thorough study of the effects of the water overburden.

Special attention should be given to the propagation of seismic
surface waves on the seafloor, because their long-duration,
oscillatory character and slow loss of energy with distance may make
them an important cause of liquefaction at relatively large
distances from earthquakes.

Finally, in order for seismologists to aid in evaluating the
potential for mass movement, it is essential that geotechnicians
tell seismologists exactly what is the relevant quantity to estimate.

ACKNOWLEDGMENTS

The authors thank Rick Adair for calculating the oceanic
seismograms, Tom Hanks and Dave Boore for reviewing this paper, and
the meeting organizers for the opportunity to contribute to this
volume.

REFERENCES

Adair, R. G., and J. A. Orcutt, 1980, Synthetic seismograms in an
    oceanic environment, EOS Trans. Am. Geophys. Un., 61:1037.

Aki, K., and P. Richards, 1980, "Quantitative Seismology," W. H.
    Freeman, San Francisco, Calif.
Archuleta, R., and S. Day, 1980, Dynamic rupture in a layered
    medium: The 1966 Parkfield earthquake, Bull. Seism. Soc. Am.,
    70:671.
Boore, D. M., W. B. Joyner, A. A. Oliver, and R. A. Page, 1978,
    "Estimation of ground motion parameters," Circular 795, U.S.
    Geol. Surv.
Boore, D. M., W. B. Joyner, A. A. Oliver, and R. A. Page, 1980, Peak
    acceleration, velocity, and displacement from strong-motion
    records, Bull. Seism. Soc. Am., 70:305.
Boore, D. M., and R. L. Porcella, 1980, Peak acceleration from
    strong-motion records: a postscript, Bull. Seism. Soc. Am.,
    70:2295.
Brady, A. G., P. Mork, V. Perez, and L. Porter, 1980, "Processed
    data from the Gilroy array and Coyote Creek records, Coyote
    Lake, California, earthquake," Open-file report 81-42, U.S.
    Geol. Surv.
Brune, J. N., 1970, Tectonic stress and the spectra of seismic shear
    waves from earthquakes, J. Geophys. Res., 75:4997.
Brune, J. N., 1971, Correction, J. Geophys. Res., 76:5002.
Dobry, R., I. Idriss, and E. Ng, 1978, Duration characteristics of
    horizontal components of strong-motion earthquake records, Bull.
    Seism. Soc. Am., 68:1487.
Fletcher, J. B., 1980, Spectra from high-dynamic range digital
    recordings of Oroville, California aftershocks and their source
    parameters, Bull. Seism. Soc. Am., 70:735.
Hanks, T., 1979, b-values and ω^{-n} source models: implications for
    tectonic stress variations along crustal fault zones and the
    estimation of high frequency strong ground motion, J. Geophys.
    Res., 84:2235.
Heaton, T., and D. Helmberger, 1977, A study of the strong ground
    motion of the Borrego Mountain, California, earthquake, Bull.
    Seism. Soc. Am., 67:315.
Heaton, T., and D. Helmberger, 1979, Generalized ray models of the
    San Fernando earthquake, Bull. Seism. Soc. Am., 69:1311.
Hersey, J., E. T. Bunce, R. Wyrick, and F. Dietz, 1959, Geophysical
    investigation of the continental margin between Cape Henry,
    Virginia and Jacksonville, Florida, Bull. Geol. Soc. Am., 70:437.
Hudson, D. E., 1976, "Strong-motion earthquake accelerograms; Index
    volume," Earthquake Eng. Res. Lab. Report EERL 76-02, Calif.
    Inst. of Tech., Pasadena, Calif.
Hussong, D., P. Edwards, S. Johnson, J. Campbell, and G. Sutton,
    1976, Structure of the Peru-Chile trench: 8°-12° S. latitude,
    in: "Geophysical Monograph 19, The Geophysics of the Pacific
    Ocean Basin and its Margins," G. Sutton, ed., American
    Geophysical Union, Washington, D.C.
Joyner, W., 1975, A method for calculating nonlinear seismic
    response in two dimensions, Bull. Seism. Soc. Am., 65:1337.

Kind, R., 1979, Extensions of the reflectivity method, <u>J. Geophys.</u>, 45:373.

McCann, M., and H. Shah, 1979, Determining strong-motion duration of earthquakes, <u>Bull. Seism. Soc. Am.</u>, 69:1253.

McGuire, R., and T. Barnhard, 1979, "Four definitions of strong motion duration: their predictability and utility for seismic hazard analysis," <u>U.S. Geol. Surv.</u>, Open-file report 79-1515.

McGuire, R., and T. Hanks, 1980, RMS accelerations and spectral amplitudes of strong ground motion during the San Fernando, California, earthquake, <u>Bull. Seism. Soc. Am.</u>, 70:1907.

Page, R. A., D. M. Boore, W. B. Joyner, and H. W. Coulter, 1972, "Ground motion values for use in the seismic design of the trans-Alaska pipeline system," <u>U.S. Geol. Surv.</u>, <u>Circular 672</u>.

Perez, V., 1980, Spectra of amplitudes sustained for a given number of cycles: An interpretation of response duration for strong-motion earthquake records, <u>Bull. Seism. Soc. Am.</u>, 70:1943.

Priestly, K., J. Orcutt, and J. Brune, 1980, Higher-mode surface waves and structure of the Great Basin of Nevada and western Utah, <u>J. Geophys. Res.</u>, 85:7166.

Schnabel, P., and H. B. Seed, 1973, Accelerations in rock for earthquakes inthe western United States, <u>Bull. Seism. Soc. Am.</u>, 63:501.

Seed, H. B., R. Murarka, J. Lysmer, and I. Idriss, 1976, Relationships of maximum acceleration, maximum velocity, distance from source, and local site conditions for moderately strong earthquakes, <u>Bull. Seism. Soc. Am.</u>, 66:1323.

Swanger, H., and D. Boore, 1978a, Simulation of strong-motion displacements using surface-wave modal superposition, <u>Bull. Seism. Soc. Am.</u>, 68:907.

Swanger, H., and D. Boore, 1978b, Importance of surface waves in strong ground motion in the period range of 1 to 10 seconds, <u>in</u>: " Proc. of the Second International Conference on Microzonation for Safer Construction - Research and Applications," M. Sherif, ed., National Science Foundation, Washington, D.C.

Thatcher, W., and T. Hanks, 1973, Source parameters of southern California earthquakes, <u>J. Geophys. Res.</u>, 78:8547.

Trifunac, M. D., 1976, Preliminary analysis of the peaks of strong earthquake ground motion - dependence of peaks on earthquake magnitude, epicentral distance, and recording site condition, <u>Bull. Seism. Soc. Am.</u>, 66:189.

Trifunac, M., and A. G. Brady, 1975, A study on the duration of strong earthquake ground motion, <u>Bull. Seism. Soc. Am.</u>, 65:581.

Vanmarcke, E., and S. Lai, 1980, Strong motion duration and rms amplitude of earthquake records, <u>Bull. Seism. Soc. Am.</u>, 70:1293.

TSUNAMI AS TRIGGERING MECHANISM OF HOMOGENITES RECORDED IN AREAS OF THE EASTERN MEDITERRANEAN CHARACTERIZED BY THE "COBBLESTONE TOPOGRAPHY"

Maria Bianca Cita, Anna Maccagni, Gilberto Pirovano

Institute of Geology
University of Milano
Milano, Italy

ABSTRACT

A several meters thick lithologic unit characterized by a very fine grained, homogeneous, structureless marl, a sharp basal contact and a rapidly fining upwards sequence was recovered in eleven piston cores of the "Cobblestone Project": 8 from the southern Calabrian Ridge (Area 4) and 3 from the western Mediterranean Ridge (Area 3). All these cores are located in the floor of flat bottomed depressions like troughs, basins, perched basins and craters.

The unit – called homogenite – is conspicuously absent on plateaus, where pelagic sedimentation occurs with oozes, sapropels and tephras, and on basin flanks, where pre-late Pleistocene sediments are consistently recorded.

Homogenite deposition is a Holocene event: the unit is over- and underlain by decimetric thick normal pelagic oozes. Sapropel S-1 (top approximately 8000 y BP) is found beneath the base of homogenites in all homogenite-bearing cores.

The tsunami induced by the collapse of the caldera of Santorini 3500 y BP (Minoan eruption) is considered the causative mechanism of homogenite.

Calculations by K.A.Kastens show that oscillating currents accompanying the tsunami were above the threshold erosion velocity, and that pressure pulse was sufficient to cause liquefaction of loose sediments draping the basin walls.

Grain-size analysis carried out on homogenite samples from two cores where the unit is approximately 500 cm thick support the theoretical calculations that a higher energy was available in the western Mediterranean Ridge Site ($\sim$500 km from Santorini) than in the southern Calabrian Ridge Site ($\sim$800 km from Santorini).

233

BACKGROUND

The Mediterranean is an enclosed basin, entirely surrounded by land masses. The connections with the Atlantic Ocean through the Gibraltar Strait - which is some 350 m deep - account for the present lack of an oceanic-type thermo-haline circulation at depth. Temperatures recorded at the bottom of the Mediterranean are consistently higher than 12.7 °C (Fairbridge, 1966; Lacombe & Tchernia, 1972).

Circulation at depth is even more restricted in the eastern Mediterranean, which is separated from the western basin by a second threshold, the Strait of Sicily. As a result of these peculiar oceanographic conditions, the eastern Mediterranean underwent cyclically repeated stagnant episodes during the ice ages, resulting in the deposition of jet black, organic C rich sapropels, which represent isochronous lithologies (Ryan & Cita, 1977).

The eastern Mediterranean contains two active volcanic arcs: one is located in southern Italy, and southern Tyrrhenian Sea; the other one in the Aegean Sea.

The late Quaternary explosive volcanic activity is materialized in deep-sea cores by numerous tephra layers, which represent isochronous lithologies (Keller et al., 1978 inter alias).

Dominant lithology in pelagic sequences is marl or ooze, minor lithologies are sapropels and tephras.

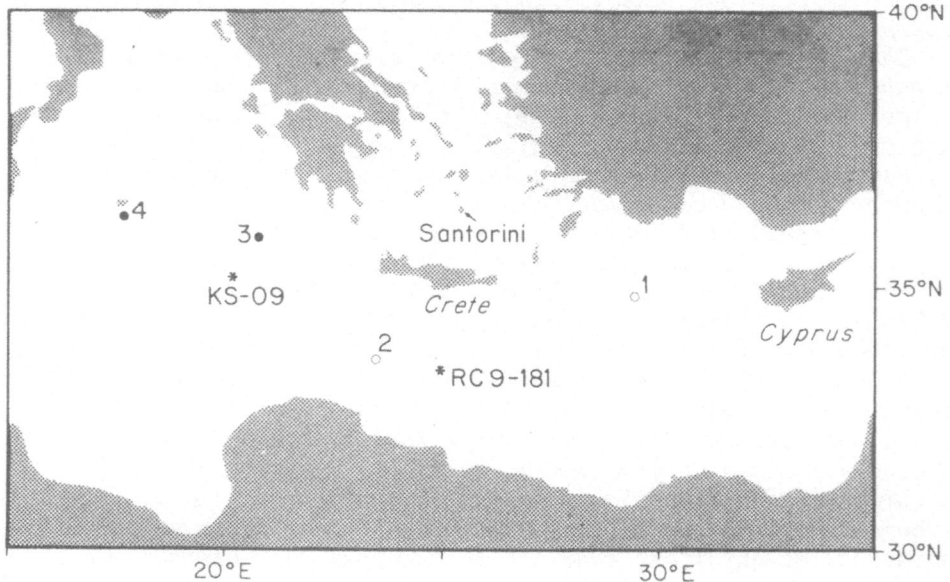

Figure 1 - Location map showing the areas explored with the deep-tow instrumentation (1-4) and other deep-sea cores discussed in the text. Solid circles indicate geological test-areas.

In other words,the deep-sea record of the eastern Mediterranean permits a high resolution stratigraphy to be applied for the late Quaternary, much higher than that worked out for the open ocean where a combination of biostratigraphy, paleomagnetic stratigraphy and isotopic stratigraphy allows the recognition and correlation of time intervals a few tens of thousand years long (Berggren et al, 1980, cum ref.).

"COBBLESTONE" PROJECT

During the summer of 1978, two oceanographic cruises were dedicated to investigate with multidisciplinary approach selected areas of the eastern Mediterranean characterized by a peculiar hummocky relief, referred to as "cobblestone topography" (Hersey, 1965). R/V MELVILLE of Scripps Institution of Oceanography explored and surveyed with deep-tow instrumentation, including echo-sounding, 4 kHz seismic profiling, side-looking sonar, and bottom photography four areas located west of Cyprus, south of Crete, west of Crete and north of the Messina abyssal plain, see Figure 1 (Spiess et al, 1978).

The last two areas, each approximately 100 km^2 wide, were subsequently visited by R/V EASTWARD of Duke University, which was positioned with a transponder-navigated system, using moored buoys left by MELVILLE. High precision bathymetric maps at the scale 1:18 000, and maps plotting the reflectivity of the sediments were made available, so that the transponder-navigated coring program performed by EASTWARD could predetermine with the greatest accuracy core location on selected physiographic features such as plateaus, domes, basin walls, slump scars, base-of-slope, craters, basin centers etcetera.

The areas explored geologically are approximately 50 km^2 for each site. Forty piston-cores were obtained overall, 23 from Area 4 (see Ryan et al, 1978), 17 for Area 3 (northern flank of the Mediterranean Ridge). For commodity of reference, we named the main physiographic features explored: after Italian Renaissance painters (basins) and sculptors (elevated areas) in the Calabrian Ridge Site; after the Greek mithology in the Mediterranean Ridge Site.

Total sediment recovery is 333 m. Although a great variety of sediment types, and ages, was obtained, especially from the basins and/or crater slopes, the large majority of sediments are late Quaternary in age, so that the high resolution stratigraphy previously mentioned could be applied (Blechschmidt et al, in press; Cita et al, in press; Mc Coy & Coughlin, in press).

DISCOVERY OF HOMOGENITES

The first homogenite-bearing core obtained is n. 7, from the Beato Angelico Trough, a several km long depression striking SW-NE,

steep walled, 1 to 2 km wide (see Table 1). We were surprised
to find a 609 cm thick unit, very monotonous and structureless, an
olive-gray marl very fine grained, over - and underlain by normal
hemipelagic marls. The unit has a sharp basal contact,and a rapidly
fining upwards sequence. Such a lithology was recorded in ten additional cores, all located in flat-bottomed basin floors.

     Homogenites are exclusive of basinal settings, and correspond
to a "transparent" acoustic layer which in the MELVILLE records is
also exclusive of basinal settings (Kastens & Cita, in press).

     The acoustically transparent layer fills the depressions and
pinches out at the base of the slopes, as a typical "ponded"facies.
Thickness of homogenites in the cores correlates with thickness
of surficial "transparent" layer in seismic records.

     Thickness of homogenites in the cores ranges from 770 to 44 cm.
Table 1 plots setting and water depth of all homogenite-bearing
cores, as well as the measured thickness of the lithologic unit.

TABLE 1

Basin name	Core n.	Thickness of homogenite	Water depth
Area 4			
Beato Angelico Trough	7	609 cm	3628 m
	9	552 cm	3625 m
	12	403 cm	3625 m
	40	770 cm	3630 m
Botticelli Basin	42	478 cm	3592 m
	44	238 cm	3489 m
Raffaello Basin	10	766 cm	3832 m
L. da Vinci Basin	8	44 cm	3535 m
Area 3			
Electra Basin	23	292 cm	3001 m
Ares Crater	31	72 cm	2967 m
Aphrodites Crater	32	504 cm	3243 m

Table 1: Distribution of homogenite in Cobblestone Area 4 (southern
Calabrian Ridge) and Area 3 (western Mediterranean Ridge).

Figure 2 and 3 show the correlation of homogenite-bearing cores from Area 4 and 3 respectively.

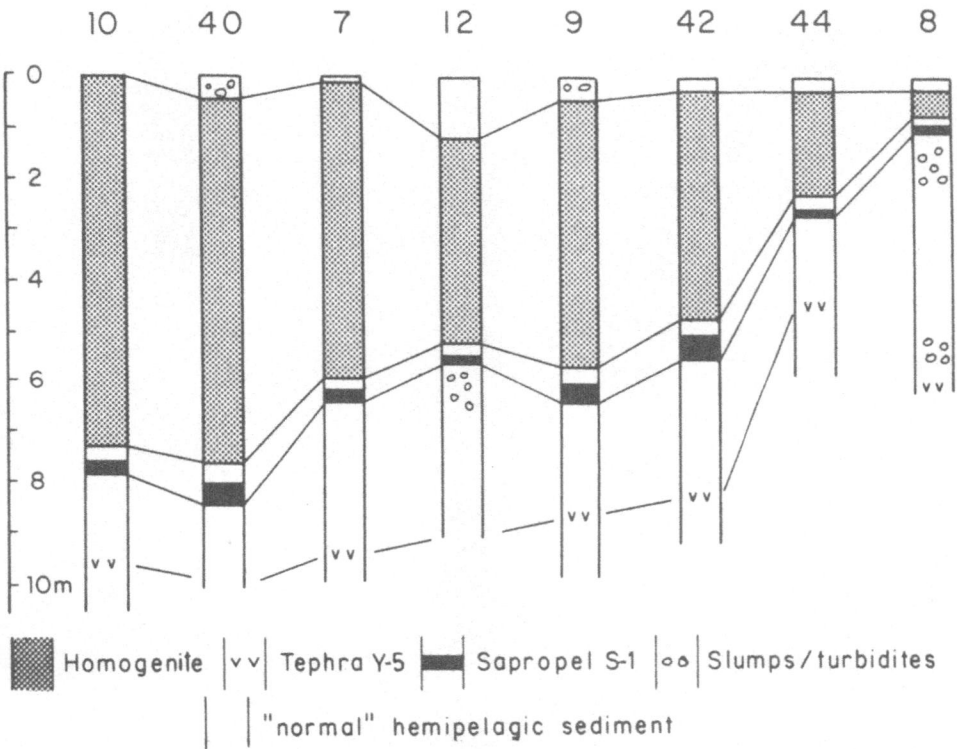

Figure 2 - Correlation of homogenite-bearing cores from Cobblestone Area 4 (southern Calabrian Ridge).

Transects of cores encompassing plateaus, slopes and basins with a total relief of approximately 100 m in Area 4 (Figure 4) and of up to 200 m in Area 3 (Figure 5) demostrate that homogenites are restricted to basin floors.

The unusually thick unit starts with a sandy layer, entirely composed of planktonic foraminiferal tests (foraminiferal sand). The base is sharp, and the unit rapidly fines upwards (Figure 6): indeed, most of the unit is devoid of sand-size sediment fraction, as proved by several dozens processings for micropaleontology(using a sieve with mesh size 63 microns, we had no residue), and by detailed studies on Cores 32 and 42.

The carbonate content also decreases upwards (Figure 7), as a result of the absence of calcitic tests of foraminifers and/or of aragonitic tests of pteropods.

The unit under discussion is called homogenite for its extremely uniform color, and structureless nature. The total lack of visually

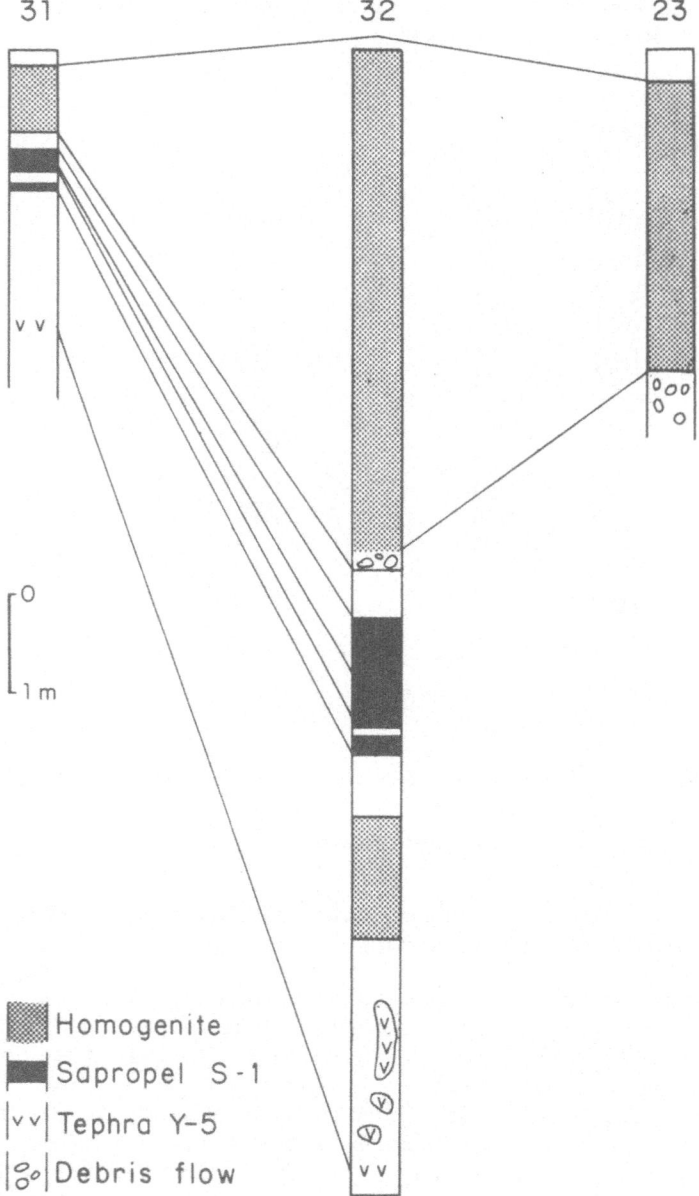

Figure 3 - Correlation of homogenite-bearing cores from Cobblestone Area 3 (western Mediterranean Ridge).

detectable internal sedimentary and organic structures has been substantiated by X-radiographs carried out on the central part of homogenites from Cores 9 and 40 (Area 4) and 23 (Area 3). Figure 8 illustrates specimens of these radiographs.

Homogenites correlate in cores from one and the same depression (for instance Cores 7,9,12 and 40, all from Beato Angelico Trough), in cores from different depressions within the same area (for instance all the above cores with Core 10 (Raffaello Basis), 42 and

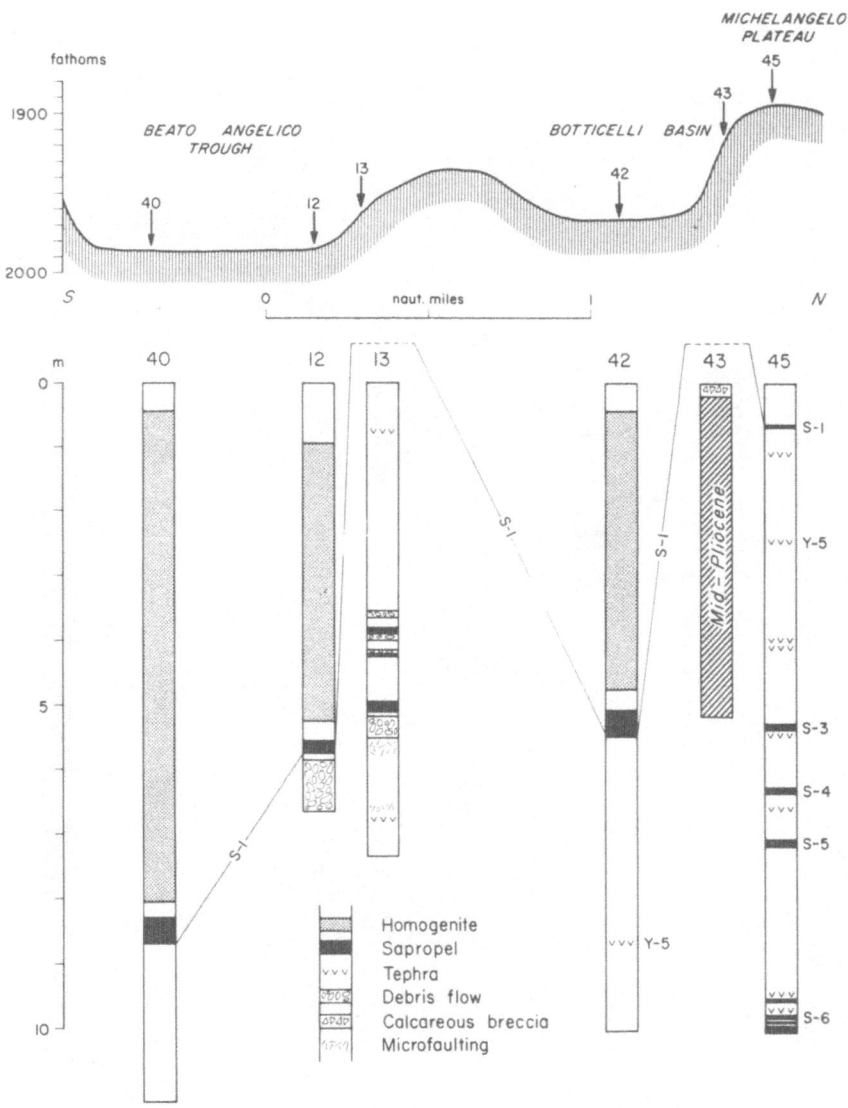

Figure 4 - Transect of cores across basins, slopes and plateaus in Cobblestone Area 4 (southern Calabrian Ridge). Total thickness of cores (excluding eventual flow-in) is considered, unlike in Figures 2 and 3.

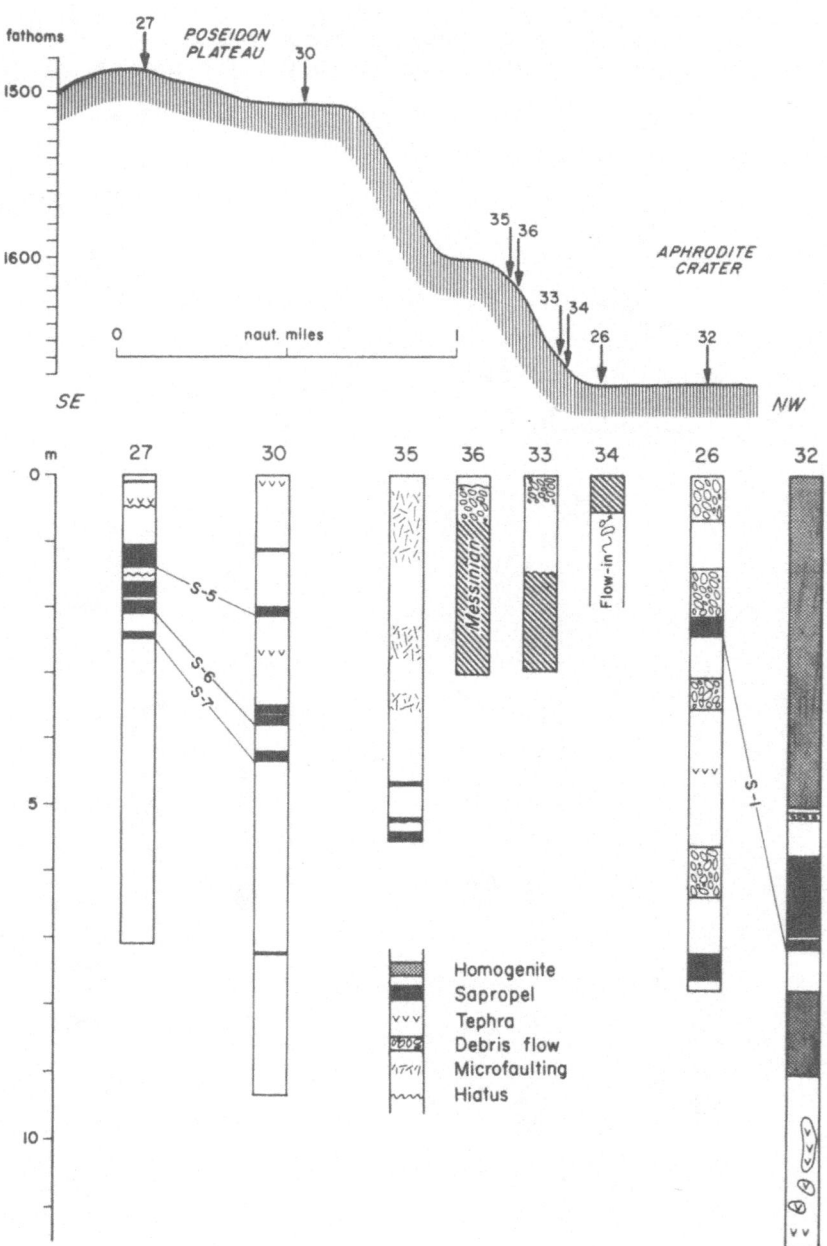

Figure 5 – Transect of cores across Poseidon Plateau and Aphrodite Crater in Cobblestone Area 3 (western Mediterranean Ridge). Total thickness of cored sediment is considered (unlike in Figure 2 and 3), but excluding eventual flow-in.

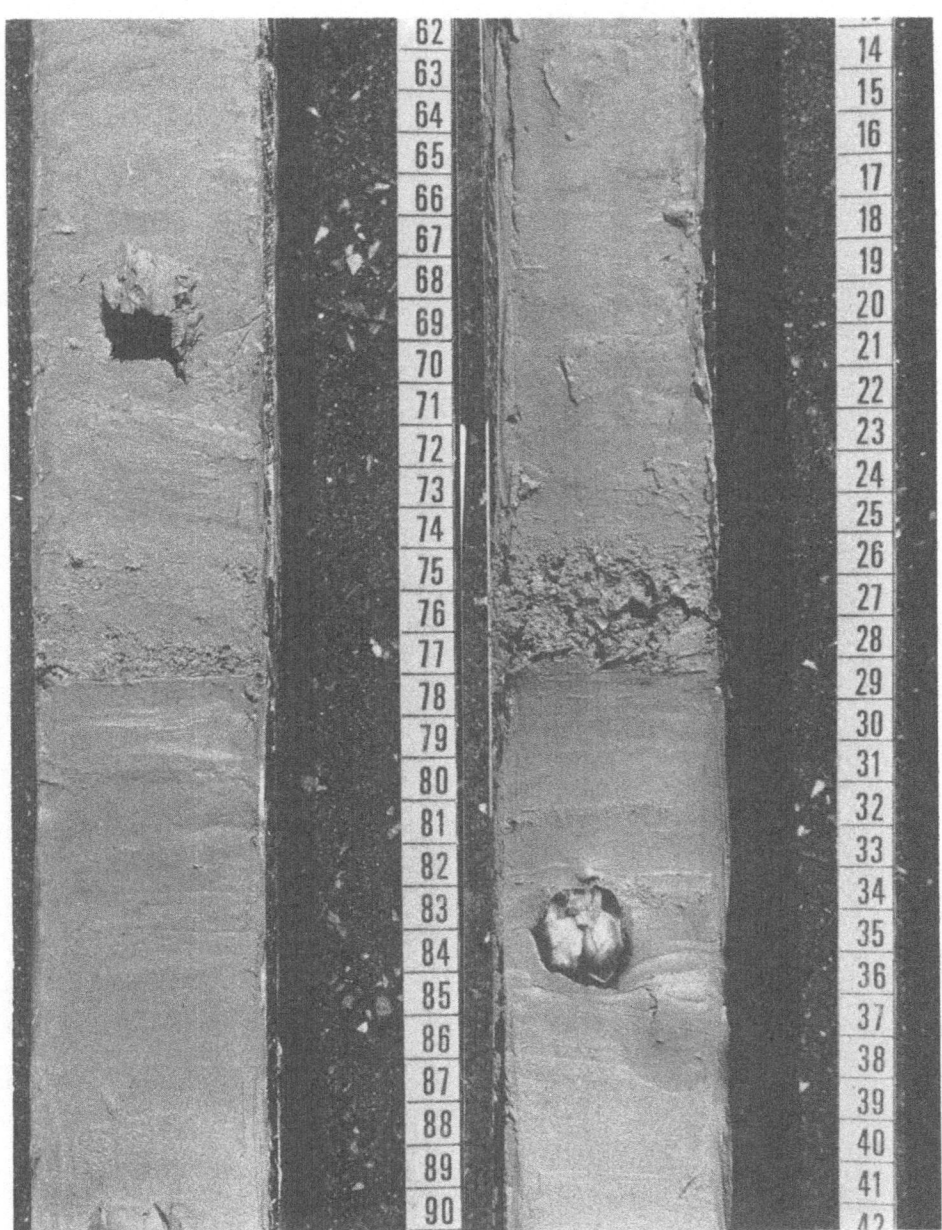

Figure 6 - Base of homogenites in Cores 40 and 9, both from Beato Angelico Trough, southern Calabrian Ridge.

43 (Botticelli Basin) and 8 (Leonardo da Vinci Basin), and also in cores from discrete basins belonging to two areas some 300 km apart (see Figure 1).

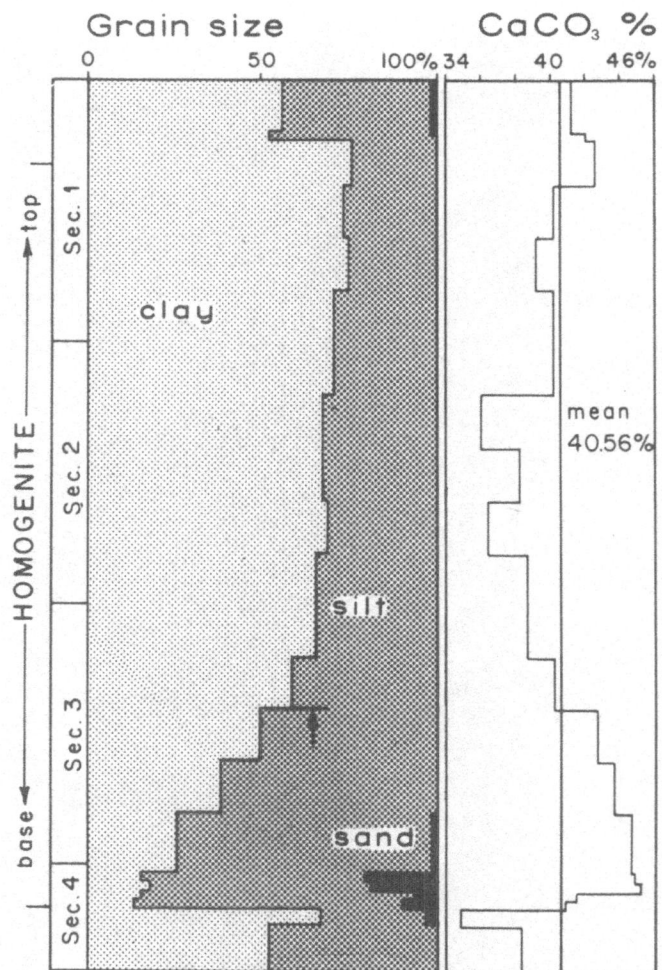

Figure 7 - Grain-size analysis (left column) and carbonate content
measured in 20 samples of Cobblestone Core 42, encompassing
homogenite. The arrow indicates the highest level where coarse
silt (greater than 31 microns) was recorded. Asterisks mark samples
whose sand-size fraction is illustrated in Figure 14. Grain-size
analysis were made with standard methods, using pipette for the
fine silt and clay fractions. Carbonate content was obtained
measuring the pressure of CO_2 developed after H Cl treatement,
corrected for temperature, using a calcimeter Mod. Pizzarelli.

STRATIGRAPHY OF HOMOGENITES

The stratigraphic position of homogenites is within the Emilia-
nia huxleyi Acme-zone (younger that 68 000 y), see Blechschmidt et al

(in press). Chronostratigraphic position is in the Holocene (Figure 9) that is younger than 10 000 y. Homogenites post date Sapropel S-1 which represents the sedimentary expression of eastern Mediterranean basis-wide stagnation induced by the post-glacial (Flandrian) trans-gression (Cita & Ryan, 1978), and has been dated several times with radiocarbon techniques, its age ranging from 9000 to 7000 y BP. Consequently homogenites are younger than approximately 8000 y BP.

Homogenites are clearly a re-sediment, as shown by the sharp basal contact and fining upwards grain-size. They strongly differ from other re-sediments encountered in the 40 transponder-navigated cores of the "Cobblestone" Project: from turbidites (recovered in basins and perched basins) for the greater thickness, and for the absence of laminations; from debris flows for the much greater thickness, and for the absence of large clasts.

Homogenites are interpreted as the result of settling on basin floor of a suspension of fine-grained loose sediment previously dra-ping the basin walls. Detailed analysis of calcareous nannofossils in the post S-1 homogenite of Core 32 by G. Blechschmidt (personal communication, 1980) shows that numerous specimens reworked from older formations are present throughout. None of these are older than mid to early Pliocene, with both Pliocene and Pleistocene forms being present. The presence of older taxa in the homogenite supports the hypothesis that these sediments are the product of slope pro-cesses.

Average sedimentation rate calculated for selected cores (Figure 10) change drastically along transects crest/slope/base-of-slope/basin (see Figures 4 and 5). The proportion is up to 1:20 from crest to basin, even more from basin to slope, where sedimen-tation rate is negative and older sediments are exposed, as a result of submarine erosion.

CONCEPT AND TIMING OF "HOMOGENITE EVENT"

The emplacement of post S-1 homogenite unit probably represents a very short geological event. Its causative mechanism has to be of more than local significance, unlike the causative mechanism of debris flows recorded in base-of-slope cores. Indeed, debris flows do not correlate even within the same basin: they are considered the result of local slope failure, whereas the ubiquitous presence of homogenites in basin floor cores requires a triggering mecha-nism effective over widely spaced areas.

First of all, we attempted to calculate with the greatest accuracy the time of emplacement of homogenites. In order to do this, we selected five cores from Area 4 and one from Area 3 in which the homogenites are both overlain and underlain by normal pelagic sediments. We measured the thickness of these pelagic sedi-ments, and assumed as boundary limits: 8000 y BP for the top of Sapropel S-1, and time 0 for core tops.

Assuming a uniform sedimentation rate through this time inter-

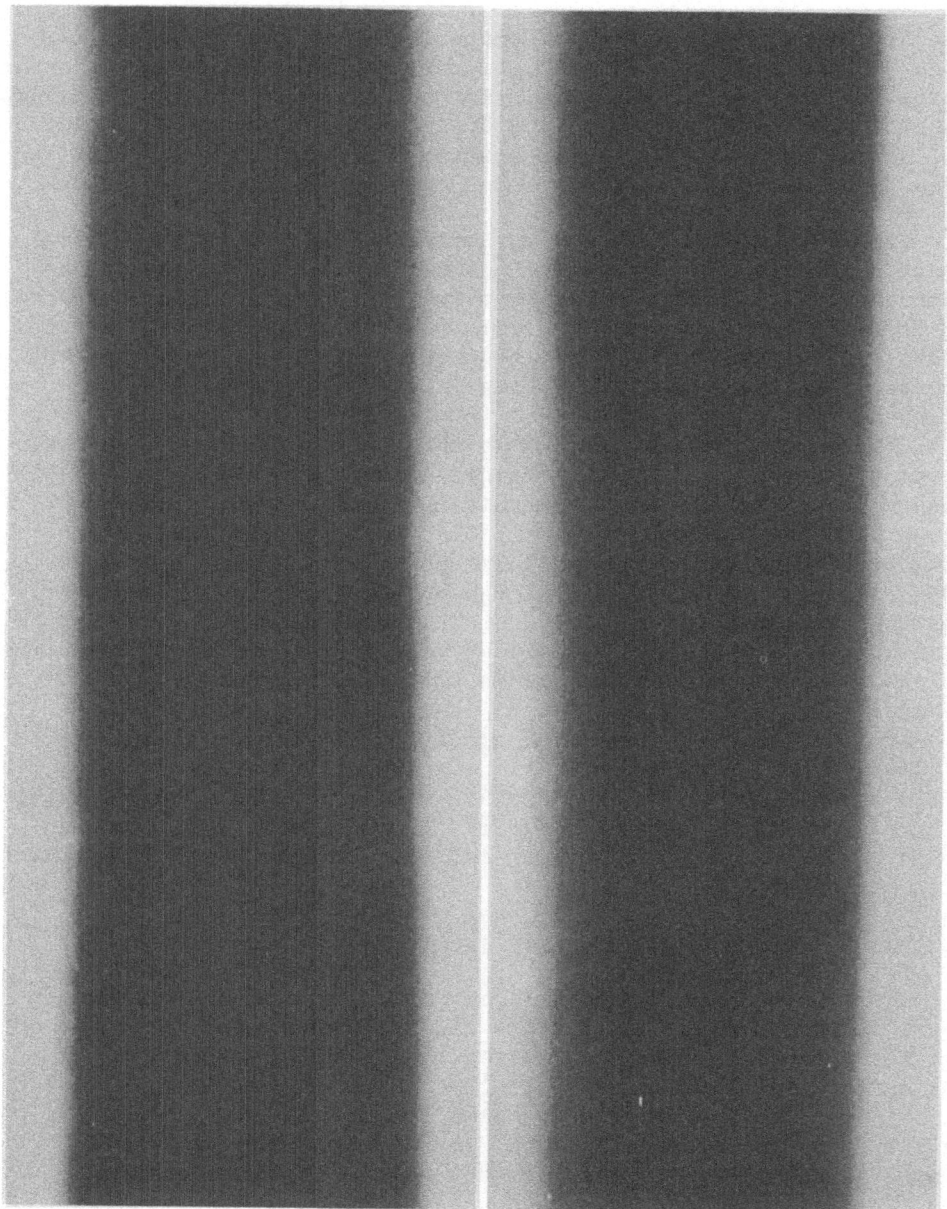

Figure 8 - Radiographs of a portion of a 40 cm interval from the central part of homogenite in Section 3 of Core 40 (left) and in Section 3 of Core 9 (right). Contact print.

val, we calculated its value by combining the thickness of the upper and lower pelagic layers, obviously excluding from our countings the thickness of homogenite itself. The resulting sedimen-

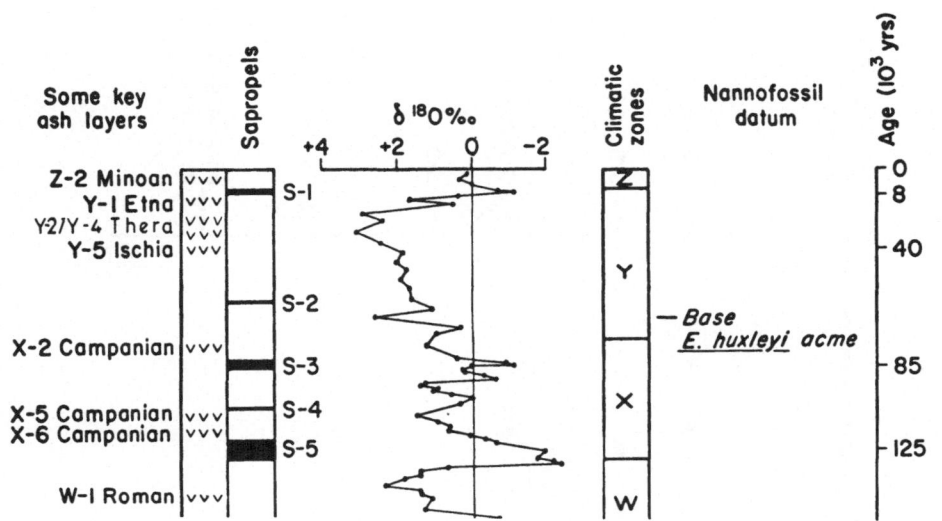

Figure 9 - Chronostratigraphy,biostratigraphy,tephrachronology,sapropel chronology of the eastern Mediterranean deep-sea record and isotopic signature of Core RC9-181(after Cita et al,1977;Vergnaud-Grazzini et al,1977;Keller et al,1978;Blechschmidt et al, in press).

tation rate was then used to compute the amount of time represented for each pelagic unit. Time of homogenite emplacement is considered virtually coincident with the lower limit of the upper pelagic unit. The results of our calculations are plotted in Table 2.

Results from Cores 40 and 42 indicate that the "Homogenite event" occurred approximately 3000 y BP, while values from Cores 9, 31 and 44 indicate that the event occurred at approximately 4300 y BP.

We consider highly improbable, however, that two discrete "Homogenite events"did occur in post-S-1 times.The best explanation we can offer is that the "Homogenite event" was indeed unique, and that its age is comprised in between 4300 and 3000 y BP. An actual age closer to the lower limit is more probable, because of the technicality of piston coring, which involves a certain amount of error when we identify core tops with time 0. The loss of a few centimeters of sediment at the top may indeed be difficult to detect. Moreover, sediment compaction affects the pelagic sediment under-lying homogenite more than the pelagic sediment above it.

The above data, calculations and arguments indicate that a sudden,short duration,major event(earthquake,catastrophic eruption or similar) should be recorded in the eastern Mediterranean <u>historic</u> record, to account for the regional "Homogenite event", at approximately 3500/4000 y BP.

We studied such record, and found a plausible event in the gigantic Bronze-age eruption which resulted in the collapse of the caldera of the Santorini volcanic island some 1500 y BC

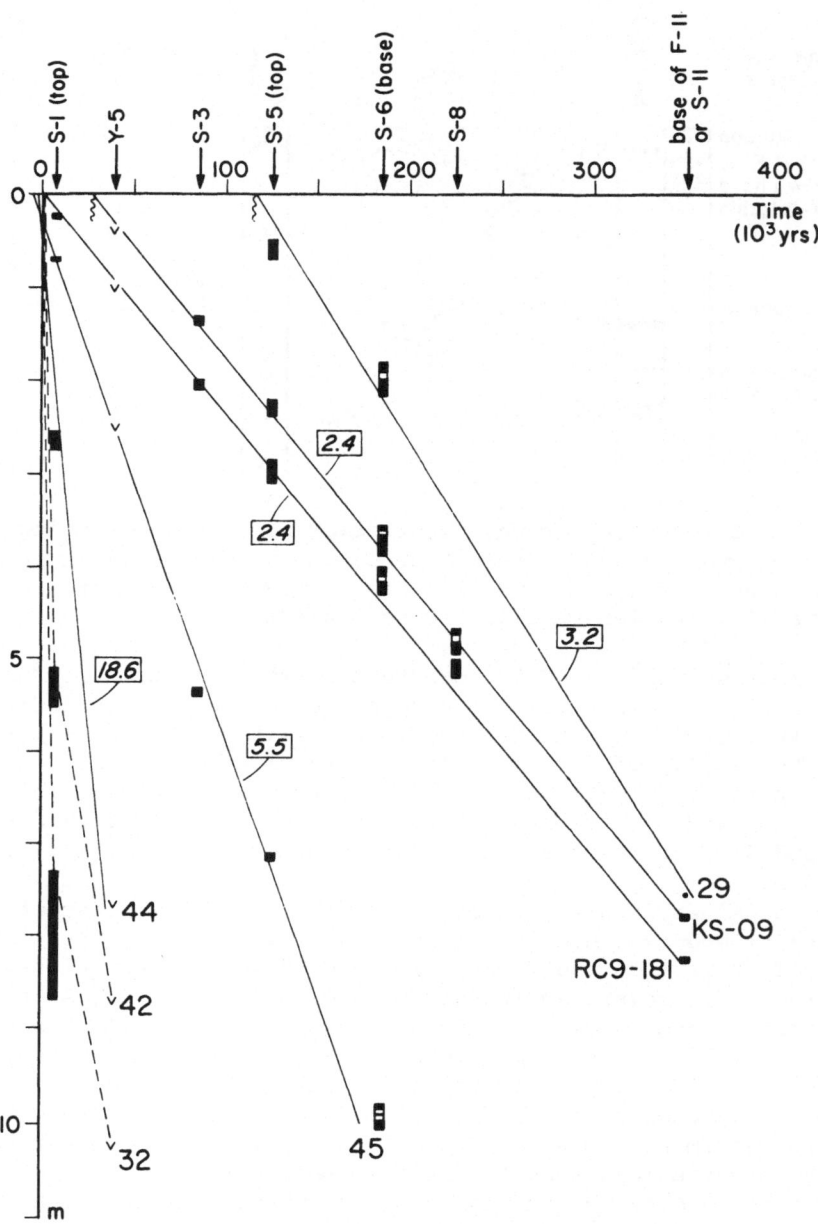

Figure 10 – Sedimentation rates calculated with the best fit method in selected cores from Cobblestone Areas 4 (Cores 42, 44, 45) and 3 (Cores 29, 32), and in Cores KS 09 and RC9-181. Chronology after Cita et al (1977) and Thunell et al (1979).

(3500 y BP) and in a large tsunami.

Reference is made to studies by Kastens & Cita (1980 and in press) for a discussion on why the tsunami, and not the earthquake itself, is considered the triggering mechanism.

TABLE 2

Core number	Thickness of pelagic seds. (cm)	Estimated sed. rate (cm/10^3y)	Duration of pelagic intervals above and below "homogenite" (y)
8	16 above 16 below	4.0	4000 y above 4000 y below
9	36 above 30 below	8.25	4360 y above 3640 y below
40	21 above 32 below	6.7	3130 y above 4850 y below
42	22 above 31 below	6.6	3300 y above 4700 y below
44	24 above 21 below	5.6	4290 y above 3750 y below
31	18 above 15 below	4.13	4360 y above 3630 y below

Table 2: Thickness and estimated time relationships of pelagic sediments bracketing the "homogenite event".

BRONZE-AGE TSUNAMI INDUCED BY THE COLLAPSE OF THE CALDERA OF SANTORINI AS TRIGGERING MECHANISM FOR HOMOGENITES.

TSUNAMI is a composite japanese word (TSU = port, NAMI = wave) which refers to gigantic waves generated by the sudden deformation of the ocean bottom. Their characters are quite unusual: wave length is usually longer than 100 km, up to 200 km. Wave height in the open ocean is of the order of 1-2 m, but when tsunami approaches the shelf, wave height rapidly increases, and may reach 30-40 m on the coast, especially when it is funnelled in narrow channels.

Tsunamis 40 m in height have been recorded against the Japanese coast in 1896, 1907 and 1933 (H.Nakagawa, personal communication, 1980). The highest tsunami ever was recorded in Alaska: all ships anchored in a port protected by a promontory were destroyed but one which was found intact on the opposite side of the promontory. Height of the promontory was 70 m above sea-level.

Velocity of shallow waves (c) is related to water depth (h) by
the following equation   c = $\sqrt{g\,h}$, consequently velocity at sea
floor can be calculated directly from bathymetry.

The origin of tsunamis is in part obscure. Apparently they
are originated by earthquakes of magnitude greater than 6.5 (Rich-
ter scale) with foci shallower than 50 km (Fairbridge, 1966). Not
all shallow-focus earthquakes produce tsunamis, however.

One of the largest and most devastating historically recorded
tsunamis was generated by the explosion of the volcanic island
Krakatoa, which belongs to an active volcanic arc, in 1883.

A direct comparison of the collapsed calderas of Santorini
and Krakatoa is at the base of Yokohama's (1978) calculations on
the tsunami caused by the Bronze-age eruption of Thera (or Santorini). Actually, the Santorini collapsed caldera is three times
bigger than the caldera formed by the Krakatoa eruption, but this
observation cannot be used to argue proportionally bigger tsunami,
since the time relationships are unknown.

Santorini is a volcanic island in the Aegean Sea which - along
with Milos, Anaphi, and many others, belong to the inner volcanic
arc bordering the Aegean plate near the northwards dipping subduction zone (Mc Kenzie,1972, Barbieri et al, 1977; Makris & Vees,
1977).

Major and trace elements determination of volcanic rocks from
Santorini permitted to reconstruct the differentiation history of
the magma (Puchelt, 1978). It is assumed that the magmas were generated in that part of the mantle which lies above the subducting
plate by partial melting induced by an aqueous phase - enriched
in incompatible elements - which was generated from the subducted
sea-floor.

Oldest radiometrically dated volcanic rocks from Santorini
are 1.5 m.y. old (Ferrara et al, 1980).

The volcanic activity of Santorini during the Bronze-age is
probably the best known in the world from that age, because of its
large cultural implications (destruction of the Minoan civiliza-
tion in Crete).

A detailed reconstruction of the mechanics of the Minoan eruption recently proposed by Pichler & Friedrich (1980) include (a)
a Plinian eruption with air-fall pumice fall (Bo 1), followed by
(b) the invasion of the magma chamber by sea water and a phreatomag-
matic eruption (Base Surge Phase Bo 2). The following (c) Ash Flow
Phase (Bo 3) with the formation of ignimbrite is accompanied by
the collapse of the caldera. Opening of the collapsed caldera is
towards SW. Surface of the caldera is 84.9 Km^2, versus 165.4 km
of the caldera plus surrounding islands.

Ashes and pumices erupted during the Plinian phase are recorded in islands of the Aegean and Levantine seas, and in deep-sea
cores lying to the east and southeast of Santorini, due to a we-
sterly wind influence (Ninkovich & Heezen, 1965; Keller et al,
1978; Watkins et al, 1978; Mc Coy, 1980).

The volcanic earthquakes preceding and accompanying the Minoan

eruptions are not likely to have been large enough to have far-
reaching effects or to cause tsunamis, but the caldera formation
is likely to have caused large tsunamis (Thorarinsson, 1978).

The Minoan tsunami of Santorini was thought to have reached
the northwest coast of Cyprus, some 680 km to the east, on the basis of field observations at Ayia Irini (Meszaros, 1978).

An argument used to estimate the strength of tsunami related
to the Minoan eruption was the occurrence of thick pumice deposits
on Anaphi island some 20 km to the east of Santorini (Marinos & Me-
lidonis, 1971), some 250 m above sea-level. However, Keller (1980)
showed that the pumice is older than the Minoan eruption(older than
18 000 y).

K.A.Kastens (in Kastens & Cita, in press) reconstructed the
ray path of wave crest orthogonals originating at 5° azimuth intervals from a point source at Santorini; she took into account
all the refractions caused by bathymetric changes (1) in order to
check if it was conceivable to consider tsunami as a causative
mechanism for slope failures resulting in the deposition of homogenites in Cobblestone Area 3 (approximately 500 km far from Santorini) and in Area 4 (approximately 800 km, see Figure 1).

Her order-of-magnitude calculations made using wave height
more conservative than those proposed by Yokohama (1978) showed
that (a) the near-bottom oscillating currents accompanying the
tsunami were above the erosion velocity of clay-sized particles,
and (b) the pressure pulse induced by P waves was sufficient to
cause liquefaction of sediments draping the slopes. Table 3 plots
the numerical values calculated for Cobblestone Area 3 (western
Mediterranean Ridge) and 4 (southern Calabrian Ridge).

COMPARISON OF WESTERN MEDITERRANEAN RIDGE AND SOUTHERN CALABRIAN RIDGE HOMOGENITES

Calculations in table 3 show that a higher energy was available in the western Mediterranean Ridge than in the southern Calabrian Ridge, the latter area being approximately 300 km farther
from Santorini. If this is the case, the sedimentologic make-up
should reflect a higher energy environment. Looking for supportive
evidence in the cores themselves, we selected for a close comparison two cores,one for each area,where the thickness of homogenites
is comparable (approximately 500 cm): Core 32 from the center of
Aphrodite Crater (see Figures 5 and 11) and Core 42 from the Botticelli Basin floor (see Figures 4 and 7).

Figure 12 compares photographs of the base of homogenite in
these two cores: a 18 cm thick debris flow, with clasts several
cm in size is present at the base of homogenite in Core 32, unlike
in Core 42.

---

(1) bathymetry affects both the direction, and the velocity of
tsunami waves, as seen before, because it is a shallow-water wave
with respect to oceanic depths.

TABLE 3

	Area 3 (Western Med. Ridge)	Area 4 (Calabrian Ridge)
Water depth	2900 m	3400 m
Wave speed	168 m/sec	182 m/sec
Wave length	151 km	182 km
Wave height	17 m	6.9 m
Near-bottom current speed	49 cm/sec	19 cm/sec
Magnitude of pressure pulse	875 g/cm^3	355 g/cm^3

Table 3: Tsunami properties according to K.A.Kastens (in Kastens and Cita, 1980, and in press).

Figure 13 compares grain-sizes measured for these two cores. In Core 32, we started our measurements from immediately above the debris flow, since grain-size analysis within this interval would have been meaningless. Piston core 32 contains homogenite till the top, whereas the normal hemipelagic sediment (a pteropod ooze) is recorded in the trigger weight.

Unlike Core 32, Core 42 measurements record both the pre-homogenite and the post-homogenite normal hemipelagic sedimentation.

The two curves are very similar, and clearly document a single sedimentary event. In the upper two thirds, they are almost identical. In both cores, coarse silt (greater than 31 microns) disappears slightly above the highest record of sand (greater than 63 microns). In both instances, clay exceeds 50% when the coarse silt disappears.

A strong difference is noticed at the base, however: unlike the similarity of the clay fraction, the sand fraction is much more abundant in Core 32 (>55%) than in Core 42 (20% maximum). This observation, and the occurrence of a distinct debris flow at the base of homogenite in Core 32, suggest that higher energy was available in the western Mediterranean Ridge.

We studied optically the sand-size fraction recorded near the base of homogenite, and documented significant changes both upcore, and between the two cores investigated. Figure 14 illustrates some SEM photographs of the sand-size fraction from selected intervals, all at the same magnification. A distinct polarity is observed, from planktonic foraminiferal tests at the base, to mica flakes, to plant fibres at the top.

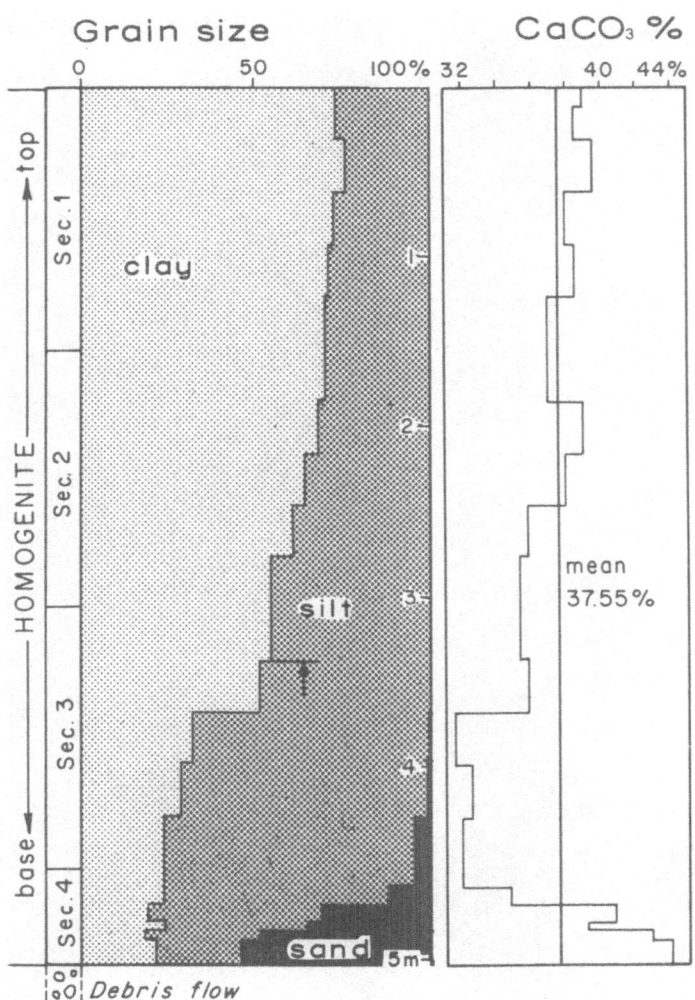

Figure 11 - Grain-size analysis (left column) and carbonate con-
tent (right column) measured in 19 samples of Cobblestone
Core 32.
Symbols and methodology as in Figure 7.

The difference in composition reflects differences in settling
velocities.
     In Core 32, unlike in Core 42, an interval very rich in
shards of volcanic glass is recorded, above an interval dominated by planktonic foraminiferal tests. Glass shards analyzed
by dr. Floyd Mc Coy and Susan Coughlin at Lamont-Doherty Geological
Laboratory in their refraction index, morphology and accompanying
minerals are referable to level Y-5 of Keller et al. (1978) eastern

Figure 12 – Base of Minoan Homogenite in Cores 32 and 42.

Mediterranean tephrachronology, a key ash-layer of Campanian
provenance with an interpolated age of approximately 40 000 y
(Cita et al, 1977; Thunell et al, 1979) see Figure 9.
     Size of planktonic foraminifers from the base of the sandy
interval is much greater in Core 32 than in Core 42, as shown by

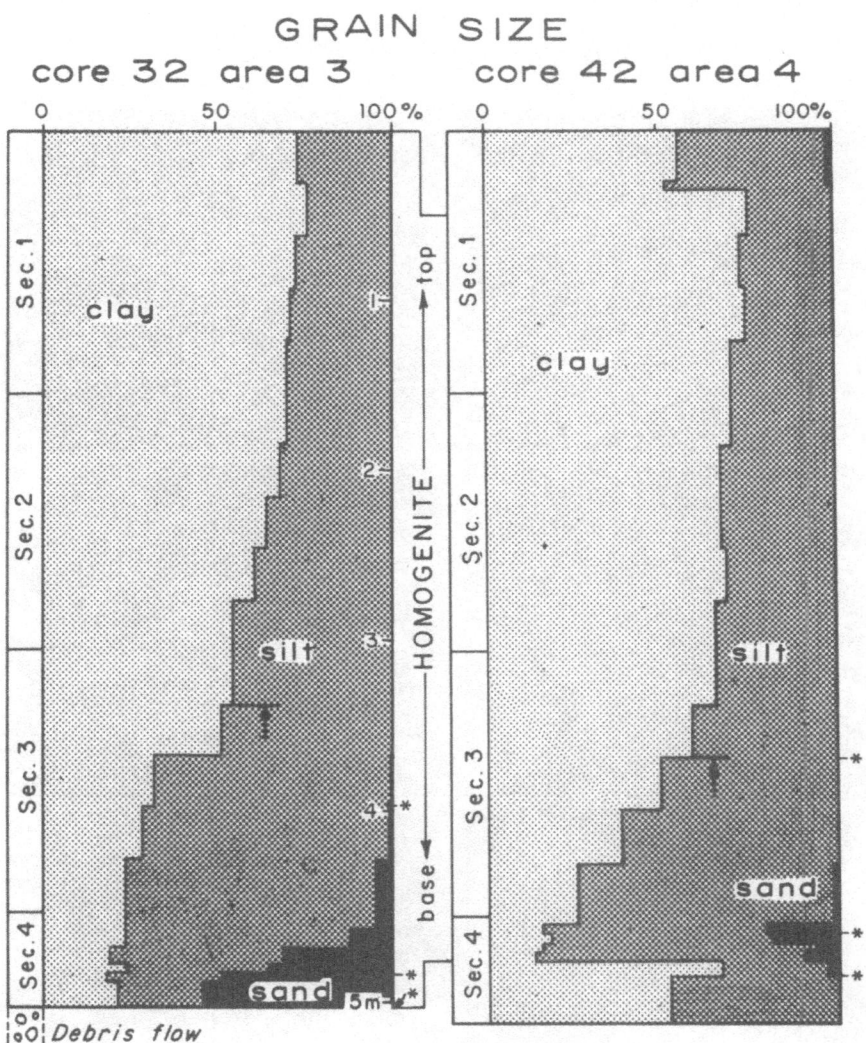

Figure 13 - Comparison of grain-size analysis obtained from the homogenites of Cores 32 and 42.

Figure 14. This observation, along with those previously discussed, supports the assumption that a higher energy was available in Area 3 than in Area 4. The occurrence of the 40 000 y BP ash Y-5 in the sand at the base of homogenite is also supportive of a high energy environment, with re-mobilization of coarse-sized tephras from the surrounding slopes.

Other differences recorded in the Area 3 homogenites versus Area 4 homogenites are plotted in Table 4. They include (a) lesser thickness and (b) presence of laminations or of debris flows at the base of the unit (1).

(1) In Area 3 cores, thickness of homogenite has been measured excluding the underlying debris flows.

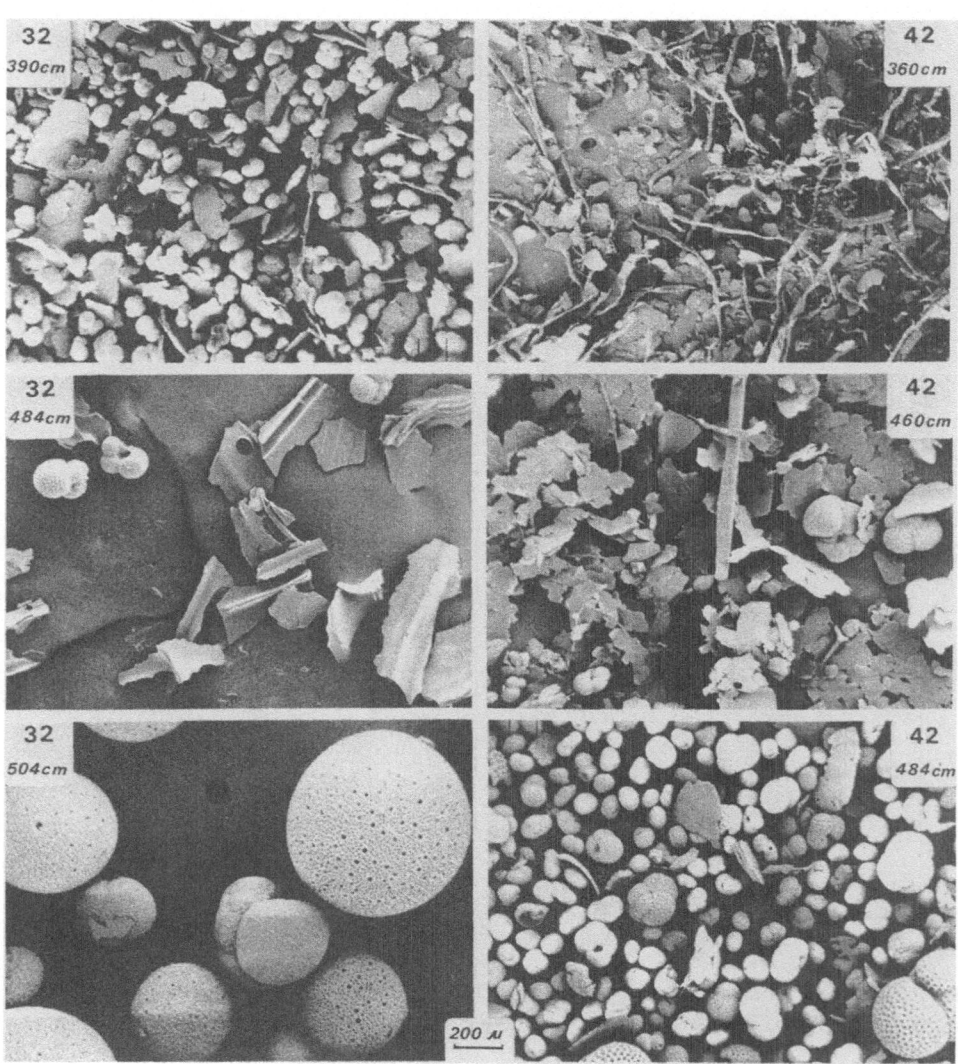

Figure 14 - Scanning Electron Microscope photomicrographs of a portion of the sand-size (greater than 63 microns) sediment fraction from three samples near the base of homogenite in Core 32 (left column, bottom to top) and from three samples in Core 42 (right column). Same magnification for all samples.

TABLE 4

	Area 3	Area 4
Distance from Santorini	500 km	800 km
Vertical relief	200 m	100 m
Maximum thickness of Minoan homogenites measured in cores	504 cm	770 cm
Mean thickness of Minoan homogenites measured in cores	289.3 cm	482.5 cm
Maximum Ø of sand-size bio-clasts at base of homogenite (*)	1 mm	0.5 mm
Percentage of sand-size fraction at base of homogenite (*)	55 %	20 %
Presence of Y-5 ash clasts at base of homogenite (*)	+	-
Presence of debris-flow and/or laminations associated with base homogenite	+	-
Presence of pre-Minoan homogenites	+	-

Table 4: Comparison of western Mediterranean Ridge (Cobblestone Area 3) versus southern Calabrian Ridge(Cobblestone Area 4)characters (*) Observations pertaining to Cores 32 and 42.

Maximum thickness recorded in Area 3 is 504 cm, versus 770 cm in Area 4. Mean thickness is 289.3 cm versus 482.5 cm. This finding is surprising, since Area 3 has a much stronger relief than Area 4, and calculations by Kastens (in Kastens & Cita, in press) have shown that thickness of homogenites in Area 4 correlates with the surface of "watersheds". We should thus expect to have more material removed from the Area 3 slopes, not less material.

The correlation of Area 3 cores (Figure 5) shows that one additional interval displaying the characters of homogenites is recorded, beneath Sapropel S-1 in Core 32, from the center of Aphrodite Crater. The explanation we offer is that the slopes surrounding Aphrodite Crater were shaken before the Minoan tsunami, and lost part of their sediment drape. This is why homogenites are thinner in Area 3 than in Area 4. One additional explanation (not mutually exclusive in respect to the former) is that the high energy resulted in suspension of fines extending well above the approximately 200 m high relief, so that they could escape in part outside the area.

Magmatic activity started in Santorini some 1.5 m.y. BP as seen before, and several explosive episodes occured in the latest part of the Quaternary. Thera ashes Y-2 and Y-4 are recorded in several deep-sea cores from the eastern Mediterranean (Keller et al, 1978; Watkins et al, 1978; Mc Coy, 1980). The stratigraphic position of these tephras (compare Figures 5 and 9) is compatible with the assumption that a tsunami of lesser proportions than the Minoan one caused slope failure and liquefaction of sediments in Cobblestone Area 3. Area 4 was too far from the source to record these older shocks.

CONCLUSIONS

Looking for a historic catastrophic event which could be accepted as causative mechanism for the "Homogenite event" recorded in the abyssal Ionian Sea, we found evidence that such event was the tsunami induced by the collapse of the Santorini caldera after the Minoan eruption some 3500 y BP. The relative timing is perfect, within the error limits of a high resolution geologic record.

Questions one might ask are: why none of the previous expeditions in the eastern Mediterranean recorded sediments similar to our homogenites? Are homogenites restricted to the Ionian Basin, or should we expect to find them also in the Levantine Basin? Are homogenites exclusive of areas characterized by Cobblestone topography?

We have no straightforward answer to these questions, but some tentative ones might be proposed.

In the E-W transect of sixteen cores from Cruise RC9 investigated and correlated by Ryan (1972), Core RC9-175 located in the Herodotus abyssal plain south of Cyprus contains a very thick Holocene sequence post-dating Sapropel S-1, which is recorded at more than 800 cm from core top. Sedimentation rate is comparable to that recorded in the Holocene of our homogenite-bearing basinal cores (Figure 10). The sediment is described as containing numerous fine grained turbidite layers, therefore it differs from that recorded in our cores, where the homogenite event was unquestionably unique in the Holocene, with one single layer up to 770 cm thick.

None of our homogenite-bearing cores is from a true abyssal plain. All were recovered from the floor of flat-bottomed depressions having sizes of the order of a few kilometers: narrow elongated troughs as Beato Angelico (Cores 7, 9, 12, 40); suboval-shaped basins as Botticelli (Cores 42, 43) or Raffaello (Core 10); small perched basins as Electra (Core 23), or subrounded craters as Aphrodite (Core 32) or Ares (Core 31). Sediments brought in suspension by the Minoam Bronze-age tsunami ponded on the basin

floor. No long distance transportation is involved in the
process, unlike in large-sized turbiditic flows accumulating on
the floor of abyssal plains after crossing wide sedimentary
aprons.

It is possible that homogenites similar to those described
here are recorded in the Hellenic Trench. In DSDP Sites 128 and
129 (Ryan, Hsü et al, 1973) homogeneous, fine-grained, thick,
structureless marls were described from the latest part of the
Quaternary, but since coring was not continuous, the stratigraphic
resolution is inadequate to prove the correlation.
The Hellenic Trench, which is much bigger than the physiographic
features explored during the Cobblestone Project, but is steep-
walled and is only a few kilometers across, could be the place of
sediment ponding similar to that discussed here.

In conclusion, homogenite deposition results from the combination of three in some way unique situations:
(a) a catastrophic tsunami,
(b) a very irregular topographic relief, and
(c) closed or semi-enclosed basins.

Moreover, a high resolution stratigraphy is necessary to recognize its synchroneity in the various cores.
One more conclusion we can draw is that without a large number of cores from limited areas, and without the high precision
navigation, which permitted to locate the cores in preselected
small physiographic features previously mapped in detail, it would
have been impossible to even conceive such a precise time/space
relationship.

ACKNOWLEDGMENTS

Cobblestone Project is part of a U.S.A.-Italy cooperative
scientific program. Italian funding was provided by Consiglio Na-
zionale delle Ricerche through grants 78.01878.66 and 79.02420.66.
American funding was provided by NSF through grant OCE 77-20047.

We gratefully acknowledge the Master of R/V EASTWARD and the
entire crew for their most cooperative attitude.
We are deeply indebted to Bill Ryan, who was a superb chief-
scientist at sea; to all the shipboard scientists of EASTWARD
Cruise E-3E-78 and E-3F-78 and especially to Gretchen Blechschmidt,
Martine Dreyfus Rawson, Elizabetta Fois, Rob Kidd, Roberto Mazzei,
Elizabeth Miller, Lucia Paggi, Gianfranco Salvatorini who had re-
sponsibility in describing and/or dating the cores.
Discussions with Kim Kastens, Bill Ryan, Gretchen Blechschmidt
and Hisao Nakagawa were of great help.
Discussion after the presentation of this contribution at the
NATO Workshop on Submarine Slides at Algarve, Portugal, and cons-
tructive criticism by P. Spudich, G. Almagor, R.B. Kidd and D.
Roberts are acknowledged.
Special thanks are extended to prof. G. Cosmacini for kindly

radiographing some cores for us, and to Floyd Mc Coy for analysing
volcanic glass recorded in Core 32.

Technical assistance was provided by G. Spezzi Bottiani,M.
Giambastiani, A. Malinverno, S. Antico, A. Rizzo and G. Chiodi.

REFERENCES

F.Barbieri, F.Innocenti, G.Marinelli and R.Mazzuoli, 1977, "Vulca-
    nesimo e tettonica a placche: esempi nell'area mediterranea",
    Mem.Soc.Geol.Ital., 13-2, 327-358, Roma.
W.A.Berggren, L.H.Burckle, M.B.Cita, H.B.S.Cooke, B.M.Funnell,
    S.Gartner, J.D.Hays, J.P.Kennett, N.D.Opdyke, L.Pastouret,·
    N.J.Shackleton and Y.Takayanagi, 1980, "Towards a Quaternary
    Time Scale", Quaternary Res., 13, 277-302, New York.
G.Blechschmidt, M.B.Cita, R.Mazzei and G. Salvatorini, in press,
    "Stratigraphy of western Mediterranean and southern Calabrian
    Ridge, eastern Mediterranean (USA/Italy Cobblestone Project)",
    Marine Micropal., Amsterdam.
M.B.Cita, C.Broglia, A.Malinverno, G.Spezzibottiani, L.Tomadin,
    D.Violanti, in press, "Pelagic sedimentation in the southern
    Calabrian Ridge and western Mediterranean Ridge, eastern Me-
    diterranean"; Marine Micropal., Amsterdam.
M.B.Cita and W.B.F.Ryan, 1978, "The deep-sea record of the eastern
    Mediterranean in the last 150 000 Years", in: Thera and the
    Aegean World I", C. Doumas ed., Thera and the Aegean World,
    45-60, London.
M.B.Cita, C.Vergnaud-Grazzini, H. Chamley, N. Ciaranfi,S.d'Onofrio,
    1977, "Paleoclimatic record of a long deep-sea core from the
    eastern Mediterranean", Quaternary Res., 8, 205-235, New
    York.
R.W.Fairbridge R.W., 1966, "The encyclopedia of oceanography",
    Reinhold Publishing Corp., 1-1021, New York.
G.Ferrara, M.Fytikas, D.Giuliani and G.Marinelli, 1980, "Age of
    the formation of the Aegean Active Volcanic Arc" in: "Thera
    and the Aegean World II", C.Doumas ed., Thera and the Aegean
    World, 37-42, London.
J.B.Hersey, 1965, "Sedimentary basins of the Mediterranean Sea",
    Colston Res.Soc., Papers, 17: 75-91, Butterworths, London.
K.A.Kastens and M.B.Cita, 1980, "Tsunami induced sediment tran-
    sport in the abyssal Mediterranean Sea", Amer. Geophys.Union,
    Fall Meeting, Abstract.
K.A.Kastens and M.B.Cita (in press), "Tsunami induced sediment
    transport in the abyssal Mediterranean Sea", Geol.Soc.America
    Bull.
J.Keller, 1980, "Prehistoric pumice tephra on Aegean islands" in:
    "Thera and the Aegean World II", C. Doumas ed., Thera and
    the Aegean World, 49-56, London.

J.Keller, W.B.F.Ryan, D.Ninkovich and R.Altherr, 1978, "Explosive
    volcanic activity in the Mediterranean over the past 200 000
    years as recorded in deep-sea sediments", Geol.Soc.America
    Bull., 89: 591-604.
H.Lacombe and P.Tchernia, 1972, "Caractères hydrologiques et circu-
    lation des eaux en Méditerranée". in: "The Mediterranean Sea",
    D.J.Stanley ed.,Dowden Hutchinson & Ross,Inc.25-36,Stroudsbourg.
J.Makris and R.Vees, 1977, "Crustal structure of the central Aegean
    Sea and the islands of Evia and Crete, Greece, obtained by
    refractional seismic experiment", J.Geophys., 42: 329-341.
G.Marinos and N.Melidonis, 1971, "On the strength of the seaquakes
    (tsunamis) during the prehistoric eruptions of Santorini",
    Acta.
F.W.Mc Coy, 1980, "The upper Thera (Minoan) ash in deep-sea sedi-
    ments: distribvution and comparison with other ash layers",
    in "Thera and Aegean World II", C.Doumas ed., Thera and Ae-
    gean World, 57-78, London.
F.W. Mc Coy and S. Coughlin, in press, "Thephrochronology of Cob-
    blestone Area 3, Mediterranean Ridge, Ionian Sea", Quaterna-
    ry Res., New York.
D.P.Mc Kenzie, 1972, "Plate tectonics of the Mediterranean region",
    Geophys.J.R.Astr.Soc., 30: 109-185.
S.Meszaros, 1978, "Some words on the Minoan Tsunami of Santorini"
    in "Thera and the Aegean World I", C. Doumas ed., Thera and
    the Aegean World, 257-262, London.
D.Ninkovich and B.C.Heezen, 1965, "Santorini tephra", Colston Res.
    Soc.Papers, 17: 413-452. Butterworths, London.
H.Pichler and W.L.Friedrich, 1980, "Mechanism of the Minoan erup-
    tion of Santorini", in: "Thera and the Aegean World II", C.
    Domumas, ed., Thera and the Aegean World, 15-30, London.
H.Puchelt, 1978, "Evolution of the volcanic rocks of Santorini"
    in: "Thera and the Aegean World I", C.Doumas ed., Thera and
    the Aegean World, 131-146, London.
W.B.F.Ryan, 1972, "Stratigraphy of late Quaternary sediments in
    the eastern Mediterranean", in: "The Mediterranean Sea",
    D.J.Stanley ed., Dowden, Hutchinson and Ross, Inc.,149-169;
    Stroudsbourg.
W.B.F.Ryan and M.B.Cita, 1977, "Ignorance concerning episodes of
    ocean-wide stagnation", Marine Geol., 23: 197-215, Amsterdam.
W.B.F.Ryan, M.B.Cita, G.Blechschmidt, D.Chayes, H.Chezar, E.Fois,
    W.Harvie, R.B.Kidd, A.Malinverno, E.L.Miller, E.Patacca, R.
    Radoicic, M.Rawson, P.Scandone, 1978, "Cobblestone topography
    in the eastern Mediterranean. Preliminary results of a tran-
    sponder-navigated coring program following a deep-tow survey,
    southern Calabrian Ridge (Area 4)". Messinian Seminar n.4,
    Abstracts. Accademia Naz.Lincei, Roma.
W.B.F.Ryan, K.J.Hsü et al, 1973, "Initial Reports of the Deep Sea
    Drilling Project", 13: 1-1455, U.S.Government Printing Offi-
    ce, Washington, D.C.

F.N.Spiess, W.B.F.Ryan, K.A.Kastens, N.Kenyon, L.Sole, A.Ibrahim, C.Evans, S.Bloomer, 1978, "Preliminary results of a deep-tow survey of Cobblestone topography on the Calabrian Ridge", Messinian Seminar n. 4, Abstracts, Accademia Naz.Lincei, Roma.

S.Thorarinsson, 1978, "Some commets on the Minoan eruption of Santorini", in: "Thera and the Aegean World I", C. Doumas ed., Thera and the Aegean World, 263-276, London.

R.C.Thunell, A.N.Federman, S.R.Sparks and D.F.Williams, 1979, "The age, origin and volcanological significance of the Y-5 Ash layer in the Mediterranean", Quaternary Res., 10, 241-253, New York.

C.Vergnaud-Grazzini, W.B.F.Ryan and M.B.Cita, 1977, "Stable isotopic frectionation, climate change and episodic stagnation in the eastern Mediterranean during the Late Quaternary", Marine Mircopal., 2, 353-370, Amsterdam.

N.D.Watkins, R.S.J.Sparks, H.Sigurdsson, T.C.Huang, H.Federmann, S.Carey and D.Ninkovich, 1978, "Volume and extent of the Minoan tephra, Santorini volcano: new evidence from deep-sea sediment cores", Nature, 271: 122-126, London.

I.Yokoyama, 1978: "The tsunami caused by the prehistoric eruption of Thera" in: "Thera and the Aegean World II I", C. Doumas ed., Thera and the Aegean World, 277-286, London.

OCEANOGRAPHY

The oceans are important for several reasons. Two
important concerns of very practical importance are the
utilization of raw material (minerals, oil, gas) and the
food resources (the bilogical food-chain) on one hand,
and the pollution question (waste, accident e.g. marine
slides) on the other hand.

It is generally believed that oceanic forces like
turbidity currents and tsunamis have been the triggering
mechanism for some marine slides' occurrence. Tidal
currents, vertical water circulation, strong surface
waves and bottom waves, among others, may also be involved in marine slides. The shape of a shoreline is determined by wave erosion, and the major erosional factor
operating along coasts is wave actions.

Not only physical oceanographic phenomena are active
in sliding and slumping, but the effect of chemical
oceanographic action which influences the deep-sea sedimentation and its instability, can be instrumental in
the sliding processes. Both physical and chemical oceanographic phenomena are therefore indirectly involved in
the marine biological life and in the marine food-chain.

In his paper on interrelationships between oceanographic events and mass wasting of the seafloor, William
Ryan discusses the various physical, chemical, and biological factors that can be responsible for subaqueous
erosion and marine slides. He concludes that because the
ocean is dynamic, and its environment is evolving and
changing through time, individual mechanisms may have
greater or lesser influence in certain geological eras.
"One thing is for certain; there is no single culprit
producing submarine landslides.".

INTERRELATIONSHIPS BETWEEN OCEANOGRAPHIC EVENTS

AND MASS WASTING OF THE SEA FLOOR

William B. F. Ryan

Lamont-Doherty Geological Observatory
of Columbia University
Palisades, New York

INTRODUCTION

The shaping of the ocean floor is dependent not
only on the local geological setting and the nature of
the substrate, but it is also influenced by the overlying water mass. The purpose of this paper is to
explore various physical, chemical, and biological
factors that can be responsible for subaqueous erosion
and landslides. Because oceanographic conditions change
in response to both global climate fluctuations and the
continuous rearrangements of the size, shape, and depth
of the ocean basins, submarine erosion has a certain
degree of time dependency. Knowledge of the so-called
"paleo-oceanographic events" is made available
primarily from the results of deep-sea drilling and
seismic reflection profiling.

CLIMATIC FACTORS

The surface circulation of the ocean is dominantly
wind driven. The lower atmosphere wind patterns are
influenced by distance from the equator and the oro-
graphic effects of mountain chains. Wind-induced
coastal currents cause upwelling and accompanying high
fertility. Regions of enhanced productivity such as
the margins of Peru-Chile, West Africa, and Western
India accumulate biogenic sediments at rates several
times faster than typical pelagic accumulation. These
sediments are enriched in siliceous skeletal tests of

micro-organisms which improve the overall permeability
and porosity of the resulting ooze. Furthermore, the
deposits under fertility belts usually have high contents of organic carbon due to very low levels of
oxygen that has been consumed by the zooplankton and
nekton. The organic-rich substrate will generate
methane upon burial and diagenesis.

The formation of in situ natural gas is a principal cause of over pressurization in the sediment layer.
The gas tends to accumulate in permeable strata such
as diatomaceous layers or well sorted grain flow
deposits such as contourites or the basal members of
turbidites. The generation of gas is time dependent
and temperature dependent. Beginning in the Late
Eocene the temperature of the deep ocean has decreased
progressively so that by Late Miocene time abyssal
temperatures were only a few degrees above freezing.
With the combination of cold bottom-water temperatures,
low crustal heat flow and moderate lithostatic
pressure in the shallow sub-surface, clathrates can
form where methane is trapped beneath compacted clays.
The impermeable nature of the clathrate itself can
further enhance the sub-adjacent pore pressures. One
of the major potential causes of mass failure of thick
and large allochthonous sheets of deep ocean sediment
may be over-pressurization and detachment at the base
of the clathrate zone.

Organic-rich sediments are also deposited in
silled basins that have entrapped bottom waters.
Examples of such situations are the present Black Sea,
the Quaternary Eastern Mediterranean and the Mid-
Cretaceous South Atlantic. Sapropellitic pebbly muds
and turbidites with plant and coaly fragments are
typically encountered in drill cores from the above
three regions.

Climatic changes shift the latitudinal distribution of deserts. The western margin of Africa is an
example where the growth and filling of submarine
canyons responds to the absence and presence of deserts.
Aeolian sediments deliver little organic matter to the
adjacent ocean, and the kerogen that is delivered is
generally refractory and not readily maturated except
at high burial temperatures. Arid margins character-
istically lack active slope gullies, and the major
sediment component of the seaward continental rise is
biogenic calcareous ooze and mud that consolidates
readily into chalk and mudstone.

One of the significant climatic factors that
affects erosion of the continental margin is glacially

induced sea-level change. Large eustatic fluctuations
will rapidly expose or drown extensive areas of the
continental shelf. At times of low stand, rivers and
deltas debauch in close proximity to the continental
slope. Progradational terraces formed by rapid local
accumulation in prodelta settings are likely to be
unstable. Sedimentation tends to be concentrated near
the heads of submarine canyons. The flushing of canyon
reservoirs is catastrophic as revealed by individual
turbidity current layers on the abyssal plains with
single bed volumes in excess of 10 exp 11 m^3.
Individual turbidites have been traced on the floors
of the Atlantic abyssal plains for distances greater
than 1000 km. Abyssal plain turbidite frequency, however, suggests that the filling of canyon reservoirs
between flushings may take thousands of years.

Eustatic sea level change will act as a hydrostatic pump to cause the lateral offshore migration
of pore waters in sub-surface aquifers beneath the
continental shelf and coastal plain. Some researchers
have speculated that sapping of fresh to brackish water
springs on canyon walls is an avalanche generating
mechanism.

BIOLOGICAL FACTORS

Living organisms have a major influence on the
physical properties of the substrate which they either
form themselves, bind together or churn up. As
example, barrier reefs with steep external escarpments
survive the onslaught of storms because of the framework structure of corals and the binding power of the
secretions of blue-green algae. Reefs, however, generate a great deal of sediment for the slope as the
result of limestone boring by molluscs, fish and
algae. The boring of the reef escarpments produces
aprons of carbonate silty mud which are unstable on
steep slopes and which are periodically flushed away
into deeper waters. The resulting grain-flow deposits
are numerous but restricted in size. In the Bahamas
the deposits form narrow linear aprons along the base
of extensively gullied slopes.

Burrowing is a common feeding attribute of
benthic creatures on the continental margin. Burrowing
is not done just by worms. Significant excavation is
also carried out by large animals such as crabs and
lobsters which construct honeycomb passages that have
been referred to as "Pueblo communities". The

pervasive tunneling weakens the canyon walls.
Selective burrowing in interlayered shales and chalks
leads to the undermining and the sudden collapsing
of chalk cliffs.

Canyon thalwegs are littered with blocks of chalk
and grainstone that have avalanched from canyon walls.
Bioerosion is selective in certain substrates and is
one of the mechanisms that is compatible with the
observed headward-directed erosion of continental
slopes. Biological disturbances may not be exclusively
those of gradual wasting. Such phenomena as whale
impacts may actually trigger mass movements. Whale
induced craters are ubiquitous in deep sea photographs
in the Eastern Mediterranean, and the pock marks are
sufficiently large to be imaged by side-looking sonar.

OCEANOGRAPHIC FACTORS

The principal agent for eroding loosely consolidated sediment is the eddy turbulent shear during the
passage of turbulent bottom currents. Deep currents
are likely to be driven by geostrophic forces and be
primarily steady state. However, along margins the
currents can be oscillatory with the diurnal tidal
cycle. Episodic near-bottom currents can also be
induced by tsunamis or gravity driven suspension flows.

Geostrophic currents have produced erosional
unconformities in the Western Atlantic and are thought
to be the major agent that created the widespread
Horizon A^u. Geostrophic circulation in Oligocene
removed more than a kilometer of sediment cover from
the West African continental slope and exposed early
Cretaceous shelf edge strata. Geostrophic currents
have the ability to locally steepen the continental
slope by back cutting the margin. Off the coast of
the West African margin seaward of Cape Bojador the
erosional surface is buried by hundreds of meters of
Miocene debris flows, indicating the inability of
subsequent sedimentation to adhere to the steeply dipping unconformity. Where the sediments alternate
between lithified beds (i. e. sandstones) and unlith-
ified beds (i. e. clays) geostrophic bottom currents
will selectively attack bedding plains. When the
resistant beds are sufficiently undermined, they
collapse as rock falls. Erosion surfaces often armor
themselves with ferromanganese or phosphatic crusts.
Strong currents will eventually breach the crusts
and winnow out steep-sided furrows or craters.
Tabular slabs of the original crusts are found, for

example, strewn across the floor of a subaqueous
ablation pits on the Blake Plateau beneath the Gulf
Stream.

Many researchers who have explored submarine
canyons with submersibles have remarked on the relatively high velocity of tidal currents in the canyon
axes. Thalweg floors are commonly composed of
rippled sand. The movement of the sand polishes and
undercuts the basal parts of the canyon walls, leading
to eventual gravity collapse of the walls.

Tsunamis have been attributed to the triggering
of massive mud flows that result in the suspension
and redeposition of pelagic sediments during brief
episodic events. In the eastern Mediterranean multi-
meter thick mud turbidites (called "homogonites") are
found ponded in intermontane basins and island arc
trenches. Such deposits are often intercalated with
pebbly muds transported in the traction carpet beneath
the suspension.

Calcium carbonate dissolves in the abyssal ocean.
The level of dissolution is known as the CCD and
migrates vertically with changing ocean chemistry,
productivity and proximity to land. Carbonate-rich
sediments which are laid down originally in relatively
shallow parts of margins or ridges will later subside
to deeper depths as the consequence of thermal con-
traction of lithosphere. Where old chalk or limestone
layers outcrop beneath the CCD, they become subject
to dissolution and disintegration. Rock falls on the
Blake and Bahama escarpments can be attributed in part
to weakening of the rock face by dissolution.
Dissolution attacks the escarpment through joints.

Semi-enclosed marginal basins such as the Red
Sea or Mediterranean Sea have enhanced salinities
because of excess solar evaporation. In these environments submarine lithification of carbonates proceeds
rapidly. Sediment outcrops become armored with
diagenetic crusts comprised of magnesium calcite. The
resulting cementations have the effect of buttressing
and strengthing the steep escarpments and thereby
diminish the likelihood of avalanches.

Massive salt layers and evaporites have accumulated in totally isolated oceanic basins (e.g.,Gulf of
Mexico, the ocean basin margins of the South and North
Atlantic, Red Sea and Mediterranean). With the
termination of the evaporitic stage and the resumption
of normal sedimentation the salt becomes susceptible
to sub-surface dissolution. A crater-like or linear
graben karst topography develops where the overlying

pelagic ooze collapses into solution cavities within
the salt layer. The walls of the craters become the
sources of debris flows that commonly form apron
shaped tongues on the crater floors. Diapirism of
the subsurface salt creates piercement structures.
Debris flows have been observed in radial patterns
around salt domes south of Cape Hatteras and in the
Sigsbee and Balearic Abyssal Plains.

PALEO-OCEANOGRAPHIC EVENTS

     The exploration of submarine canyons with sub-
mersibles and their careful stratigraphic sampling
has shown that canyon cutting is intermittent. In
fact, there can be interim stages when previously
dissected slopes become smoothed and former canyons
become buried (e.g., the margin of West Africa off
the Sahara desert). On the east coast North American
margin canyon cutting is observed in pre-Maastrichtian
time, again in the pre-Late Eocene and extensively
during the Oligocene. The later phases of erosion
truncate and expose the canyon fill which has been
deposited during prior periods of net deposition.
Canyon fill is recognized by its grain flow facies,
allochthonous mudstone conglomerates and displaced
shallow water fauna. There is a first hand correlation between episodes of mass wasting in New England
submarine canyons and episodes of defacement along
the Bahama Escarpment. The most likely agents of
canyon incision which can vary with time are geostrophic circulation, carbonate dissolution, natural
gas generation and eustatic sea level change.
     Although tidal currents have been considered to
be a major erosive agent we know that tides cannot
have been significant in small marginal seas where
submarine canyons also exist. It is hard to imagine
deep ocean tides which can be turned on and off during
geological history. There is the possibility that
bioerosion might also have a time dependency. The
degree of burrowing and boring shows some relationship
with the nature of the substrate and the amount of
organic matter (food) present. For example, beneath
nutrient poor regions such as the Sargasso Sea or the
Mediterranean bioturbation is not at all well developed
in deep-sea cores. It is therefore possible that
periods of canyon fill might accompany either ocean
conditions conducive to low productivity and, or,
atmospheric conditions leading to coastal aridity and
high sea level stand. Bioerosion can be attenuated

with the development of abyssal pavements and hard
grounds.

CONCLUSIONS

There are many factors that can be correlated with
the phenomena of submarine erosion, mass wasting and
avalanching. Because the ocean is dynamic and its
environment is evolving and changing through time,
individual mechanisms may have greater or lesser
influence in certain geological eras. Conditions can
change drastically from tranquil settings with modest
sedimentation to dissolution or to total desiccation
and subaerial denudation. One thing is for certain;
there is no single culprit producing submarine
landslides.

MARINE SEDIMENTOLOGY

The implications of plate-tectonic theory for continental shelf development have been assuming increasing
interest during the last 20-30 years. Continental shelves
- more or less deformed tectonically - are characteristic
of continental margins that coincide with the leading
edges of lithospheric plates. Sediments are accumulated
in these continental shelves, which are abruptly termina-
ted by the continental slopes. The sediments of most
slopes are muds and silt, derived from continental erosion, or originating from inclined layers that have been
deformed by slumping or sliding down the incline. Some
of the problems in marine sedimentology are the recognition and stratigraphic consequences of slide deposits on
modern as well as ancient sediments.

Hill et al. in their contribution study the thin-
bedded debris-flows and the mechanism of deposition of
such deposits. Based on some 20 piston cores from a small
area of the Nova Scotian Slope off eastern Canada, they
conclude that such thin-bedded debris-flows are an important constituent of submarine slope sequences.

Long-range sidescan sonar technique has for several
years been of importance in marine sedimentology studies,
and Kidd in his paper concentrates on the applicability
of the GLORIA system on small scale features with several
examples from the Atlantic. He concludes that individual
sediment slides show sufficient downslope changes in a-
coustic character to allow some inferences to be made of
the competence of the flow at varying stages in its movement.

Maria Cita et al. present an unusual debris flow
deposits from the base of the Malta escarpment in the
eastern Mediterranean. Based on cores, the lithology and
ages as well as the components of the major minerals and
clay minerals are determined. It is concluded that the
age of the matrix (latest Pleistocene) indicates that the

271

the mass movement occurred during the last glaciation,
when low sea-level stands accompanied by enhanced thermo-haline circulation at depth resulted in frequent
slope failures.

THE DEPOSITION OF THIN BEDDED SUBSQUEOUS DEBRIS FLOW DEPOSITS

Philip R. Hill, Ali E. Aksu and David J.W. Piper

Department of Geology
Dalhousie University
Halifax, Nova Scotia, Canada B3H 3J5

INTRODUCTION

Although submarine slides, slumps and debris flows have long been recognised as major features of continental and basin slopes, there have been few attempts to incorporate them into sedimentological facies models applicable to ancient rocks. A major problem is that of scale: most slumps, slides and flows identified from surveys of present-day submarine slopes have areal extents of several kilometres (e.g. Embley, 1976; Jacobi, 1976), so that only small parts of an individual slump body are sampled by either cores or most land outcrops. Thin-bedded debris-flow deposits are an important constituent facies in the slope environment. In this paper we analyse the mechanism of deposition of these thin-bedded deposits.

EXAMPLES FROM MODERN ENVIRONMENTS

(a) Nova Scotian Slope

About twenty (5-metre long) piston cores have been obtained from a small area of the Nova Scotian Slope off eastern Canada (Fig. 1). No deep canyons cross the area, but long-range sidescan (GLORIA II) and high resolution seismic reflection profiles show meso-relief on a scale of tens of metres, which can be resolved into a dendritic channel system developed in what is interpreted as an old slump-scar (Fig. 1; Hill, 1981). Most cores were taken prior to this interpretation and are not strategically placed. Those described in this paper are located downslope of the main channeled area between 600 and 1000 metres water depth (Fig. 1).

273

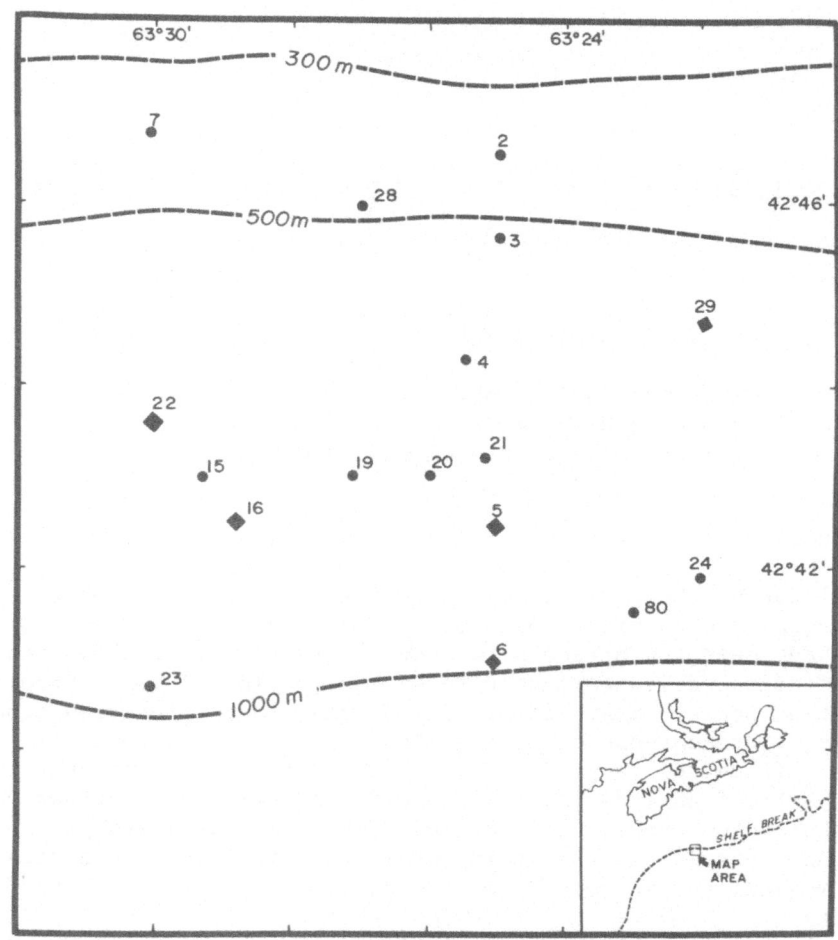

Fig. 1.   Location of Nova Scotian Slope piston cores. Diamonds
          indicate cores in fig. 2. Mesotopography interpreted
          from GLORIA II and high resolution seismic records (Hill,
          1981).

        Poorly sorted gravelly sandy mud beds occur in the lower parts
of almost all the cores. They are pre-Holocene, based on C^{14}
dating and microfossil abundance (Fig. 2) and occur within sequences of hemipelagic and turbidite muds with occasional sand beds.

        Gravelly sandy mud beds are interpreted as mass-flow deposits
and not ice-rafted (paratill) deposits using the criteria (Table 1)
of Kurtz and Anderson (1979) and Aksu (1980). Five beds can be
identified positively as debris-flow deposits. They occur in
widely separated cores (Fig. 1) illustrated in Figure 2. It is
unlikely that they represent the same depositional event. Three
of these beds occur at the bottom of the cores with thicknesses of

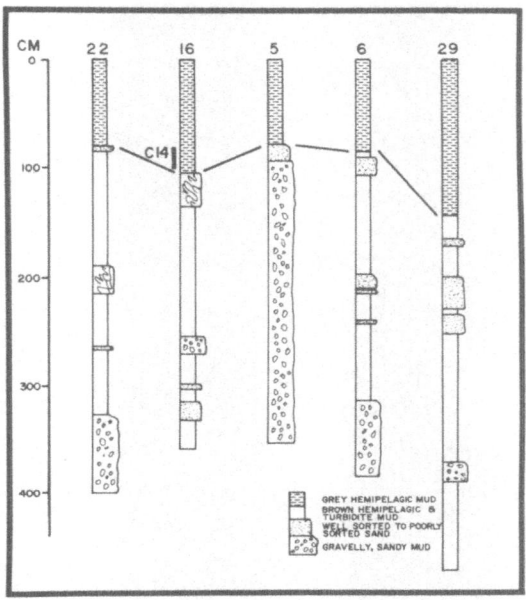

Fig. 2. Descriptive logs of piston cores from Nova Scotian Slope
showing occurrence of late Pleistocene gravelly sandy mud.
C^{14} date is 18,860± 860 yr. B.P. (GX 7452).

Table 1. Most Useful Criteria for Distinguishing Debris Flows and
Paratills (modified from Kurtz and Anderson, 1979, and
Aksu, 1980)

	Criterion	Debris Flows	Paratills
1.	Contacts	Sharp	Gradational
2.	Faunal content	Rare, or if present, displaced fauna	Normal in situ assemblage
3.	Fabric	Imbricated or sub-horizontally oriented clast	Randomly to horizontally oriented clasts
4.	Texture	Single population	Multiple populations

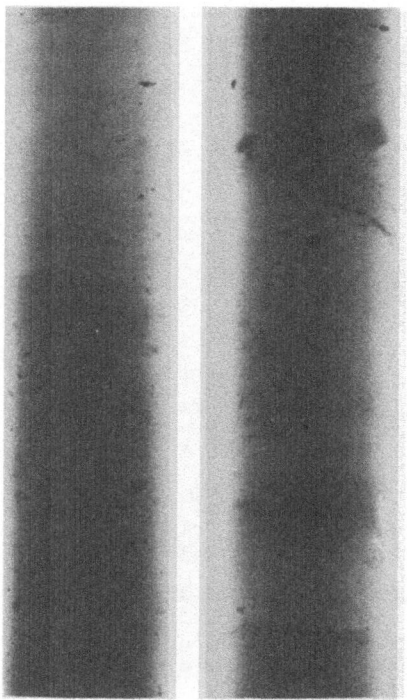

Fig. 3.   X-radiographs of 20 cm thick gravelly sandy mud bed from
          core 29, Nova Scotian Slope.

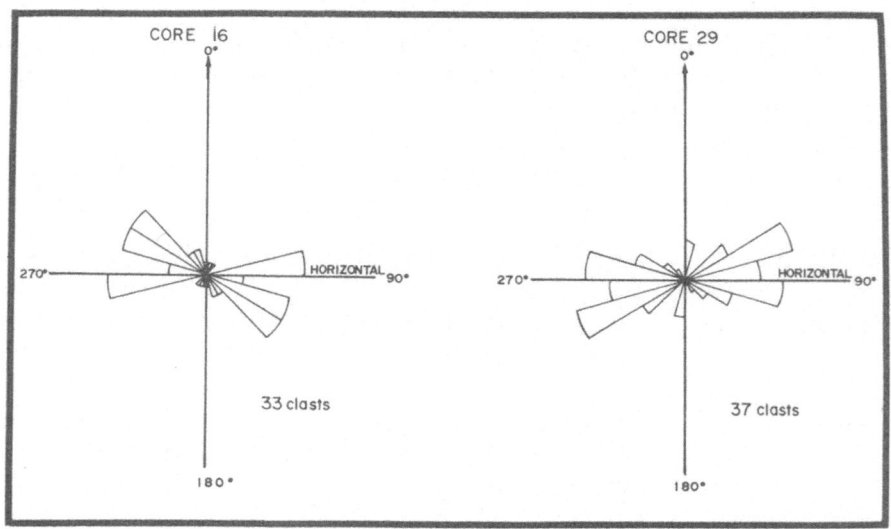

Fig. 4.   Orientation of elongate clasts >-1ϕ in vertical section
          of thin gravelly sandy mud beds from cores 16 and 29,
          Nova Scotian Slope.

>50 cm (cores 6 and 22) to >200 cm (core 5). However, the other
two beds of the same lithology (cores 16 and 29) have thicknesses
less than 20 cm. Most of the following discussion will pertain to
these two beds.

    Both beds are massive, gravelly, sandy muds with sharp tops
and bases (Fig. 3). They show well-defined pebble fabrics (Fig. 4),
in which dominant long axis orientations are sub-horizontal, but
apparently lack imbrication. Underlying and overlying muds contain
relatively low abundances of both planktonic and benthonic foraminifera, which may be reworked. The gravelly, sandy mud lithology
contains significantly fewer foraminifera per gram and almost lacks
planktonic species. This suggests a reworked fauna, and is supported by the lack of juveniles, the poor state of preservation of
the less robust tests and the apparent size sorting of many forms
with different test-geometry (M.A. Williamson, pers. comm.). Grain
size distribution is rather uniform within the same bed (Fig. 5a).
Single size distributions (Fig. 5b) closely approximate a normal
population (and thus plot as a straight line on log-probability
plots c.f. Visher, 1969).

(b) Baffin Bay

    Ten cores from the northern basin-slope and basin-plain of
Baffin Bay (Fig. 6) contain a distinctive lithological unit of
yellow-brown to brown gravelly, sandy mud (Aksu, 1980). The unit
is uniform in composition, but variable in texture and sedimentary
structures. Aksu (1980) gave detailed descriptions and suggested
that they represent progressive modification of a debris-flow to a
turbidity current. This paper will consider only the thin-bedded,
massive debris-flow beds that occur in the most northerly (proximal) cores (901, 903, Fig. 6). These beds have an average thickness of 40 cm and from extrapolation of C^{14} dates, are all older
than 25,000 yrs. B.P.

    These beds show similar characteristics to the Nova Scotian
Slope examples. They have sharp upper and lower contacts
a moderately well developed fabric (Fig. 7), single population
size distributions (Fig. 8) and they contain less than 2 foraminiferal tests per gram (Aksu, 1980).

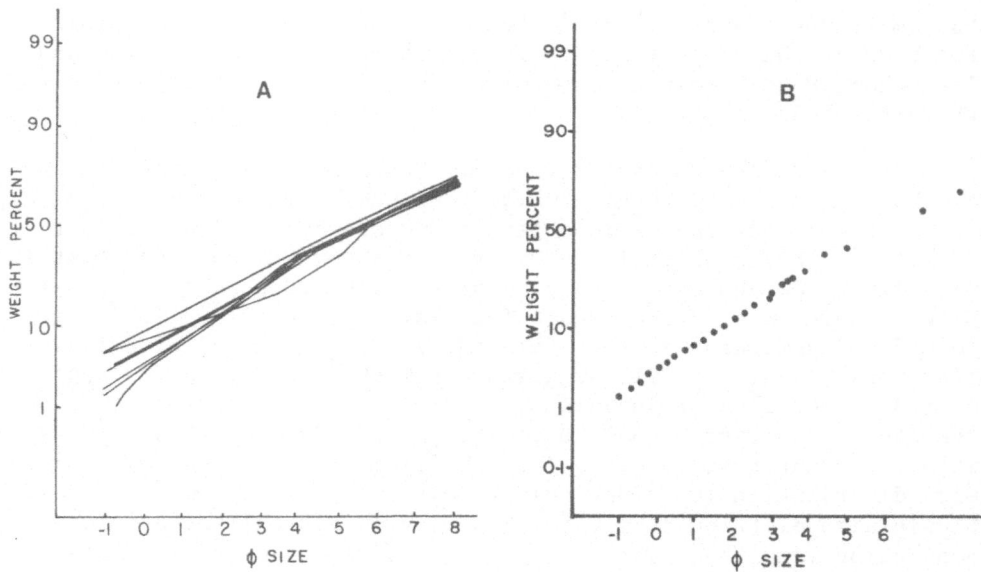

Fig. 5. Cumulative grain size distributions of gravelly sandy muds
from Nova Scotian Slope plotted on probability scale
A: Seven analyses from one bed in core 29.
B: Single analysis from same bed.

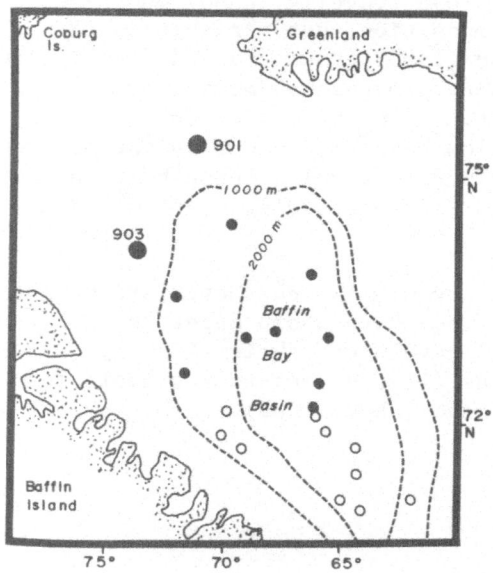

Fig. 6. (left) Location of Baffin Bay piston cores. Large solid
circles indicate thin bedded gravelly sandy muds described
in text; small solid circles indicate distal equivalents
(Aksu, 1980).

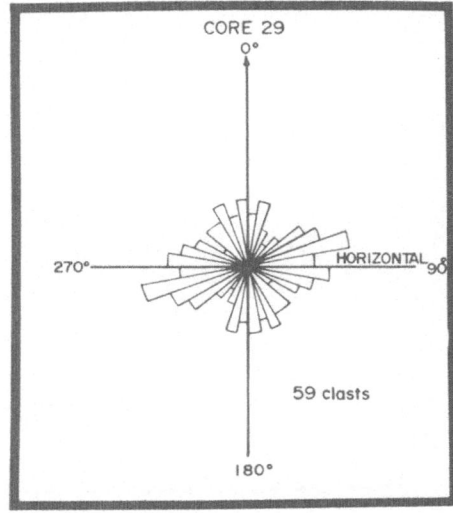

Fig. 7. Orientation of elongate clasts >∅ in vertical section of massive gravelly sandy muds from Baffin Bay (from Aksu, 1980).

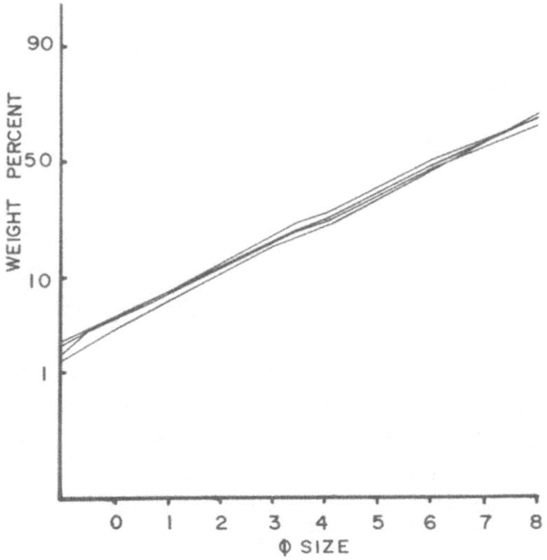

Fig. 8. Cumulative grain-size distributions of massive gravelly sandy muds from Baffin Bay.

Fig. 9. Lateral termination of limestone breccia bed with convex top and flat base, Cow Head, Newfoundland.

Fig. 10. Two thin lenses of limestone breccia, Cow Head, Newfoundland. Note lateral continuation of lower bed (arrow)

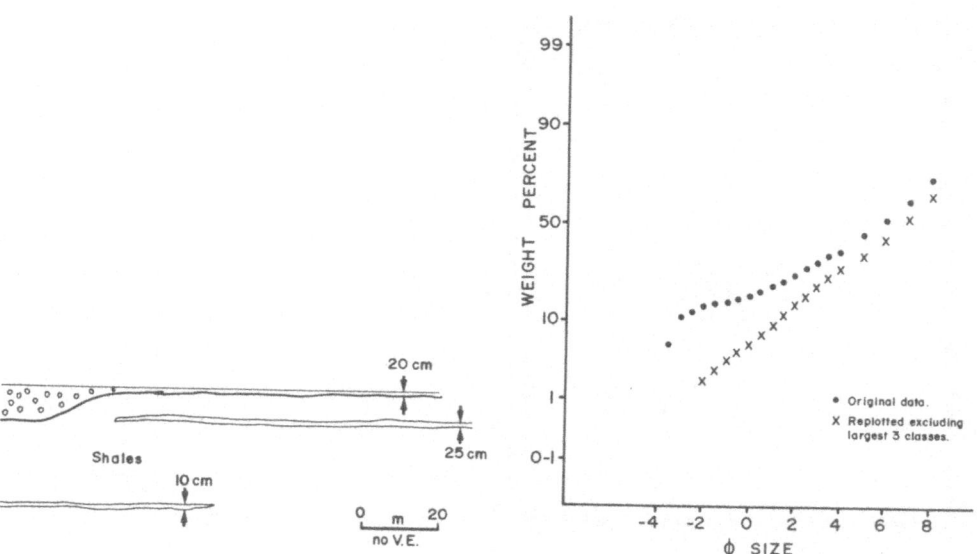

Fig. 11. Field skectch from
Miocene Potamoula Shale, Greece,
showing three thin pebbly mudstone beds in shale sequence.

Fig. 12. Cumulative grain-size
distribution of a sample from
uppermost bed in Fig. 11 and
replotted curve excluding two
largest clast sizes.

EXAMPLES FROM ANCIENT SEQUENCES

(a)   The Cow Head Breccia

The Cambro-Ordovician Cow Head Group of western Newfoundland
(Hubert et al., 1977; James et al., 1980), is famous for its megabreccia beds which have thicknesses of tens of metres and contain
huge blocks of limestone and contorted shale beds. However, between
the megabreccia units stratigraphically above Bed 6 of James et al.
(1980), there is a wide variety of sedimentary facies, including
some thin-bedded limestone breccia beds. These are presented as
analogues to the Pleistocene beds described above; the carbonate
environment precludes the possibility of confusion with paratills.

The thin-bedded breccias vary in thickness and are generally
made up of a fine carbonate matrix containing tabular clasts of
limestone, which are lithologically similar to the surrounding
autochthonous slope deposits of lime mudstone (James et al., 1980).
In many beds, the long-axes of clasts are aligned subhorizontally.
Most thin beds can be traced laterally across outcrop, but some
terminate laterally, showing either convex tops and flat bases
(Fig. 9) or flat to convex tops and convex bases (Fig. 10). Some
of the latter type are seen to recur several times along the same
bedding plane, sometimes connected by a very thin (2 to 5 cm) bed
of similar lithology.

(b) Miocene Flysch, Western Greece

     Piper et al. (1978) describe a flysch sequence of Miocene age
in western Greece, which they interpret as a deep-sea fan sequence.
Three pebbly mudstone beds of 10 to 25 cm thickness are described
from a proximal sequence (the Potamoula Shale, near Klima, see
Piper et al., 1978, Fig. 2). Although probably not a slope
sequence, this example demonstrates that thin pebbly mudstone beds
can have considerable lateral extent. Two beds can be traced over
at least 100 metres (Fig. 11), and one extends laterally from a
much thicker (7 metre) pebbly mudstone bed which fills a large
channel. A grain-size distribution of a sample from the uppermost
bed is shown in Figure 12.

CHARACTERISTICS OF THIN BEDDED MASS-FLOWS

     From these examples, the following appear to be the important
characteristics of thin-bedded, subaqueous debris-flow deposits:

(1) Texture

     Beds are massive, showing little variation in grain-size
distribution within the same bed, except in the gravel portions of
the distributions. The size distributions of any one sample,
although showing poor sorting, consists of a single population
(Figs. 5, 8 and 12), similar to distributions compiled by Pierson
(1980) for subaerial flow deposits. We find that the coarse end
of many size distributions deviate from straight-line plots as a
result of statistically inadequate sampling of gravel clasts (Fig.
12).

(2) Fabric

     Most beds studied here show some degree of clast orientation
(Figs. 4 and 7) sub-parallel to the bedding plane. There are no
distinctive changes in fabric or clast size through a bed.

(3) Bed geometries

     Thin-bedded debris-flow deposits assume varying geometrical
configurations as summarised in Fig. 13. Four types are recognised:
(1) sheet; (2) sheet associated with channel; (3) lensoid; (4)
lensoid connected to other lensoid. These may not constitute a
complete range, but the geometries suggest that both channelised
and sheet deposits are important. Types 1 and 2 are similar to the
"Finer Rudite Sheets" described by Cook et al. (1972), who also
described larger scale channeled debris-flow deposits.

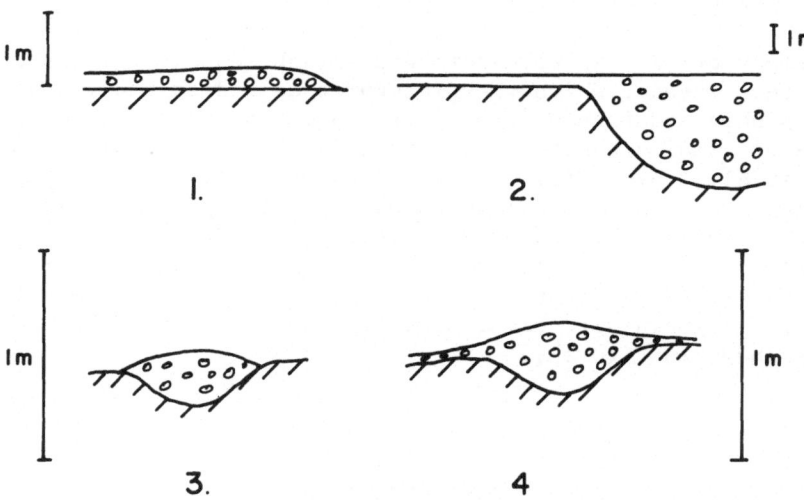

Fig. 13. Four debris-flow bed geometries recognised from ancient
          deposits. See text for explanation.

DISCUSSION

     The mechanics of submarine debris flows and their deposition
are described by Hampton (1970) and Middleton and Hampton (1973,
1976), based on the terrestial field observations and experimental
work of Johnson (1970) and Hampton (1970). Since then, there has
been a general acceptance of their model, but a lack of further
published experimental work. There seems to have been an uncon-
scious acceptance that all debris flows are the same and conform to
the Hampton model. This was certainly not the contention of the
authors, who stressed that in many cases transitions with other
kinds of gravity-flow are probably important. It is likely therefore that there should be variability within the grouping of
"debris-flows", controlled by factors such as grain-sizes available,
water content of flow and clay composition.

     This contention is supported by the differing observations of
authors with regard to fabric development (clast orientation)
within debris flows e.g. Cook et al. (1972), Crowell (1957), Aalto
(1972) and Cook (1979). Some flows seem to show good long-axis
alignment (Cook 1979), others show no fabric development at all
(Crowell, 1957). This might be explained by the work of Lindsay
(1968) who demonstrated that fabric development would be cyclical
and dependant on viscosity in a purely viscous, laminar flow. If
applied to the hybrid plastico-viscous model of Hampton (1970), a
better developed fabric might be expected in a less dense ("fluid")

flow than in a more viscous flow (Reineck and Singh, 1975). However, since grain size distribution partly controls viscosity, it also has a large effect on fabric development.

The debris flow models of Johnson (1970) deals mainly with the gross features of flows and is of little use in interpretation of single samples of a deposit in a core. Hampton's (1970) experimental work, however, can be applied to core samples by use of parameters such as the maximum clast-size carried by a flow D_{max}, given by:

$$D_{max} = \frac{8.8 \, k}{g(\rho_s - \rho_f)} \qquad (1)$$

where k = yield strength, ρ_s = density of the clast, ρ_f = density of the fluid matrix (Hampton, 1975); and the critical bed thickness for flow, yc, given by

$$y_c = \frac{k}{\gamma \sin \sigma} \qquad (2)$$

where γ = unit weight of debris and σ = slope angle (Hampton, 1970). The yield strength of the debris, k, is highly sensitive to water content and is not independent of the density (Hampton, 1970). Any use of these equations must therefore assume a water content. We have made some calculations based on data from Hampton (1970, p. 167, Table G3) to determine whether the Hampton model can be used to interpret the beds in Scotian Slope cores. The proportions of sand, silt and clay in our samples closely approximate those used in the experiment, but the dominant clay mineral in our samples is illite, whereas pure kaolinite was used in Hampton's experiment.

Values obtained using equation (1) and (2) were: D_{max} = 4.3 cm, y_c = 5.2 cm for a water/(water + clay) content of 60% and a slope angle of 4°. The maximum clast size observed in X-radiographs of the beds is 2 cm and the bed thicknesses are of the order of 20 cm. This indicates that, on a gentle slope, even with a fairly watery consistency such as this (a) there would have been sufficient strength to transport the observed clasts and (b) the flow thickness would have been considerably greater than the rigid plug thickness. While there are obvious weaknesses in this kind of treatment, it does indicate that it is possible that these thin beds are the products of complete and discrete debris flows. Detailed use of Hampton's theory would require experimental work on real samples.

Hampton's model for subaqueous debris-flow was based on observations of subaerial flows (Johnson, 1970), which were largely confined to channels by the deposition of levee-like lateral deposits. The overall deposit of such a flow has a linear, channelised

geometry (Johnson, 1970) rather than a sheet-like geometry as
observed here. Several mechanisms can be suggested for producing
sheet debris-flow deposits:

(1) A channel may become overfilled and the flow spill over
the channel margins. If substantial overfilling occurs, a sheet-
like geometry may result, connected to a channel (Figure 11).

(2) Bypassing of a flow and deposition of a lag can occur if
the debris flow separates from the source of the debris. Such a
separation has been observed and described by Hampton (1970); also
Middleton and Hampton, 1973), in experimental situations, but in
these cases, the separated mass slid rigidly after separation. The
bypassed area was characterised by erosional features such as parallel grooves, similar to large-scale features observed by Ryan (this
volume) on the eastern US continental slope, but not by a uniform
lag deposit. It is conceivable however, that separation could
occur by necking of the flow, as a result of velocity increase at
the head, with a maintenance of the flow rather than the transition to a slide. Surges in the flow (Johnson, 1970; Pierson, 1980)
are a common feature of debris flows and could produce a relatively
thin deposit on bypassing. However, these have been observed only
in channel flows.

(3) Sheet deposits could represent final deposition at the
end of a channelised flow. Some subaerial flows are observed to
produce fan-shaped, low relief lobes or sheets when they emerge
from valleys (Shelton, 1966). Presumably, submarine flows would
behave similarly. We know of no documented examples of this,
although with some of the recent technological developments of
sidescan sonar (e.g. this volume), there may soon be some. Deposition of larger clasts during the earlier phase of channelised
flow (Blatt et al., 1980, p. 188) may produce a relatively fine-
grained sheet deposit.

(4) If a flow is very fluid, so that the strength is sufficiently reduced, there is no reason why the flow should not spread
out from the source in a sheet-like manner rather than become channelised. If strength is regarded as the primary support mechanism
for debris flow, (Middleton and Hampton, 1970), then this type of
flow may no longer fall in the debris-flow category and a purely
viscous model may be more appropriate. This is indicated by equation (2) based on the experimental results of Hampton (1970),
Figs. 17-21) where the strength of a slurry only increases significantly if the water content is reduced beyond a certain value.

The Hampton model does not simply explain lenticular beds of
Type 3 and 4 (Fig. 13). Many of these small lenses cannot be
small channel deposits, as illustrated in Figure 10, where the
lower bed is conformable on, not incised into, the underlying shale

and thin limestone beds. Although not enough information is available in a single outcrop to properly interpret this situation, we
present it to illustrate that the actual mechanics of a debris-
flow moving across a soft mud bottom are not known. For example,
there would be large density contrasts between the flow material
and a watery mud bottom. Could a loading effect be important in
debris-flow? If so, to what extent? Such problems might also be
addressed by experimental work.

CONCLUSIONS

    1. Thin-bedded debris-flows are an important constituent of
submarine slope sequences.

    2. Debris flow deposits can be distinguished from paratills
by evaluation of a number of criteria, of which the most important
are: contacts with adjacent beds, faunal content, fabric and texture. No single criterion is diagnostic, but if several are considered together, objective and unambiguous distinctions can usually
be made.

    3. The Johnson-Hampton models for debris-flow are useful for
explaining the mechanics of thin-flows, but less useful in explaining observed bed geometries.

    4. Experimental work is needed to investigate the behaviour
of thin debris flows of varying viscosities and strengths, particularly in terms of fabric development, behaviour of the bed of
the flow and the mechanics of channelised versus sheet flow.

ACKNOWLEDGEMENTS

    Cores were obtained through the co-operation of the captains
and crews of CSS Dawson and CSS Hudson under the auspices of
Bedford Institute of Oceanography. Noel James and Bob Stevens are
gratefully acknowledged for discussion in the field and permission
to use the material on the Cow Head Breccias. Mark Williamson
kindly advised on the foraminiferal assemblages. This work was
supported by NSERC (Canada) and Imperial Oil grants to Piper. Noel
James and Brian Pratt read early drafts of this manuscript.

REFERENCES

Aalto, K.R., 1972, Flysch pebble conglomerate of the Cap-des-Rosiers
      Formation (Ordovician), Gaspé Peninsula, Quebec, J. Sed. Pet.,
      42:922-926.
Aksu, A.E., 1980, Late Quaternary Stratigraphy, Paleoenvironmento-
      logy and Sedimentation History of Baffin Bay and Davis Strait,
      Unpubl. Ph.D. thesis, Dalhousie University, Halifax, N.S.

Blatt, H., Middleton, G., and Murray, R., 1980, "Origin of Sedimentary
    Rocks", 2 ed., Prentice-Hall, New Jersey.
Cook, H.E., 1979, Ancient continental slope sequences and their
    value in understanding modern slope development, in: "Geology
    of Continental Slopes", Doyle, L.J. and Pikley, O.H., ed.,
    SEPM Special Publ., 27:287-305.
Cook, H.E., McDaniel, P.N., Mountjoy, E.W., and Pray, L.C., 1972,
    Allochtonous carbonate debris flows at Devonian Bank ('reef')
    margins, Alberta, Canada, Bull. Canadian Petrol. Geol.,
    20:439-497.
Crowell, J.C., 1957, Origin of pebbly mudstones, Geol. Soc. Am.
    Bull., 68:993-1009.
Embley, R.W., 1976, New evidence for the occurrence of debris flow
    deposits in the deep sea, Geology, 4:371-374.
Hampton, M.A., 1970, Subaqueous debris flow and generation of tur-
    bidity currents, Unpubl. Ph.D. thesis, Stanford University,
    Stanford, Calif.
Hampton, M.A., 1975, Competence of fine-grained debris flows, J.
    Sed. Pet., 45:834-844.
Hill, P.R., 1981, Erosion of an old slump-scar on the Nova Scotian
    Slope and a possible mid-slope depositional lobe, (abstr.)
    AAPG-SEPM Annual Meeting, San Francisco, June 1981.
Hubert, J.F., Sucheki, R.K., and Callahan, R.K.M., 1977, The Cow
    Head Breccia: Sedimentoloty of the Cambro-Ordovician Conti-
    nental Margin, Newfoundland, in: "Deep Water Carbonate Envi-
    ronments", Cook, H.E. and Enos, P., eds., SEPM Special Publ.,
    25:125-154.
Jacobi, R.D., 1976, Sediment slides on the northwestern continental
    margin of Africa, Mar. Geol., 22:167-174.
James, N.P., Klappa, C.F., Skevington, D., and Stevens, R.K., 1980,
    Cambro-Ordovician of West Newfoundland - Sediments and Faunas,
    Geol. Assoc. Canada, Annual Meeting, Halifax, 1980, Field Trip
    Guidebook No. 13.
Jeffrey, G.B., 1922, The motion of ellipsoidal particles immersed
    in a viscous fluid, Proc. Roy. Soc. London, Ser. A, 102:161-
    179.
Johnson, A.M., 1970, "Physical Processes in Geology", Freeman Cooper
    and Co., San Francisco, Calif.
Kurtz, D.B., and Anderson, J.B., 1979, Recognition and sedimento-
    logic description of recent debris flow deposits from the Ross
    and Weddell Seas, Antarctica, J. Sed. Pet., 49:1159-1170.
Lindsay, J.F., 1968, The development of clast fabric in mudflows,
    J. Sed. Pet., 38:1242-1253.
Middleton, G.V., and Hampton, M.A., 1976, Subaqueous sediment trans-
    port and deposition by sediment gravity flows, in: "Marine
    Sediment Transport and Environmental Management", Stanley, D.J.
    and Swift, D.J.P., eds., Wiley Interscience, N.Y., 197-218.
Pierson, T.C., 1980, Erosion and deposition by debris flows at Mt.
    Thomas, North Canterbury, New Zealand, Earth Surface Processes,
    5:227-247.

Piper, D.J.W., Panagos, A.G., and Pe, G.G., 1978, Conglomerate Mio-
     cene Flysch, Western Greece, J. Sed. Pet., 48:117-126.
Reineck, H.-E., and Singh, I.B., 1975,"Depositional Sedimentary
     Environments", Springer-Verlag, New York.
Shelton, J.S., 1966, "Geology Illustrated", W.H. Freeman & Co.,
     San Francisco.
Visher, G.S., 1969, Grain size distributions and depositional Pro-
     cesses, J. Sed. Pet., 39:1074-1106.

LONG-RANGE SIDESCAN SONAR STUDIES OF SEDIMENT SLIDES AND THE EFFECTS OF SLOPE MASS SEDIMENT MOVEMENT ON ABYSSAL PLAIN SEDIMENTATION

Robert B. Kidd

Institute of Oceanographic Sciences
Wormley
Godalming
Surrey.   U.K.

INTRODUCTION

The long-range sidescan sonar technique has been used to great effect in surveys aimed at understanding processes responsible for the large-scale morphology of the deep ocean floor.   Areas of thin sediment cover over igneous basement such as mid-ocean ridges (Laughton et al., 1979) and fracture zones (Searle, 1979), and areas of major compressive deformation (Stride et al., 1977) have been the prime targets of interest.   Since 1977, a second phase of development of the GLORIA (Geological Long Range Inclined Asdic) tool, used by the Institute of Oceanographic Sciences, has provided the impetus to examine further the more subtle sedimentary features that had been recognised previously on continental margins and rises (Kenyon & Belderson, 1973).

Roberts (this volume) presents results of sidescan sonar surveys which mosaic entire sections of the passive continental margins from the continental shelf edge to the abyssal plains. The approach has been to use the long-range sonar in conjunction with both deep airgun and high resolution seismic reflection profiling to build up three dimensional models of continental margin areas.   Within such models large-scale submarine slides, such as the Grand Banks slide, are clearly delineated and conclusions may be drawn on slide morphology and internal geometry. In this paper, I plan to concentrate on slide features of somewhat smaller scale and show how an understanding of the more subtle identification capabilities of the long-range sidescan sonar technique has been built up over the past three years.

THE GLORIA SYSTEM

The Mk II version of the GLORIA device is a dual-scan towed sonar operating at frequencies of 6.3 - 6.7 kHz (Somers et al., 1978). The vehicle is towed around 300 metres behind the ship at a depth of about 50 metres. Surveys are generally run at between 8 and 10 knots: speeds which provide for optimum vehicle stability. Sound pulses are transmitted at rates of 20 or 40m sec. Beam width is about 2.7° in the azimuth and 10° in the vertical. The data are tape-recorded and then photographically anamorphosed to adjust the short-range scale. This process results in a sonograph correctly scaled in horizontal range, with the exception of its near-field. Typical horizontal ranges are around 30km on the continental rises and abyssal plains, shortening considerably on the uppermost continental slopes because of depth dependent variations in sound propagation. Thus in abyssal regions over 12,000km^2 of seafloor might be surveyed in a single day. In practice, 'whole-margin' surveys such as that of the Iberian Margin (Roberts and Kidd, in press) are conducted such that parallel track spacings overlap by around 10km. As slopes increase the constraint of range limitation brings the line spacings closer together.

During shipboard operations, sonographs resulting from the satellite-navigated GLORIA surveys become available for pasting along plotted track lines about 24 hours after their initial record- ing. One to one-quarter million and one to one million scales are most frequently used. First interpretations aboard ship are traced on overlay sheets, using for control the simultaneously recorded 12 kHz echo sounder and 2 kHz or 3.5 kHz high resolution seismic profiles. Subsequent, usually shore-based, interpretation is made using all available echo sounder and high resolution seismic profiles that cross the GLORIA coverage.

MEDIUM SCALE SEDIMENT SLIDES

A number of sediment slides have been recognised since surveys began with GLORIA Mk II vehicle in 1977 during studies either of continental margin sedimentation as a whole or the effects of deep ocean circulation.

A. The Rockall Trough

The first deployment of the new vehicle was west of Ireland in the Rockall Trough. The object of the survey was an examination of the surface of a major abyssal sediment drift, the Feni Ridge (Roberts & Kidd, 1979). The 60km wide sonograph coverage showed a complex array of sediment wave fields. Individual waves were typically 25 to 50 metres in height, 1 to 4km in wavelength and were traceable for distances of up to 26km.

The Feni Ridge sediment wave fields were disrupted in the
northern part of the survey area by a sediment slide that originated
on the flanks of Rockall Bank (Roberts, 1972). The sonographs
(Figure 1) depict the slide as an area of irregular topography,
while it is represented on the seismic profiles as a region of
hyperbolae. The feature is bounded by a circular southward-facing
scarp that on the 3.5 kHz profile shows a relief of 11 meters
(Figure 2). The hyperbolic echo returns in this area had previously
been thought to result from further mudwave fields. The sonographs
however provide a plan view that suggests that the irregular hummocky
relief here is oriented in a pattern concentric with the outer edge
of the slide.

A DEEPTOW and transponder-navigated coring survey, conducted
in the same year over part of the outer edge of the slide, showed
that the scarp represents the edge of a large debris flow that was
transported from the eastern margin of Rockall Bank some 15,000 -
16,000 years B.P. (Flood et al., 1979). Cores from the debris
flow display a chaotic mixture of marls, chalks, mud balls, sands
and pebbles. Those further south are typically current-deposited
clays and silty clays with silt and sand layers. Between, at the
foot of the small scarp, is a wedge of acoustically transparent
sediment that remains unsampled but may represent less competent
sediment that travelled at, and was deposited as homogenized
sediment near, the 'nose' of the debris flow.

The survey demonstrated the need for high resolution seismic
profiling data, both along track and as crossings, to get some idea
of sub-bottom acoustic character within 100 meters of the seabed
below the ship and in the farfield of the sonographs.

B.   The South African Margin

During a short GLORIA survey of the eastern margin of South
Africa in 1979, a sediment slide was crossed that had been recognised
previously from seismic profiling. As mapped by Dingle (1977) this
slide, "the Agulhas Slump", has an areal extent of some 79,488km^2.
The recorded GLORIA data was affected by interference problems.
Nevertheless, a head region adjacent to the Agulhas Fracture Zone
is clearly seen, and at around 36°S, 26°E a lobate feature similar
to the Feni Ridge slide is visible on the lower continental rise to
the port side of our track. This lobe of the slide represents a
feature approximately 15km long and 10km wide, which appears detached
from the main slide.

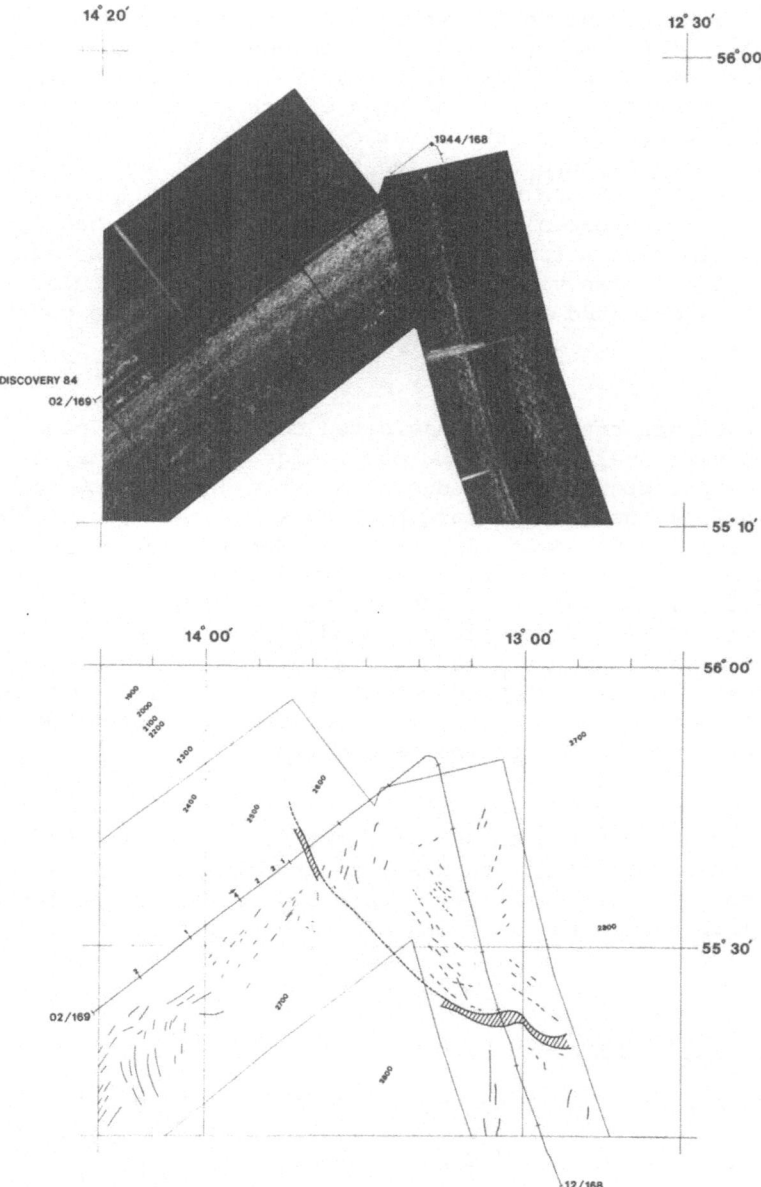

Fig. 1.   Sediment slide in Rockall Trough - Sonographs and
           interpretation.   Bathymetry in fathoms;   times and day
           numbers, sediment wave heights and migration directions
           shown along track.   Shaded is the transparent sediment
           wedge shown in Figure 2.

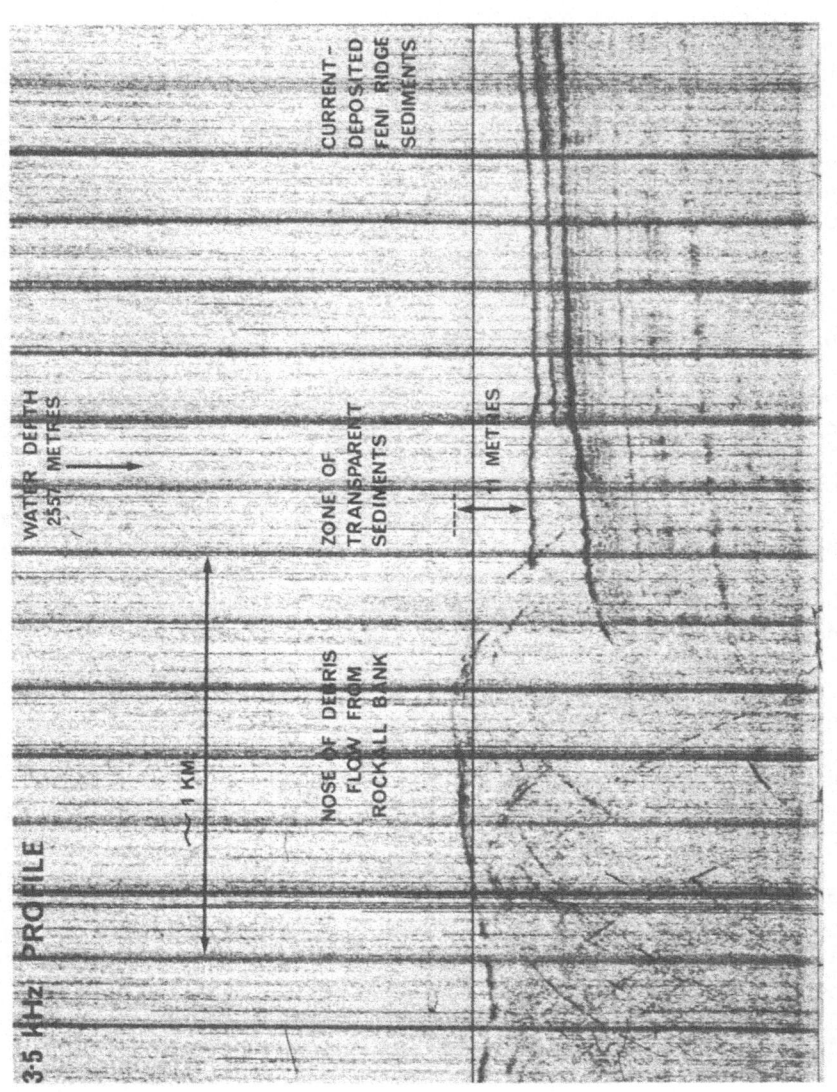

Fig. 2.   3.5 kHz High Resolution seismic profile record during the second crossing by "Discovery" of the edge of the Rockall sediment slide near 13°45'W.   For location see Figure 1.

C.   The Blake-Bahama Region

An opportunity presented itself to make short surveys during transit of the Blake-Bahama region with GLORIA aboard the charter ship "Starella" late in 1979 and aboard "Discovery" in 1980, the latter with a newly-developed 2 kHz high resolution seismic profiler. This area (Figure 3) has been the subject of intensive survey by American ships using 3.5 kHz profiles and a great deal of additional information of geological significance was available with which to interpret features observed in the farfield of our sonograph coverage. This included data from coring, current metering, DEEPTOW and submersible programs (Flood, 1978). The main aim of our surveys was to derive large-scale plan-views over the flanks of the Blake and Bahama Outer Ridges, two major sediment drifts developed by the Western Boundary Undercurrent (Flood and Hollister, 1974).

Figure 4 is an illustration from a paper being prepared jointly with R.D. Flood and D.G. Roberts in which all available profiling data has been used to interpret the sonograph coverage. Trends of the major targets visible on the sonograph are traced on the interpretive diagram. These include the crest to the north (heavy dashed lines), the crests of major sediment waves developed on the flanks of the drift (curving lines), trend lines representing abyssal sediment furrows (thinner straight lines, only some shown). High resolution seismic profile tracks which cross these features (thinnest lines, terminating in a record of cruise number and times) have noted along them the heights of individual sediment waves and also the direction of wave migration as deduced from sub-bottom reflectors. Below is an example of a typical 3.5 kHz seismic profile over the sediment waves. This illustration represents the level to which interpretation of long range sonar data has progressed to date. Clearly a combination of this technique with high resolution seismic profiling should provide a powerful tool for regional surveys of large scale sediment slides.

At the end of the "Discovery" survey of the Blake-Bahama area the edge of a major submarine slide was crossed, which traverses the northern Blake Outer Ridge. Figure 5 has been interpreted in the same way as the preceding figure, with the addition of a moated feature and the slide edge itself. The 2 kHz seismic profile illustrated below displays an acoustically-transparent layer, up to 50 metres in thickness, masking pre-existing topography. The sonograph shows a coincident change in surface topography which becomes increasingly rough and irregular northwards. The edge of the transparent layer is interpreted as a debris flow. It can be traced from the sonograph through a number of 3.5 kHz seismic profiles, which display the same acoustically transparent layer. One interesting point about the edge of this slide is that it contrasts with the Rockall slide illustrated earlier in that no concentric targets are visible here. Indeed, sediment waves appear

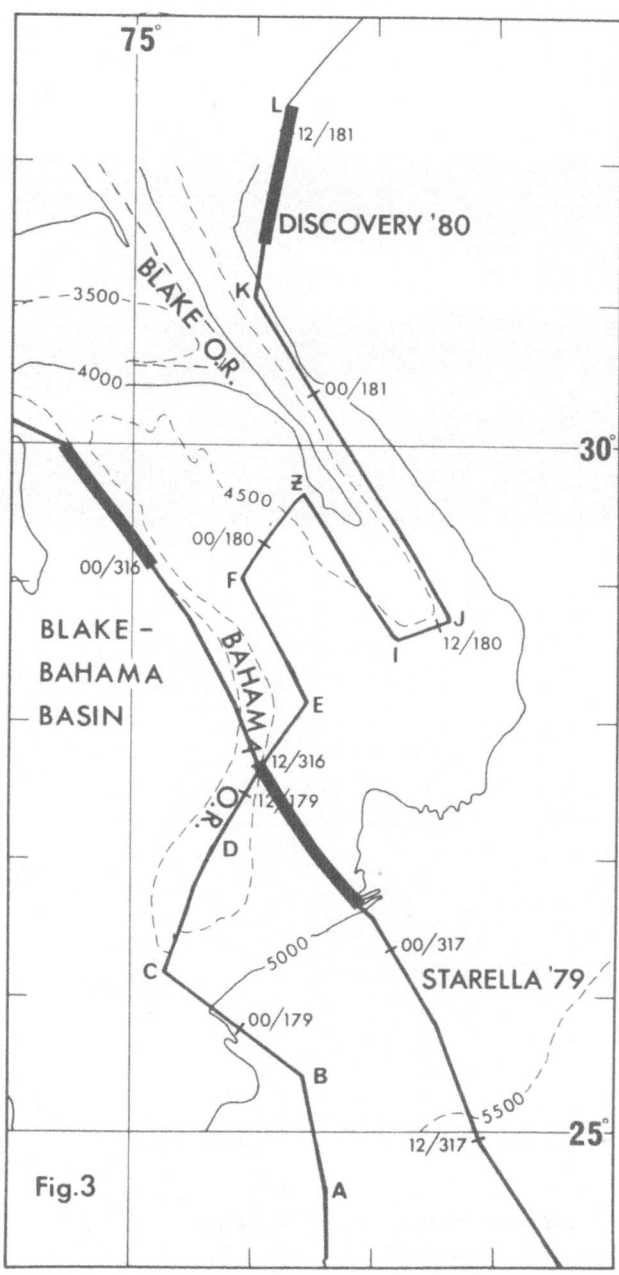

Fig. 3. "DISCOVERY" and "STARELLA" tracks through the Blake-Bahama region. Thickened parts of the track show the location of the sonographs in Figures 3, 4 and 5. Bathymetry in meters. Times and day numbers shown along track.

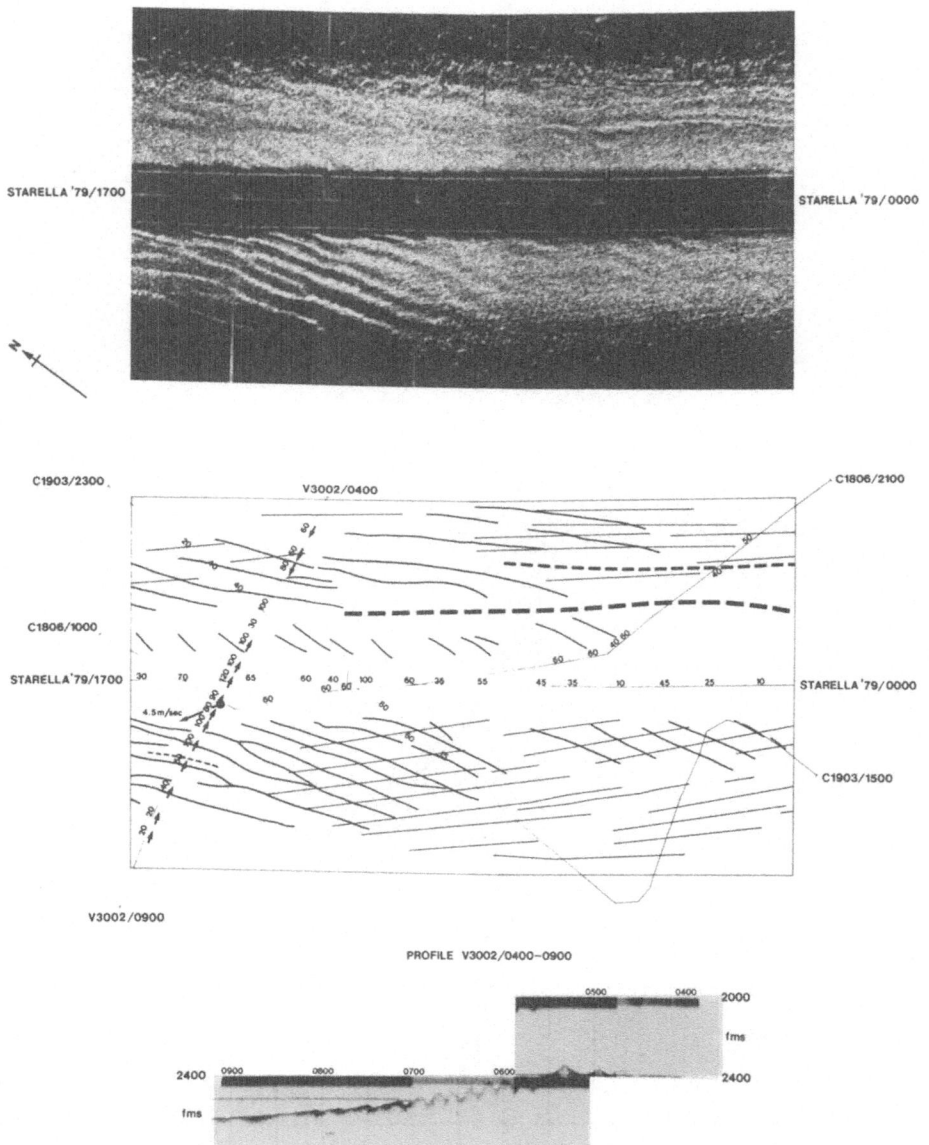

Fig. 4.   Sediment features on the flanks of the northern Bahama
           Outer Ridge.   Sonograph coverage and interpretation.
           For location see Figure 3;   for explanation see text.

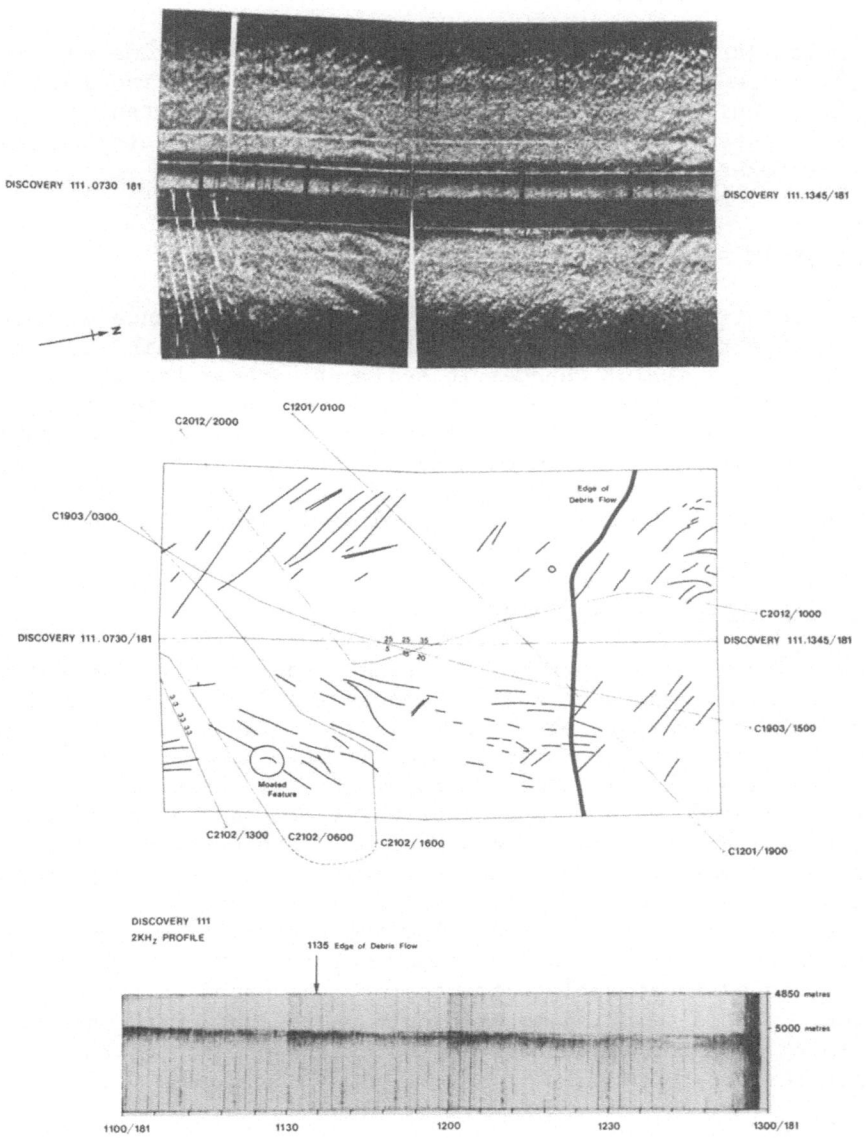

Fig. 5. Edge of a submarine slide on the flanks of the northern Blake Outer Ridge: sonograph coverage, interpretation and 2 kHz seismic profile; compare Figure 4 and text for explanation.

to have developed across the slide since its emplacement. The
acoustic nature of the slide material also suggests that the material
deposited was less competent than that of the body of the Rockall
slide. Thus this extensive edge zone is the equivalent of the
"transparent" wedge on the Feni Ridge.

Embley (this volume) describes this particular slide in its
entirety mapped on the basis of 3.5 kHz seismic profiling and coring
data (his Figure 16). He estimates its total areal extent at
25,000km², extending from the upper continental slope off Charleston
to the Hatteras Abyssal Plain at 5,400 metres.

ABYSSAL PLAIN SEDIMENT DISTRIBUTION

One relatively recent discovery made from long range sidescan
sonar surveys that have extended on to the abyssal plains is the
possible recognition of changes in sediment type or bed roughness
below the normal target resolution of the device. A number of
crossings of abyssal plains have displayed an appreciable acoustic
texture, which is thought to be a backscattering effect caused by
bed roughness changes. Wavelength considerations would suggest
that resolution in this case could be down to 25cm.

A.   The Hatteras Abyssal Plain

An example of part of the "Starella" transit with GLORIA from
the Blake-Bahama region to Puerto Rico is illustrated in Figure 6.
The sonograph shows part of the southern Hatteras Abyssal Plain near
its progression into the Nares Abyssal Plain. Echo sounder profiles
through the area show little topography at all. Despite this,
lineated and wavy features are clearly visible on the sonograph
oriented in a general downslope direction. The profile on the
diagram shows one crossing of these features with their approximate
positions arrowed. No relief is indicated, indicating that any
small relief variations must be below the resolution of the echo
sounder (± 1m). Changes in the seabed multiple below might however
suggest changes in acoustic character of the seabed at these
locations. It is tempting to attribute these changes to changes
in sediment type and/or texture caused by the passage of turbidity
currents through the area.

B.   The Biscay Abyssal Plain

Similar sedimentary features were discovered during earlier
Iberian Margin surveys in 1978 that extended on to the Biscay Abyssal
Plain (Figures 7 and 8). The coverage extends from north of Spain
westwards and southwards to the Theta Gap, which is a gorge-like

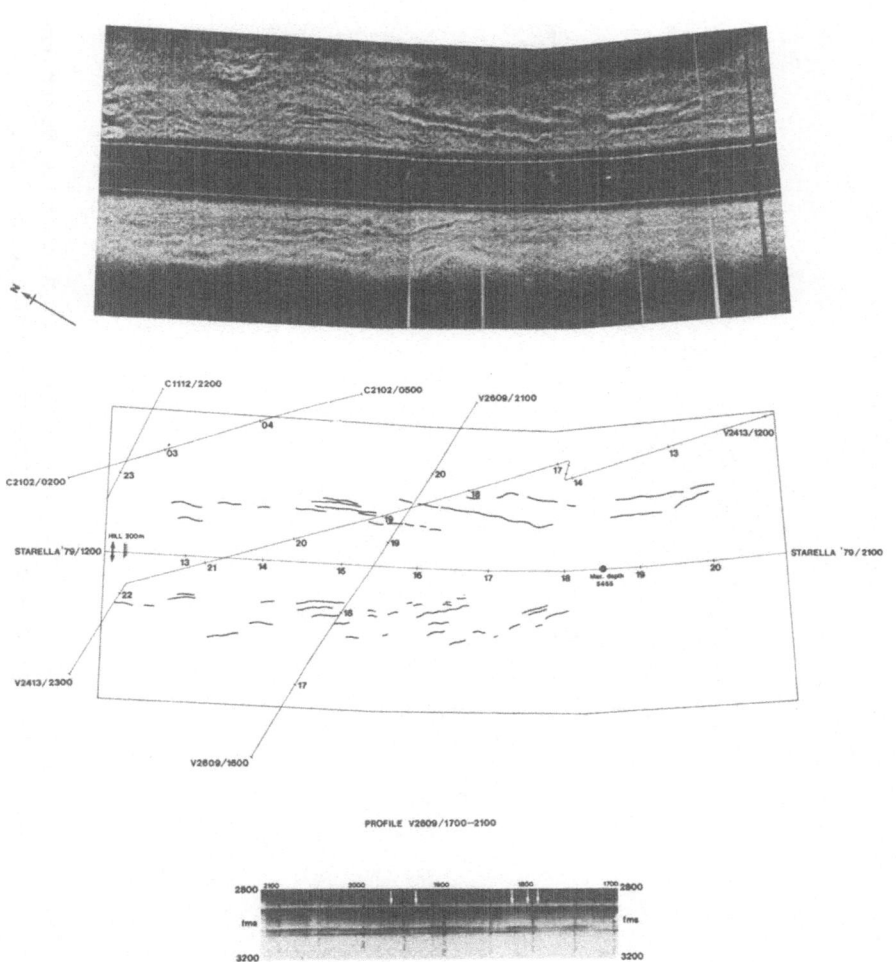

Fig. 6.   Sediment features on the southern Hatteras abyssal plain:
sonograph coverage, interpretation and profile.   Hour
times shown along track, other notations as for Figure 4.

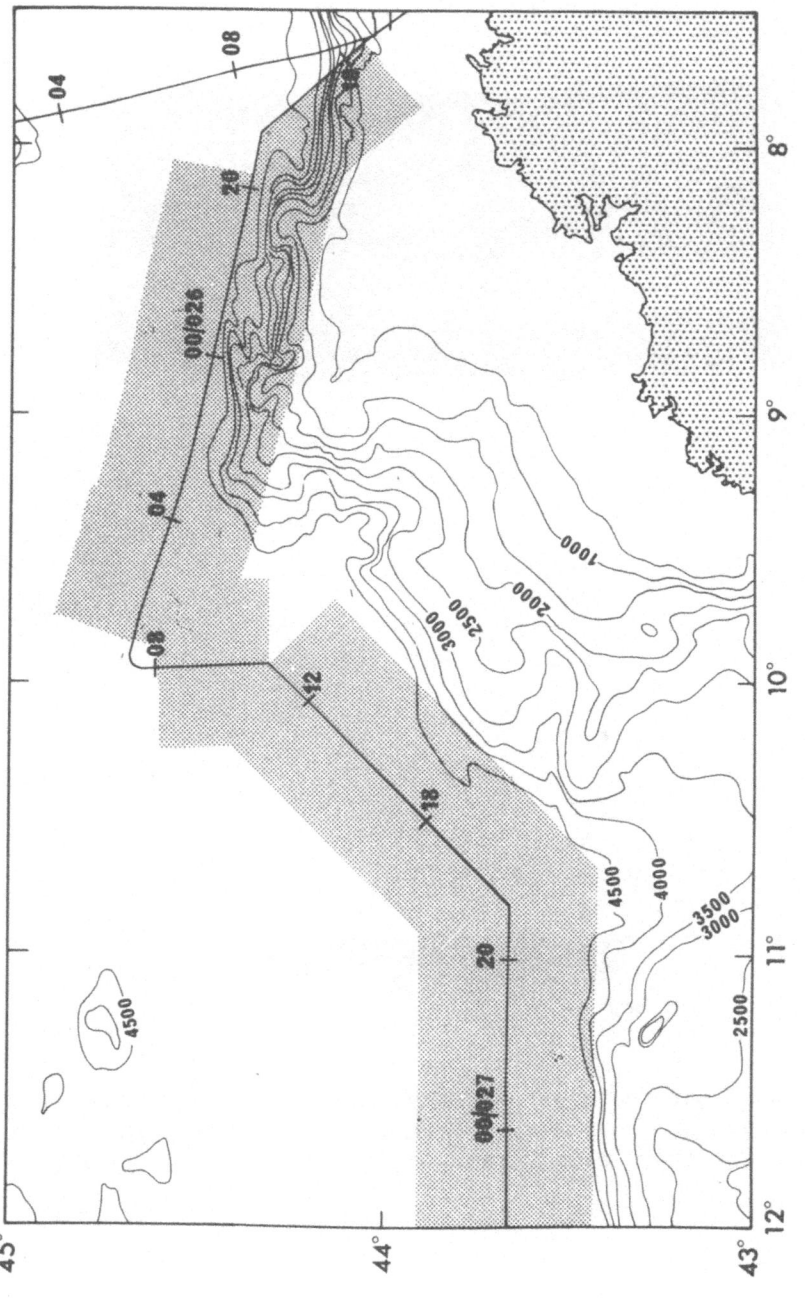

Fig. 7.   GLORIA coverage on the Biscay Abyssal plain.   Dotted land area is Northern Spain, bathymetry in meters.   Hour marks and day numbers shown along track.

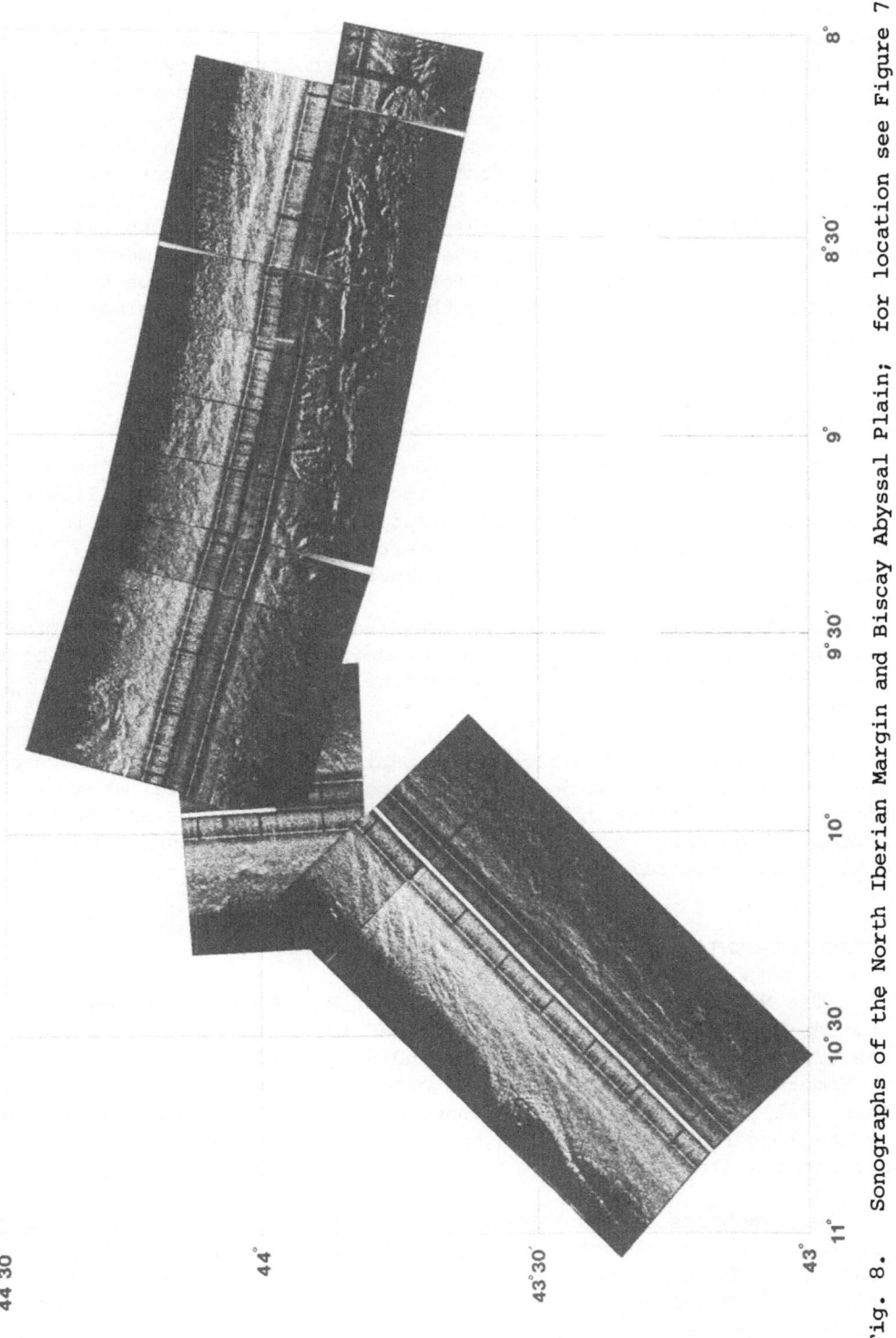

Fig. 8.   Sonographs of the North Iberian Margin and Biscay Abyssal Plain;   for location see Figure 7.

connection to the deeper Iberian Abyssal Plain.   Laughton (1960) had
mapped from echo sounder records a channel system close to the Theta
Gap which extends eastwards onto the western Biscay Abyssal Plain.
The sonographs, on the other hand, show clear evidence of lineations
extending far beyond any topographic expression of these channels to
areas north of Spain that are considered from echo sounding "flat"
abyssal plains.   In these northern areas, the lineations appear to
connect with canyons dissecting the continental slope.   If the
acoustic changes do indeed represent an expression of turbidity
current flow direction, rather than changes such as sediment ripple
fields or zones of contrasting bioturbation, the technique could
provide a means of investigating the distal effects of submarine
sediment slides.

CONCLUSIONS

The GLORIA system provides a tool that allows the mapping of
large scale features indicative of regional sediment instability.
Individual sediment slides show sufficient downslope changes in
acoustic character to allow some inferences to be made of the
competence of the flow at varying stages in its movement.

It becomes essential in the detailed investigation of individual
slides to incorporate high-resolution seismic profiling data at both
the time of the survey and in subsequent interpretation of the
records.

The device may also have the capability to trace the effects of
sediment sliding on the continental margins to the distal abyssal
plains.   A programme of transponder-navigated sampling and near-
bottom photography, profiling and sidescanning is necessary to
confirm this capability.

ACKNOWLEDGMENTS

The Masters, crews and scientists of R.R.S. Discovery and the
charter ship "Starella" are acknowledged for their support during
the cruises which resulted in the findings reported above.
M. Somers, J. Revie, B. Barrow and other members of the GLORIA
technical support team were responsible for the successful operation
of the long range sonar system and D. Bishop for the seismic
profiling systems.

D.G. Roberts is thanked for a critical review of this manuscript
and for continued co-operation in all aspects of these long-term
studies.   The Department of Energy and Department of the Environment are acknowledged for support of parts of these studies under
commissioned research projects.

REFERENCES

Dingle, R.V., 1977, The anatomy of a large submarine slump on a sheared continental margin (S.E. Africa), Journ. Geol. Soc. Lond., 134, 293-310.

Flood, R.D., 1978, Studies of deep-sea sedimentary microtopography in the North Atlantic Ocean, Doctoral dissertation, Woods Hole Oceanographic Institution, unpublished manuscript, 395pp.

Flood, R.D. & Hollister, C.D., 1974, Current-controlled topography on the continental margin off the Eastern United States, in: "The Geology of Continental Margins", C.A. Burk & C.L. Drake, editors, Springer-Verlag, New York, pp. 197-205.

Flood, R.D., Hollister, C.D., & Lonsdale, P., 1979, Disruption of the Feni sediment drift by debris flows from Rockall Bank, Mar. Geol., 32, 311-334.

Laughton, A.S., 1960, An interplain deep-sea channel system, Deep Sea Research, 7, 75-78.

Laughton, A.S., Searle, R.C., & Roberts, D.G., 1979, The Reykjanes Ridge Crest and the transition between its rifted and non-rifted regions, Tectonophysics, 55, 173-177

Roberts, D.G., 1972, Slumping on the Eastern Margin of the Rockall Bank, North Atlantic Ocean, Mar. Geol., 13, 225-237

Roberts, D.G., & Kidd, R.B., 1979, Abyssal sediment wave fields on the Feni Ridge, Rockall Trough; long-range sonar studies, Mar. Geol., 33, 175-191.

Roberts, D.G., & Kidd, R.B., (in press), Sedimentary and structural patterns on the Iberian Continental margin; an alternative view of continental margin sedimentation, Geology.

Searle, R.C., 1979, Side-scan sonar studies of North Atlantic fracture zones, Jour. Geol. Soc. Lond., 136, 283-292.

Somers, M.L., Carson, R.M., Revie, J.A. Edge, R.H., Barrow, B.J., & Andrews, A.G., 1978, GLORIA II - an improved long range sidescan sonar, Proc. IEE/IERE Sub-conference on Offshore Instrumentation & Communications, Oceanology International 1978, B.P.S. Publ. Ltd., London, pp16-24.

Stride, A.H., Belderson, R.H. & Kenyon, N.H., 1977, Evolving miogeanticlines of the East Mediterranean (Hellenic, Calabrian & Cyprus) Outer Ridges, Phil. Trans. Roy. Soc. Lond., A, 284, 255-285.

UNUSUAL DEBRIS FLOW DEPOSITS FROM THE BASE OF THE MALTA ESCARPMENT

(EASTERN MEDITERRANEAN)

Maria Bianca Cita (1), Francesco Benelli (2), Biagio
Bigioggero (3), Alessandro Bossio (4), Cristina Broglia
(1), Henry Chezar (5), Georges Clauzon (6), Annita Co-
lombo (3), Massimo Giambastiani (1), Alberto Malinverno
(1), Elizabeth L. Miller (7), Elisabetta Parisi (1),
Gianfranco Salvatorini (4), Piero Vercesi (8)

(1) Istitute of Geology, University of Milano, Italy
(2) AGIP-SGEL, Dan Donato, Milano, Italy
(3) Institute of Mineralogy, University of Milano,Italy
(4) Institute of Geology, University of Pisa, Italy
(5) Lamont-Doherty Geological Observatory of Columbia
    University, Palisades, N.Y., USA
(6) Laboratoire de Géographie, Université d'Aix-Marseil-
    le, France
(7) Department of Geology, Stanford University,Stanford,
    Calif., USA
(8) Institute of Geology, University of Pavia, Italy

GEOLOGICAL SETTING

The more than 3000 m high Malta Escarpment forms the western boundary of the eastern Mediterranean. The escarpment has an average
dip of about 20°, however the lower part of the escarpment is steeper and has a slope of approximately 60° (Figure 1). Extensive exposures of late Triassic-lower Liassic shallow water limestone occurs along this lower part of the escarpment (Chayes et al, 1980;
Scandone et al, in press; Cita et al, in press).
     Canyons and gullies incise the front of the Malta Escarpment.
In places, canyons and gullies originating on the middle part of the
slope coalesce into small round (1 km diameter) basins that occur
as major reentrants or indentations along the base of the escarpment. These small basins are separated from the bathymetrically
deeper abyssal plain to the east by a slight 100 m high ridge or
low hill (Figures 2 and 3).
     A gravity core from the center of one such basin and one from

305

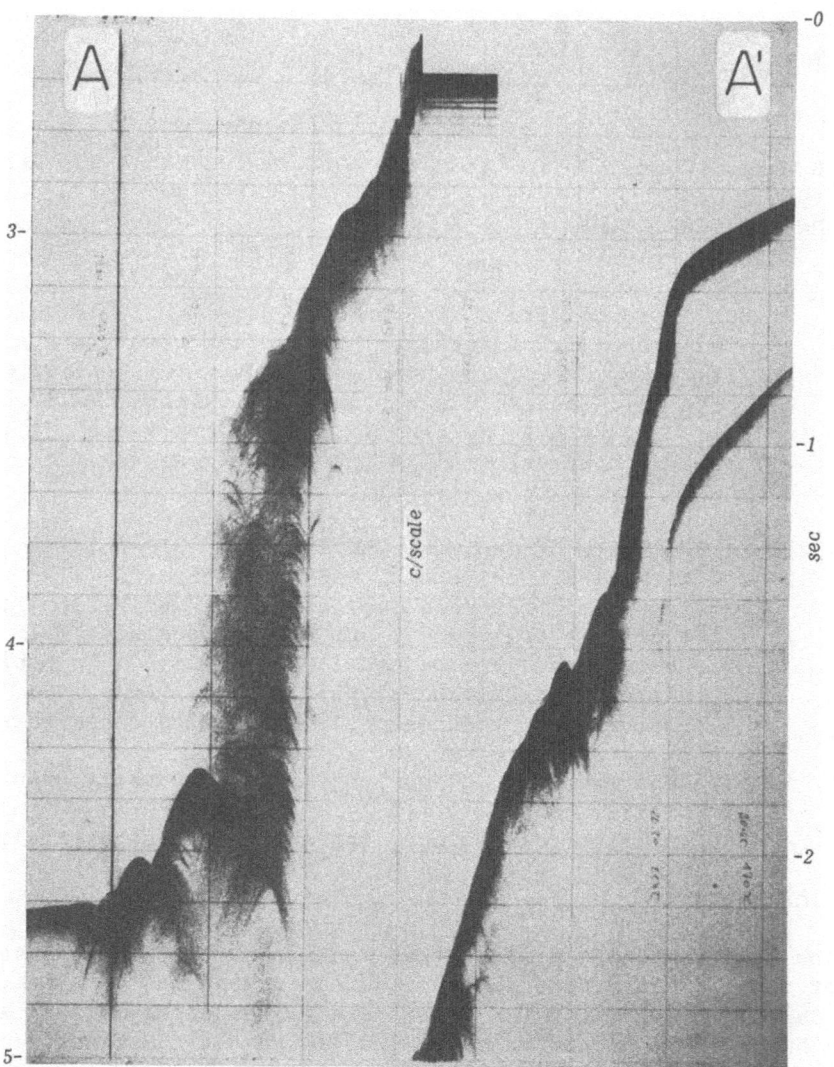

Figure 1 - BANNOCK PDR profile across the Malta Escarpment. See profile location A-A' in Figure 2.

a hill or ridge that separates this basin from the abyssal plain contains debris flow deposits overlain by an undisturbed cover of Holocene pelagic marl.

These debris flow deposits contain cobbles and pebbles of many lithologies we have dredged from the escarpment itself, plus others previously unknown. Thus this single deposit we sampled represents an excellent overview of the rocks present on the escarp-

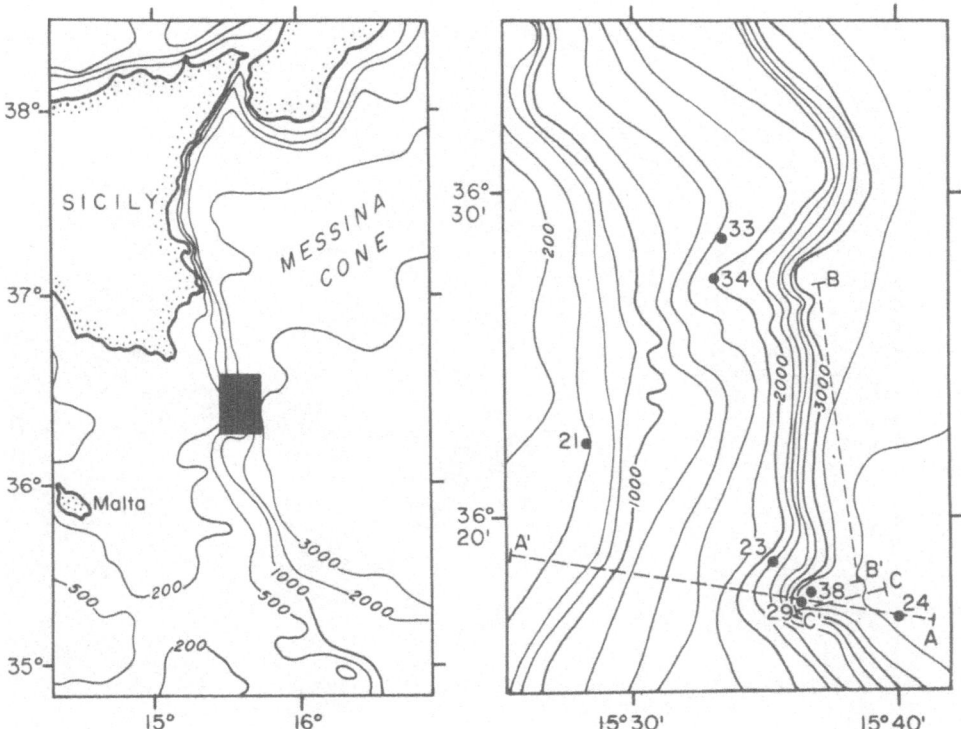

Figure 2 - Location of the area investigated, Malta Escarpment. Bathymetry after Ist. Geogr. della Marina, slightly modified. Also shown are the position of the cores discussed in the text, and that of the acoustic profiles illustrated in Figures 1, 3 and 9.

ment and provides us with additional information on the ages and environments of these older rocks.

THE DATA SET

Figure 4 shows the visual correlation of the six gravity cores and the single piston core investigated; Table 1 plots their location. Cores form three transects:

(1) an approximately W-E running transect across the escarpment includes cores 21 (shelf), 23 (intermediate slope) and 24 (abyssal plain);
(2) a N-S transect along the intermediate slope includes cores 23 (intercanyon), 34 (canyon axis) and 33 (intercanyon);
(3) a W-E transect across the base of the escarpment includes cores 29 (center of small basin), 38 (hill or ridge) and 24 (abyssal plain).
Sediments recovered are Holocene and late Pleistocene in age. The Pleistocene/Holocene boundary is tentatively identified on the

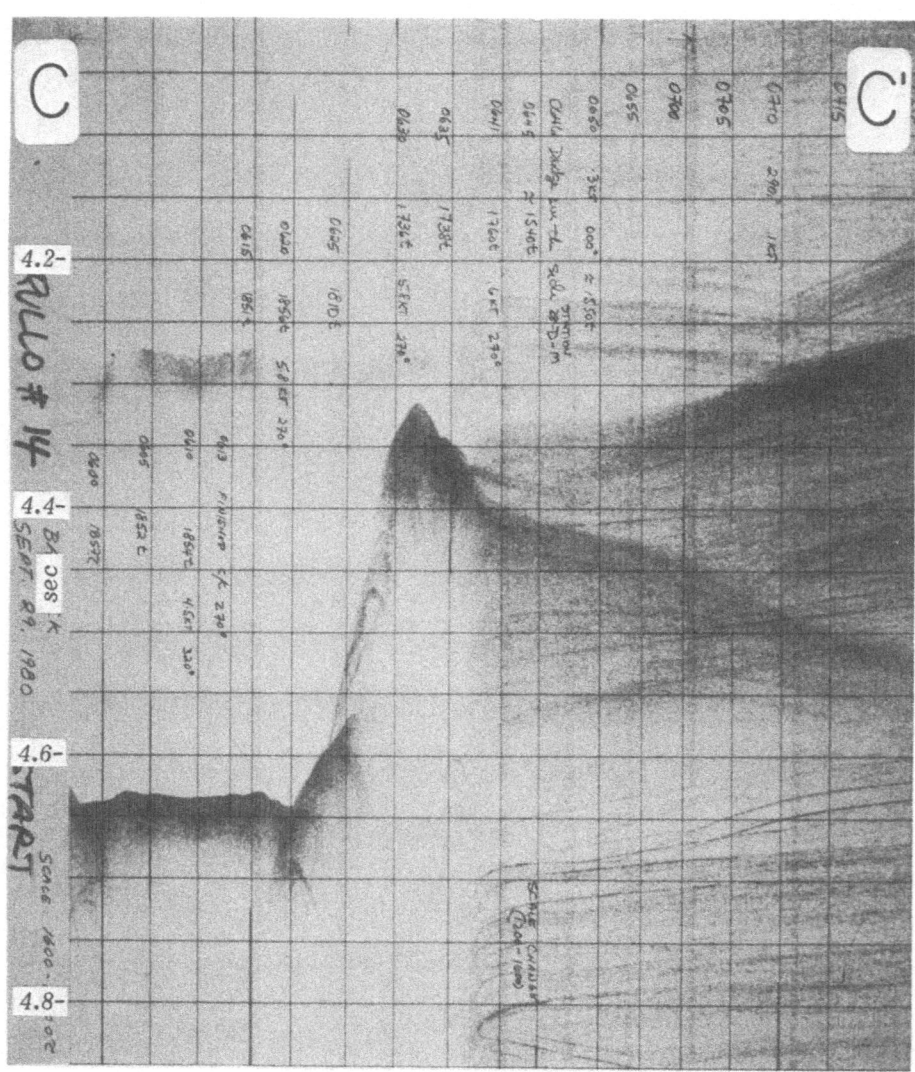

Figure 3 - BANNOCK PDR profile across the small basin and the ridge
where Cores 29 and 38 respectively were located. See location of
profile C-C' in Figure 2. Vertical exaggeration is 17:1.

basis of planktonic foraminifereal assemblages. The Holocene fauna
is characterized by warm-water species, including <u>Globigerina cali-
da</u>, <u>G.digitata</u>, <u>G.rubescens</u>, <u>Globigerinoides ruber</u>, <u>G.quadrilobatus</u>,
<u>G.sacculifer</u>, <u>Globorotalia truncatulinoides</u>, <u>Hastigerina siphonife-
ra</u>, <u>H.pelagica</u>, <u>Orbulina universa</u>. Foraminiferal assemblages referable to the latest Pleistocene are dominated by cool water species
like <u>Globoratalia inflata</u>, <u>G.scitula</u>, <u>Globigerina bulloides</u>, <u>G.
pachyderma</u>, <u>G.quinqueloba</u>.The distinction of the two faunas is quite

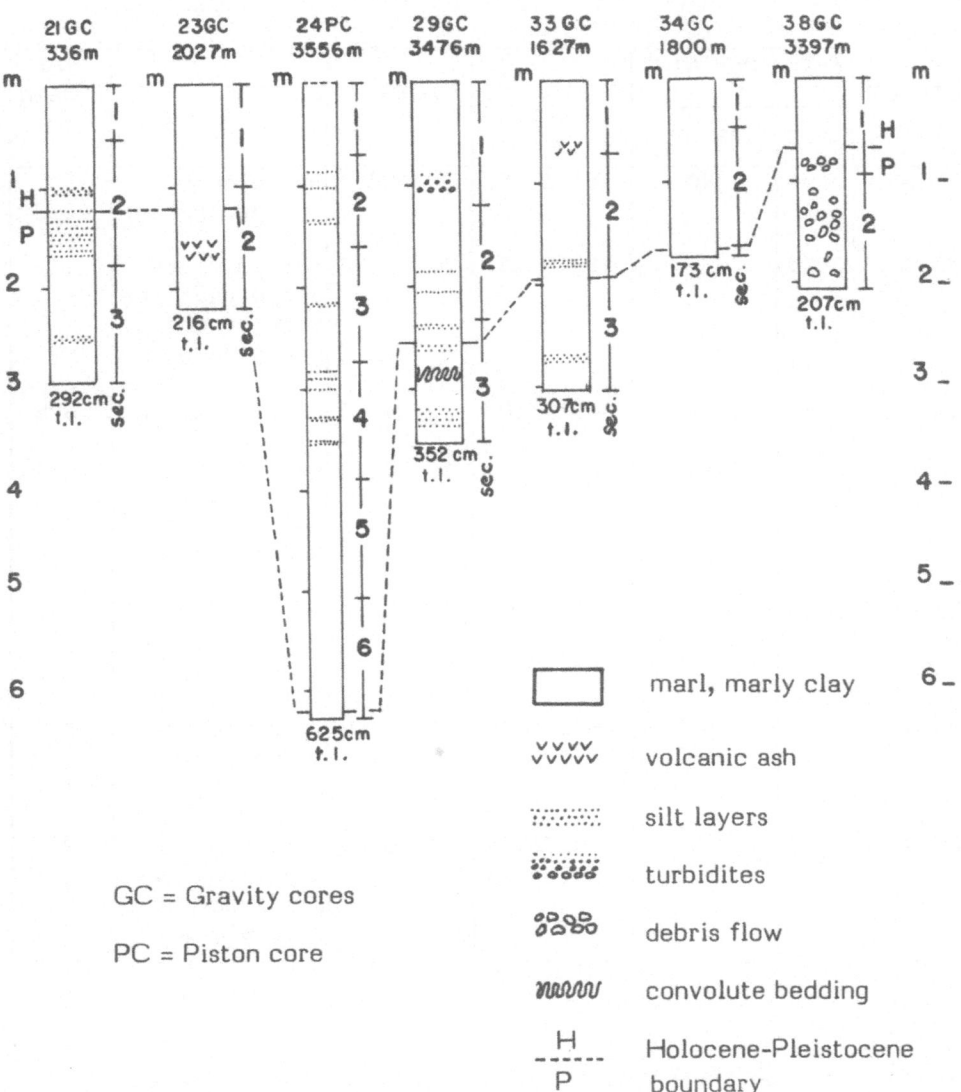

Figure 4 - Correlation of the cores raised from the Malta Escarpment during Cruise BAN-80. The Pleistocene/Holocene boundary is based on climatic change recorded in planktonic foraminiferal faunas.

easy. In most cases a color change, from brown to olive brown (in the Holocene) to olive gray (in the latest Pleistocene) is noticed, when sedimentation is hemipelagic.

CORE 29

Core 29, 352 cm long, was raised from the center of the small

TABLE 1 - Location of the cores discussed in the text.

STATION n.	WATER DEPTH (m)		LAT. N.	LONG. E	SETTING	TOTAL LENGTH (cm)
	PDR	WIRE OUT				
21 GC	336	340	36°22'.5	15°28'.1	slope above the escarpment	292
23 GC	2029	2069	36°18'.8	15°35'.0	intermediate slope near the base of the up-per escarpment	216
24 PC	3556	3548	36°17'.0	15°39'.9	abyssal plain	625
29 GC	3476	3498	36°17'.6	15°36'.2	small basin in the center of the southern amphitheater	352
33 GC	1643	1668	36°28'.7	15°33'.4	intermediate slope, inter-ca-nyon area	307
34 GC	1800	1849	36°27'.4	15°33'.0	intermediate slope, canyon axis (?)	173
38 GC	3296	3390	36°17'.9	15°36'.5	relief separa-ting small ba-sin from the abyssal plain	207

basin of the southern amphitheater explored, at a water depth of 3476 m. It consists of brown, olive and olive gray clayey marls with silt layers and decimetric thick turbidites (Figure 5). One such turbidite is rich in volcanic glass. The lower one, shown in the insert, has a sharp basal contact, fining upwards sequence, and is directly underlain by a 18 cm thich chaotic interval, with in-dications of convolute bedding.

X-Ray analysis was carried out on eight samples from Section 3 (see Table 2). The level at 37-40 cm from a sand turbidite (see insert of Figure 5) displays the greatest abundance of terrige-nous minerals (quartz plus feldspars) and the least abundance of clay. Amont clay minerals, the most common is illite, followed by chlorite, kayolinite and smectite.

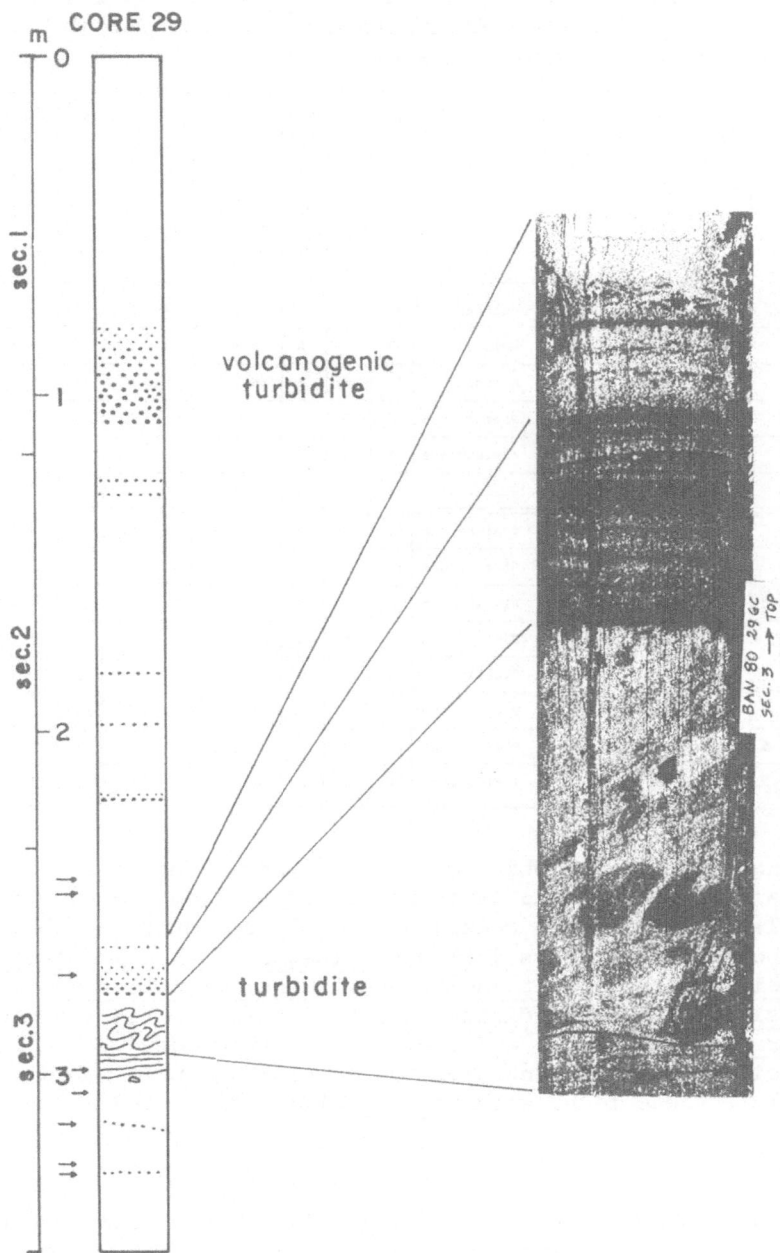

Figure 5 - Columnar log of Core 29. The photographic insert shows a portion of Section 3 with a turbidite layer and a chaotic interval with convolute bedding. Arrows indicate the position of samples analyzed for Clay Mineralogy.

TABLE 2 - Clay Mineralogy of six samples from Core 38 and eight samples from Core 29. X-Ray doffractometry by Geochemical Laboratory of AGIP-SGEL, S. Donato Milanese.

CORE / SECTION	cm from top of section	LITHOLOGY and AGE		PERCENT MAJOR MINERALS COMPONETS						RELATIVE PERCENT CLAY MINERALS				
				QUARTZ	ALBITE	ORTHOCLASE	CALCITE	DOLOMITE	CLAY MINERALS	SMECTITE	ILLITE	CHLORITE	KAOLINITE	MIXED LAYERS
38-1	14-18	marly clay	Holocene	15	6	4	11	-	64	17	36	19	28	-
38-1	42-45	marly clay	Holocene	17	6	8	11	-	58	13	29	19	39	-
38-1	66-70	marly clay	Holocene	19	5	5	14	-	47	12	43	19	26	-
38-1	83-86	marly clay	Pleistocene	16	5	4	15	-	60	16	35	18	31	-
38-2	37-41	marly clay -matrix-	Pleistocene	18	5	4	15	-	58	11	39	23	27	-
38-2	94-98	marly clay -matrix-	Pleistocene	25	5	4	15	-	51	32	38	11	19	-
29-3	8-11	marly clay	Pleistocene	16	8	4	18	-	54	14	39	21	26	-
29-3	12-15	marly clay	Pleistocene	15	5	5	18	-	57	25	32	17	26	-
29-3	37-40.8	marly clay -turbidite-	Pleistocene	23	13	10	19	1	34	15	41	26	18	-
29-3	63-66	marly clay	Pleistocene	23	5	5	17	-	50	22	35	19	24	-
29-3	74.5-77	marly clay	Pleistocene	24	10	6	19	-	41	17	39	23	21	-
29-3	78-81	marly clay	Pleistocene	16	6	5	20	-	53	14	38	22	26	-
29-3	90-92.5	marly clay	Pleistocene	23	9	7	13	1	47	13	29	32	26	-
29-3	93.5-96.5	marly clay	Pleistocene	17	6	6	16	-	53	18	42	20	20	-

An angular clast of white micritic limestone has been recorded at the base of the cahotic interval, at 67 cm from the top of Section 3 (1). The limestone contains a fairly abundant assemblage of planktonic foraminifers including Praeglobotruncana delrioensis, Rotalipora spp., Whiteinella spp., Globigerinelloides cf. bentonensis. The fauna indicates a late Cenomanian age (probably Rotalipora cushmani Zone) and a pelagic environment.

The olive-gray marl in which the angular clast is embedded yields a late Pleistocene fauna with Globorotalia inflata and accompanying species.

CORE 38

Core 38, which is only 207 cm long, is located almost at the top of the abyssal hill shown in Fig. 3. From top to bottom, it consists of:

---

(1) The 2 cm long clast is from the "working half" of the core, therefore it is not visible in Figure 5, which shows the "archive half". No other clasts were recorded within the entire thickness of the core, notwithstanding a careful three-dimensional search. The finding is thus unique.

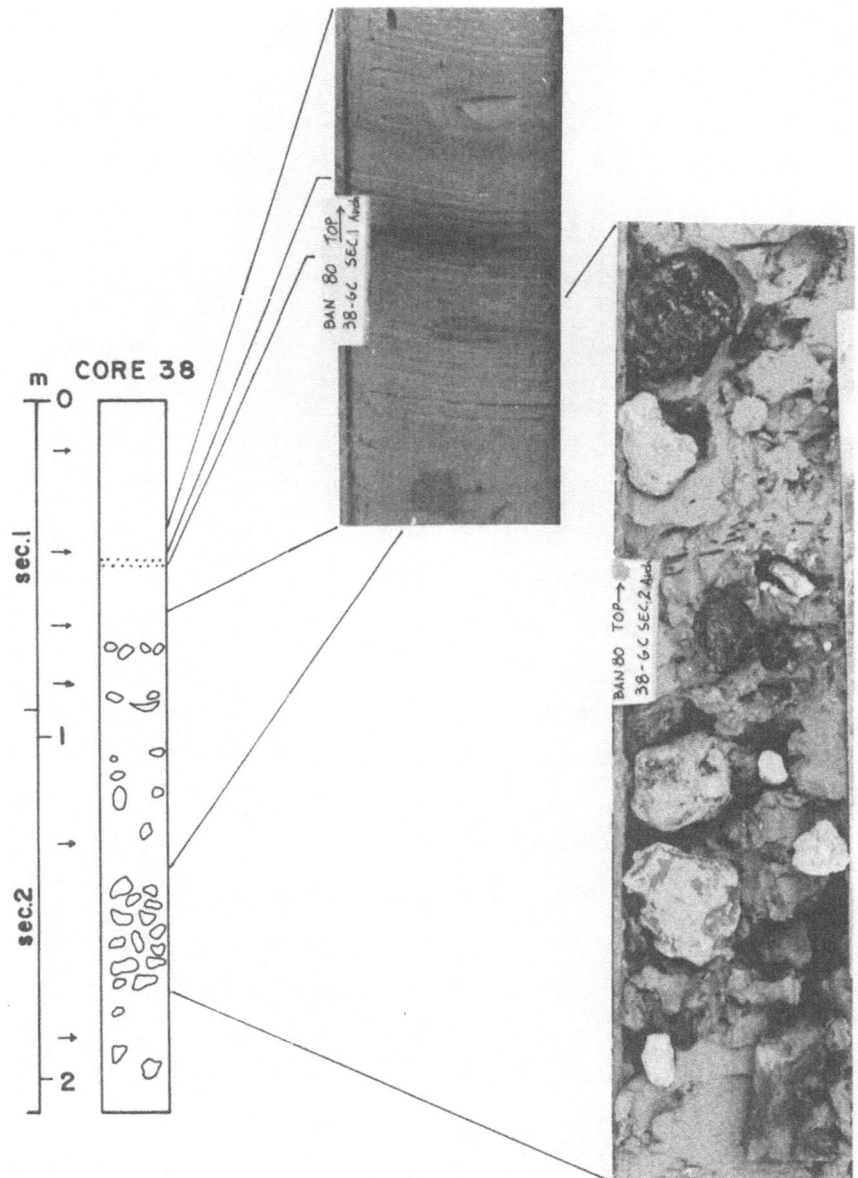

Figure 6 – Columnar log of Core 38. The photographic insert shows the interval of the debris flow deposit where the greatest concentration of lithic clasts is observed. Arrows indicate the position of samples analyzed for Clay Mineralogy.

0-74 cm: succession of clays and clayey marls of prevailingly brown color, with well preserved sedimentary structures and undeformed

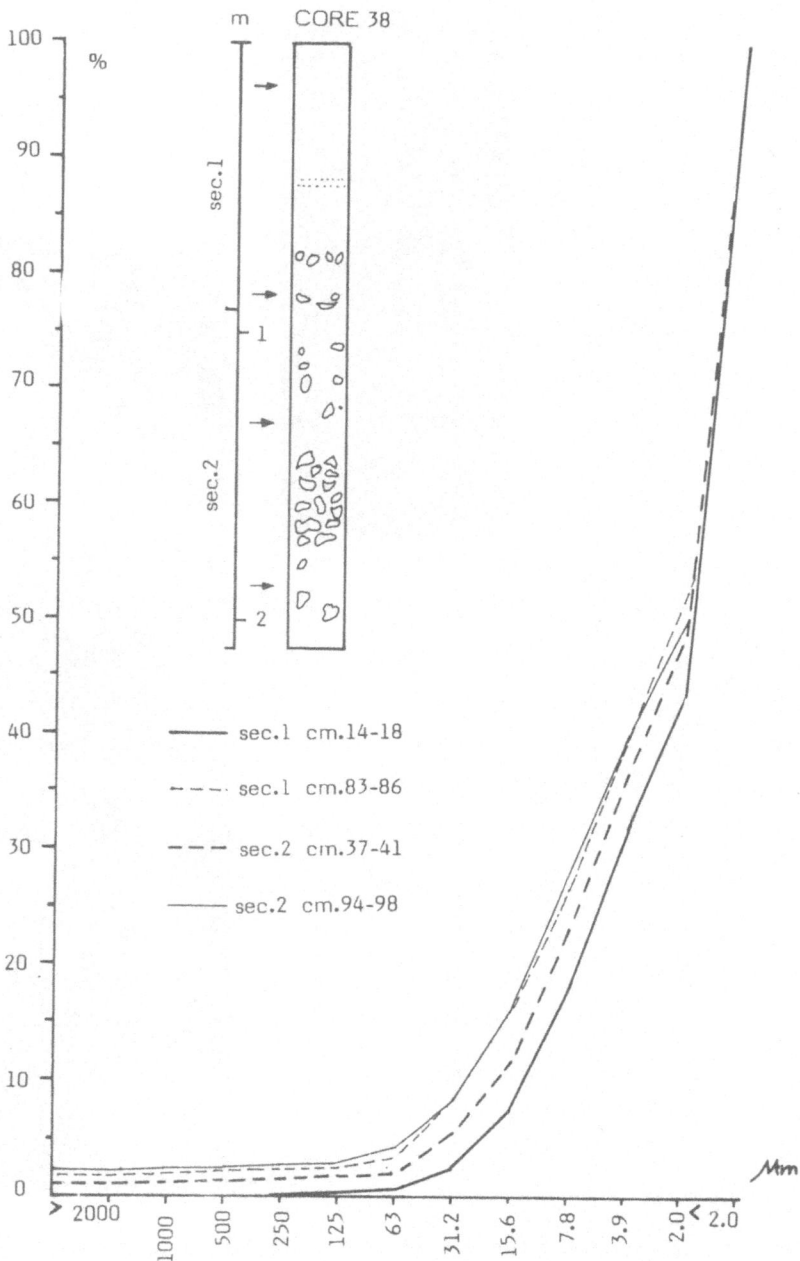

Figure 7 - Grain-size curves obtained from four samples analyzed from Core 38. Location of the samples is indicated by arrows in the columnar log to the left. Scale 1 Ø. Notice the absence of the greater than 250 μm is the sediment from 14-18 cm in Section 1, taken from above the debris flow deposit.

bedding planes (see insert in Figure 6). The fauna indicates a
Holocene age for this upper section. Illite and kaolinite are
the dominant clay minerals (Table 2); terrigenous minerals are
represented by quartz, albite, orthoclase and calcite.
Grain-size analysis (Figure 7) show a strong dominance of clay
fraction (60%) and of fine silt (23.63%).

74-77 cm: interval with centimetric clasts, including a large bio-
clast (flat fragment of <u>Ostrea</u> shell).

77-80 cm: normal sedimentation as from 0 to 74 cm.

80-92 cm (Section 1), plus the entire Section 2: lithic fragments
of different nature and size in a clayey matrix of olive gray color.
Clasts are subcentimetric from 80 to 92 cm(Section 1)where they
represent less than 5% (vol) of the sediment. Mineralogical
composition of the matrix is similar to that of the overlying
Holocene (Table 2).Clasts are mostly subcentimetric,but include
some elements up to 3 cm in size from 0 to 60 cm in Section 2
with clasts representing 5-10% (vol) of the sediment. The matrix displays chaotic structures with some indication of convolute laminations. Clasts are biggest and most frequent in a 30
cm thick interval from 60 to 90 cm Section 2 (see insert of
Figure 6):largest dimension measured is 7 cm, and several clasts
are 4 cm or more across.They represent approximately 70%(vol)
of the sediment.
The lowermost part of the core from 90 to 120 cm in Section 2
contains less clasts than the immediately overlying interval;
clasts are smaller, and represent approximately 10% (vol) of
the sediment. Unlike the interval above the greatest concentration of lithic clasts, the matrix does not show here any internal structure.
Six samples from the matrix were investigated for micropaleon-
tology.All yield the characteristic latest Pleistocene cool-water
foraminiferal fauna. Associated with the planktonic species,
benthic forms indicative of a shelf or upper slope environment
are recorded (i.e.<u>Asterigerinata mamilla</u>, <u>A.planorbis</u>, <u>Ammonia
beccarii</u>, <u>Elphidium</u> spp.),which indicate downslope displacement.
Grain-size analysis carried out on 4 samples from the matrix
below, within and above the interval rich in large clasts
 and   in the normal sediment recorded at the top of the core
indicate a drastic decrease in grain-size upcore (Figure 7).
X-Ray diffraction analysis on two samples of the matrix from
above and respectively beneath the interval with large clasts
shows a marked difference in composition: the upper sample is
similar to those of the overlying intervals, whereas the lower
sample contains more quartz,more smectite (32%), less chlorite
and less kaolinite.

The lithologic and paleontologic composition of several dozens
discrete clasts has been checked. Soft and/or semi-indurated clasts
were washed (see inventory in Table 3); lithified ones were thin
sectioned. Figure 8 shows the microfacies of some of the limestones.

TABLE 3 - Inventory of lithologies, ages and environments documented by small marly clasts from the debris flow deposit. Most of the centimetric marly lumps are from above the interval with large lithic clasts (60-90 cm in Sec. 2); only one is from beneath it.

cm from top of section 2	lithology	age	zone	environment
13	white marl	Zanclean MPl 3 G.margaritae		pelagic
17	white calcareous marl	G.puncticulata Burdigalian N 7		pelagic
43	gray marl	Maastrichtian		pelagic
43	white calcareous marl	Campanian		pelagic
50-53	white marl	late Paleocene P 4 (G.pseudomenardii) early Eocene P 8 (G.aragonensis)		pelagic
50,55,62	three small lumps	late Paleocene P 4 (G.pseudomenardii) middle Eocene P 11 (G.subconglobata) middle Eocene P 14 (T.rohri) late Oligocene P 21(?)-22		pelagic
57	white calcareous marl	Campanian		pelagic
65	white calcareous marl	Campanian		pelagic
68	white marl	middle Eocene P 11 (G.subconglobata)		pelagic
73	whitish marly limestone	Liassic(?)		bathyal open marine
91	white marl	Aquitanian N 4		pelagic

We noticed the fresh nature of the clasts, the angular shape of limestones, which represent the dominant lithology of large-sized clasts, in contrast with the rounded or subrounded shape of basalt. None of the rocks was coated with Mn-oxydes.

Basalts are represented by few, generally subrounded pebbles. The largest measured dimension is 7 cm. There are also two angular clasts of volcanic breccia with basaltic fragments embedded in a

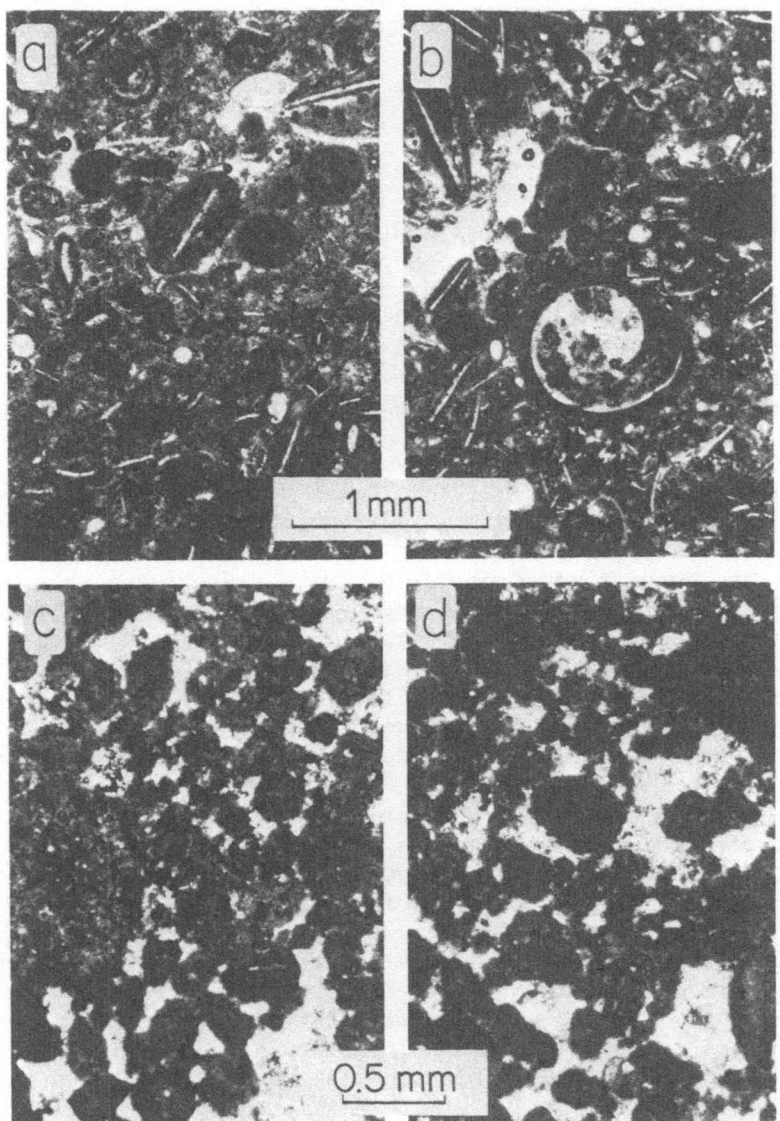

Figure 8 - Microfacies of some limestone clasts from the interval
60-90 cm in Section 2, Core 38. a= oolitic packestone with filaments. The nucleus of some oolites is represented by filaments. Mi-
crite is slightly recrystallized. High energy environment.
b = packstone with coated grains, filaments and embryonic ammonites.
c = packstone with intraclasts, strongly recrystallized, with evident fenestral structures. d = similar to c. Shallow water carbonate
platform. Age of a and b is late Jurassic. Age of c and d is late
Triassic(?) - early Jurassic(?).

carbonatic matrix. Basalts are aphiric alkali-olivine basalts with
intersertal to intergranular texture; vesicles are very abundant
(40%), filled by calcite.
  Age of fossiliferous rocks range from middle Liassic (?) to
early Pliocene.

DISCUSSION

  The fact that these cores contain deposits with unsorted angular to rounded clasts that are supported by a matrix composed of
structureless marl suggests that these sediments were deposited by
a debris-flow mechanism such as described and reviewed by Middleton and Hampton (1976) and by Rupke in Reading (1978). Reverse gra-
ding, such as seen at the base of the deposit in Core 38, has been
reported in rocks deposited by debris flow mechanism (Fischer,
1971).
  Debris flows possess a high viscosity and their mode of flow is
laminar (Rupke in Reading, 1978). Convolutions and/or flow folds
in the matrix of the deposits we cored probably represent the result of velocity gradients and/or local perturbations in laminar
flow as the deposit moved downslope. Debris flows can take place on
gentle slopes of less than one degree and are well known from recent deep-sea environment where they apparently move across areas
of many thousands of square kilometers (Embley, 1976; Moore, Curray
and Emmel, 1976; Embley, 1980). Flow velocity is variable, but is
assumed to be faster than soil creep (Rupke, 1978). Thus it is not
unusual that debris flows have occurred down the much steeper
slopes of the Malta Escarpment
  The age for the matrix (latest Pleistocene) indicates that the
mass movement occurred during the last glaciation, when low sea-
level stands accompanied by enhanced thermo-haline circulation at
depth resulted in frequent slope failures (Embley, 1980).
  The debris flow deposit that we sampled occurs at the base of
the Malta Escarpment and contains a variety of older lithologies
in a Pleistocene matrix. Most of the pre-Quaternary lithologies
correspond to lithologies identified in various dredges taken along
the escarpment from 3450 to 2400 m depth (Scandone et al., in
press; Cita et al., 1980). Lithologies previously unknown from the
Malta Escarpment include:

Jurassic: the marly limestone with Gastropods, small Ammonites,
    radiolarians, siliceous sponge spicules, ostracodes (Bairdia
    sp., Bairdia cf.molesta, Bythocypris, Paracypris), foraminifers
    (Spirillina, Conicospirillina, Dentalina cf.terquemi) recorded
    at 73 cm in Section 2 (Core 38).
    Similar faunal assemblages are known from the later part of the
    Villagonia Formation (Barbieri, 1964) of Domerian-Toarcian age.
    The packstone with oolitized filaments illustrated in Figure
    8 (a and b). Microfacies with filaments are a common record
    from the Middle Jurassic of southern Sicily (Rigo and Barbieri,

1959; Patacca et al, 1980) but the present one, which is associated with oolites, suggests a high energy environment
usually foreign to the pelagic realm.
Cretaceous: the single clast in Core 29 documents for the first
    time the existence of pelagic deposition during Cenomanian
    time along the Malta Escarpment. Previous findings of Creta-
    ceous pelagic facies were of Albian and Campanian age (Scan-
    done et al, in press; Cita et al, 1980).
Cenozoic: chronostratigraphic intervals documented by well data-
    ble sediments, all of them characterized by deep-sea pelagic
    facies include late Paleocene, early Eocene, middle Eocene,
    late Oligocene, early Miocene (Aquitanian and Burdigalian),
    all from marly clasts recovered in Core 38.

    It is difficult to say whether this debris flow deposit
originated at the top of the escarpment and picked up pebbles and
cobbles that had already accumulated at certain places along the
escarpment front (due to submarine erosion, downslope movement of
debris in canyons, and/or accumulation of talus and conglomerate
during subaerial exposure in the Messinian). The presence of Cretaceous, Eocene, Miocene and Pliocene age clasts of unconsolidated marl in the debris flow deposits suggest that at least these
lithologies were probably plucked from their original site of deposition as the debris flow moved across their exposure.
    The hypothesis suggested by some that the accumulation of
clasts in Core 38 is the result of individual rock falls is contradicted by the absence of bedding planes in the matrix, by the
coarsening upwards sequence, by the occurrence of cahotic intervals.
    The hypothesis suggested by others that the deposit might
result from the activity of northerly currents originating from
the Straits of Messina is contradicted by the W-E directed pattern
of erosional channels well visible in profile B-B', running parallel to the base of the escarpment, as shown by Figure 9 (see location in Figure 2).
    In conclusion, the debris flow deposits we sampled provide
us with an excellent and convenient cross-sectional sample of
what is exposed or was exposed along the front of the Malta Escarpment.

ACKNOWLEDGMENTS

    The cores discussed in this contribution were raised during
Cruise BAN-80 as part of the activities of Progetto Finalizzato
Oceanografia e Fondi Marini of CNR, funded by Contract 80.00728.88.
    We thank the Master of R/V BANNOCK Piazza, the officers and
the crew for their supportive attitude.
    During the cruise we had the opportunity to use a detailed
bathymetric map of the Malta Escarpment constructed with the SEA-
BEAM method by CNEXO-IFP. We thank B.Biju-Duval for generously

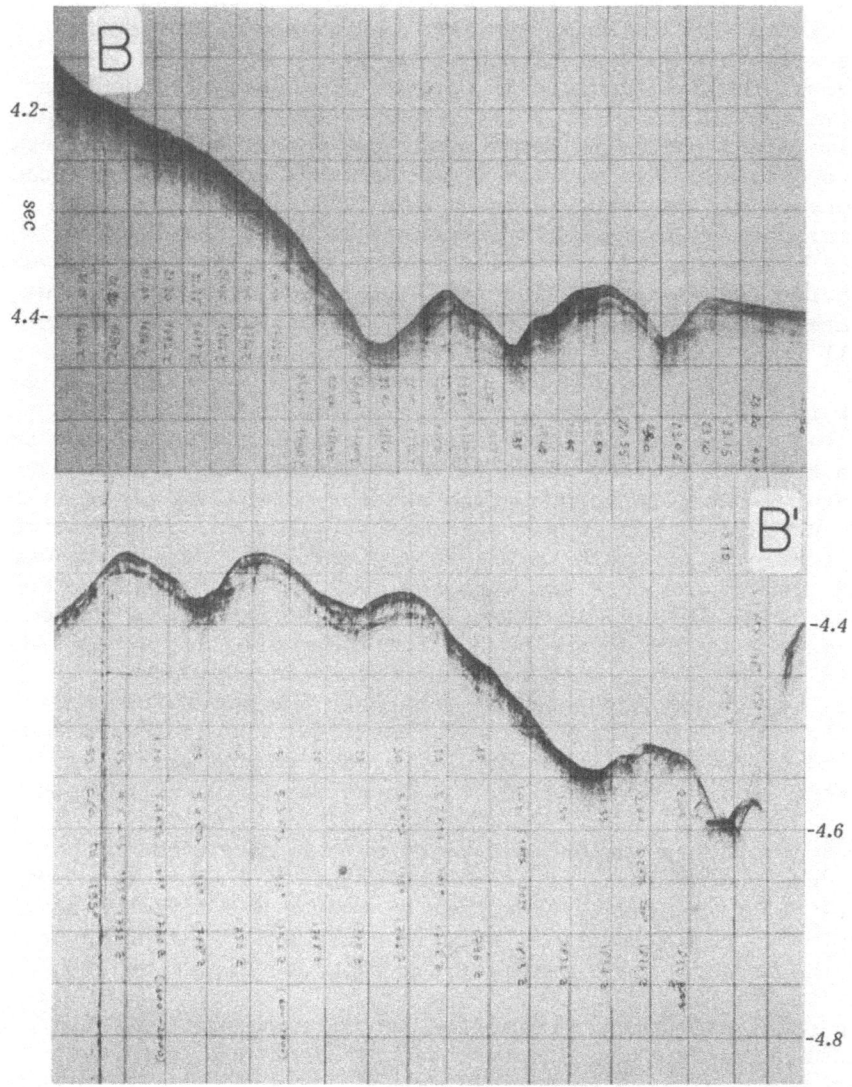

Figure 9 - BANNOCK PDR profile running parallel to the slope at the foot of the Malta Escarpment (see location in Figure 2). Several erosional channels are visible, which document the existence of active erosion across the escarpment.

providing us with this unpublished document.

Discussion with participants to the NATO Workshop on Submarine Slides R.W. Embley, W.B.F.Ryan, R.B.Kidd, G.Almagor, and during a post-cruise meeting on Cores 29 and 38 with M.Gnaccolini and

A.Rizzini were of stimulus to our research.

We benefited of the advise of E. Bellini and I.Premoli Silva for the identification of some microfaunas.

We are indebted to prof. C.Morelli of AGIP-SGEL for X-Ray analysis promptly made available.

Technical assistance was provided by G.Spezzibottiani, A. Camerlenghi and G.Chiodi.

REFERENCES

F.Barbieri, 1964, "Micropaleontologia del Lias e Dogger del Pozzo Ragusa 1 (Sicilia)", Riv.It.Patoleont., 70(4), 709-830, Milano.

D.Chayes, H.Chezar, M.B.Cita, G.Giunta, E.Miller, E.Patacca, R. Radoicic, M.Rawson, M.D.Rawson, S.Rossi, W.B.F.Ryan and P. Scandone, 1979, "Preliminary results of dredgings on the Malta-Siracusa, Apulia and Cephalonia Escarpments ("EASTWARD" Legs E-3E-78 and E-3F-78)", Rend.Soc.Geol.It., 2, 3-4.

M.B.Cita, F.Benelli, B.Bigioggero, H.Chezar, A.Colombo, N.Fantini Sestini, R.Freeman-Lynde, S.Iaccarino, F.Jadoul, E.Legnani, A.Malinverno, P.Massiotta, L.Paggi and I.Premoli Silva, 1980, "Contribution to the geological Exploration of the Malta Escarpment (Eastern Mediterranean)", Riv.It.Paleont., 86(2), 317-356, Milano.

M.B.Cita, B.Bigioggero, A.Bossio, C.Broglia, H.Chezar, G.Clauzon, A.Colombo, M.Giambastiani, L.Lecca, R.Lenaz, A.Malinverno, E.Miller, E.Parisi, A.Rossi, G.Salvatorini, P.Vercesi, 1980, "Preliminary results of a dredging and coring campaign on the Malta Escarpment (BAN 80 Cruise)", Paper presented at the 27th C.I.E.S.M. Congress, Cagliari, 9-18 oct. 1980.

R.W.Embley, 1976, "New evidence for occurrence of debris flow deposits in the deep sea", Geology, 4, 371-374.

R.W.Embley, 1980, "The role of mass transport in the distribution and character of deep-ocean sediments with special reference to the north Atlantic", Marine Geology, 38, 23-50.

R.V.Fisher, 1971, "Features of coarse-grained, hi-concentration fluids and their deposits", Journal of Sedimentary Petrology, 41, 916-927.

G.V.Middleton and M.A.Hampton, 1976, "Subaqueous sediment tran sport and deposition by sediment gravity flows", Marine sediment transport and environmental management, D.J.Stanley and D.J.P.Swift eds., 197-218.

D.G.Moore, J.R.Curray and F.J.Emmel, 1976, "Large submarine slide (olistostrome) associated with Sunda Arc subduction zone, Northeast Indian Ocean", Marine Geology, 21, 211-226.

E.Patacca, P.Scandone, G.Giunta and V.Liguori, 1979, "Mesozoic paleotectonic evolution of the Ragusa zone (Southeastern Sicily)", Geologica Romana, 18, 331-369.

M.Rigo, F.Barbieri, 1959, "Stratigrafia practica applicata in
    Sicilia", Boll. Serv. Geol. Ital., 80, 1-92.
N.A.Rupke, 1978, "Deep clastic seas", H.G.Reading ed., Sedimen-
    tary Environments and Facies, Blackwell Scientific Publica-
    tions, London, 377-379.
P.Scandone, E.Patacca, R.Radoicic, W.B.F.Ryan, M.B.Cita, M.Rawson,
    E.Miller, C.Chezar, S.Rossi and J.McKenzie (in press), "Meso
    zoic and Cenozoic rocks from the Malta Escarpment (Eastern
    Mediterranean). Results of a dredging campaign", A.A.P.G.Bull.,
    Tulsa,

MARINE BIOLOGY

What are the consequences of submarine mass movements on ancient and modern marine life? The first
question is mainly of academic interest, and let us put
it this way:
Mass movements of sediments take all forms, from transport of sediments (fossils) by turbidity currents to
marine slides which displace large volumes of sediments
(fossils) from shallow water environments to deeper water
environments. The problem is then to recognize the sediments (fossils) in the cores from the ocean.

Winterer (1980) has stated that planktonic fossils
are commonly redeposited in their proper settling velocity order in the turbidites. Some turbidites may consist almost entirely of redeposited pelagic materials,
and thus special care should be given to the vertical
sequential properties of all cyclic-looking beds, even
those without much terrigenous material. The redeposited
fossils most commonly are of the same age as the sediments enclosing them, suggesting continual stripping of
the continental slope.

The second question is of vital interest to mankind,
because it concerns the oceanic food-chain. Will the
different populations survive after a marine slide or
after a much weaker displacement?

Almaça analyses in his exposé the possibility that
marine slides and other mass movements may function as
isolating barriers for the marine animal populations and
thus participate in eventual speciation processes by
mentioning for which types of sea animals marine slides
could be efficient as isolating barriers.

He concludes that more research is needed to en-
lighten the problem whether marine slides and other mass
movements may participate in allopatric speciation processes in the sea.

Barbara Hecker has divided her contribution into two
parts, the first dealing with the contribution of biological activity to slope instability, and the second with
regard to the biological consequences of submarine slope
failure.

The first section where biologically induced erosion
is discussed is illustrated by a flow diagram of the
activities by which benthic organisms destabilize substrata and lead to slope instability. The second section,
where the effect of submarine slides on benthic fauna
is dealt with, is highlighted by another flow diagram
of the factors that may affect the biological recovery
of areas that have been disturbed by marine slides.

Ref.: Winterer, E.L., 1980, Sedimentary facies on the
      rises and slopes of passive continental margins,
      Phil. Trans. R. Soc. London, A 294: 169-176.

MARINE SLIDES AND ALLOPATRIC SPECIATION

Carlos Almaça

Laboratório de Zoologia
Centro de Fauna Portuguesa (INIC), Faculdade de Ciências
Rua da Escola Politécnica - 1294 Lisboa Codex

INTRODUCTION

Speciation, i.e. the formation of new species, deals with factors that are intrinsic and extrinsic to the populations and which cause the interruption of the gene flow amongst them or amongst groups of individuals of the same population.

Such interruption may lead to reproductive isolation amongst fractions which have been submitted to it, and then new species will be formed. It is thought to be possible to establish reproductive isolation from one generation to the next when certain chromosomal rearrangements take place. In this case, it is a quesion of an instantaneous speciation. But, in the majority of cases, the acquirement of reproductive isolation will be a gradual process arising as a consequence of the interruption of the gene flow during several generations. Meanwhile, during that period, the fractions of the population would have diverged genetically and acquired isolation mechanisms, i.e., intrinsic characteristics which prevent hybridization amongst them.

The number of generations in which the interruption of the gene flow is necessary in order to obtain reproductive isolation seems to depend a lot on the genetic system and on the reproductive strategy of each population. Therefore, there are populations which have acquired reproductive isolation in a comparatively short time, while others need many millions of years for that to take place. In fact,

the speciation period depends on many variables, amongst which and
besides the ones already mentioned, the mutation rates, selection
pressures, average duration of the generations, etc., are important.
Moreover, many speciation processes are conditioned by phenomena
which are extrinsic to the organisms, as, for example, the establish-
ment of spatial barriers, which act as obstacles to the free exchange
of genes amongst populations or fractions of populations.

In this work we will endeavour to analyze the possibility that
marine slides and other mass movements may function as isolating
barriers for the marine animal populations and thus participate in
eventual speciation processes.

SPECIATION IN THE SEA

Whatever may be the speciation process followed by a fraction
of a population, the beginning of the process is always characterized
by a reduction of the genetic variability. The extent of this reduction will depend, amongst other factors, on the dimension of the unit
involved in the process. Also the compatibility of this reduction
with the future viability of the speciating fraction will depend on
the genetic system and on the reproductive strategy of this fraction.
These facts led me to consider (Almaca, 1979) that the speciation of
aquatic animals is a process characterized, in sequence, by:

1. Amount of genetic variability lost by the fraction which
   starts the speciation in relation to the variability of the
   original population.
2. Adaptability of the genetic system and of the reproductive
   strategy of the speciating fration to 1.
3. Nature of the spatial barriers which will eventually isolate
   this fraction.

In fact, amongst animals, there are genetic systems which are
able to bear great reductions of genetic variability, while other
quickly become extinct when limited to a small number of individuals
(Ford, 1971) and, thus, when the variability of the whole is small.
This means that there are populations adjusted to endogamy, while
others are typically exogamous. Obviously, the survival of the latter
will be greatly jeopardized if their whole is divided in small isolated groups.

For sometime, due to the large divulgation made by Mayr (1942,
1965, 1970), the previous spatial isolation of populations by physical barriers that, for several generations, prevented the gene flow
amongst such populations, was considered as a necessary condition

for the success of speciation. This conception was based, mainly, on speciation studies of large Birds and Mammals, i.e. in K-selected species. As the speciation studies were extended to other animal groups, particularly to r-selected species, it was found out that spatial isolation is not a necessary condition for the success of all the speciation phenomena. In fact, Bush (1975) was already considering three essencial ways of speciation: allopatric (with spatial isolation), parapatric (in spatial contiguity) and sympatric (separation of one or more subpopulations within one population). Other modern authors, for example White (1978)and Almaça (1979), maintain and describe in detail these three forms of speciation.

The speciation of sea animals has been little studied, particularly that of species living at great depths, because of obvious difficulties. Although admitting that little is known about this subject, Mayr (1970) is, however, of the opinion that, also in the oceans, speciation is allopatric and relatively slow. In a Symposium about "Speciation in the sea" (Harding and Tebble, ed., 1963), the various authors concede, in general, great importance to allopatric speciation, without denying, however, the possibility that other forms of speciation may occur. Day (1963), referring to the benthic communities, admits the possibility of existence of a correlation between the speciation rate and the complexity of the biotic environment. This means that the greater the diversity in species, the more opportunities exist for speciation, or, in terms of biotic environment, the greater the environmental heterogeneity, the greater the number of new species that will be formed. In this hypothesis, suggested particularly by the diversity of the carnivores of the benthic communities, a progressive specialization of the trophic niche is implicit, being itself dependent on the new entities which speciation is adding to the community. However, the hypothesis of a progressive specialization implies, in itself, the existence of populations with a considerable genetic variability. This was precisely demonstrated by Gooch and Schopf (1972) in several species caught between the depths of 1,033 and 2,080 metres. The genetic variability found in these species living in the depths is of the same order as that of the terrestrial species and of the species living in fresh water, in spite of the fact that the sea depths constitute one of the most stable and homogeneous habitats. Amongst the hypothesis suggested by Gooch and Schopf (1972) to explain the high genetic variability not correlated with the environmental variability, are the following ones: (1) the neutrality of alleles, (2) physiological heterosis and (3) diversifying selection operating over a pressure gradient. I believe that another explanatory factor, and by no means less important,will be precisely, and in accordance with what has been elaborated above, the diversifying selection due to the biotic environmental heterogeneity of such depths.

MARINE SLIDES AS SPATIAL BARRIERS

A great majority of the marine species are r-selected, reason why the action of the barriers of spatial isolation will not always be a necessary condition for speciation (Almaça, 1979). However, it is natural that in many cases, if not in all of them, the speciation processes be accelerated by the existence of spatial barriers. Therefore, to this extent, the intervention of the marine slides and other mass movements might play an important role in sea speciation, besides the various processes of allopatric speciation in relation to which such intervention will be essential.

In the allopatric speciation, or speciation preceded by spatial isolation, several ways are considered at present (Bush, 1975; White, 1978; Almaça, 1979):

(1) Strict allopatry without a narrow population bottleneck.
(2) Founder principle.
    a) Marginal colonisation.
    b) Bottleneck effect.
    c) Habitat fragmentation.
(3) Extinction of intermediate populations in a chain of races.

While the types of speciation by founder principle only seem possible in r-selected populations, the ones described in (1) and (3) are applicable to both r-selected and K-selected populations (Almaça, 1979, 1980). Besides, any one of these ways of speciation seems to have possibilities of occurring in the sea and the effect eventually exerted on them by the marine slides and other mass movements as physical isolating barriers will probably be very important.

FOR WHICH TYPES OF SEA ANIMALS COULD THE MARINE SLIDES
BE EFFICIENT AS ISOLATING BARRIERS?

As regards the efficiency of the marine slides as obstacles to the gene flow amongst populations, it is necessary to distinguish, in general, two groups of sea animals: (1) the pelagic and the benthic with pelagic phases; and (2) the benthic with benthic larvae or direct development. For (1) a barrier of this type could only by chance secure spatial isolation for a period long enough for the speciation to be successfull. With the exception perhaps of some fishes and other bathypelagic animals without superfical larvae and with low vagility, which seem to be easy to isolate in deep basins (Marshall, 1963).

However, in relation to (2) the problem is different. Populations of benthic animals, even with external fecundation (if the viability of the eggs and/or spermatozoa is not excessively long) and without pelagic larvae, can be isolated for a long time by marine slides or similar incidents. In these circumstances, it is very likely that the allopatric speciation of these populations will take place successfully.

Without intending to be exhaustive, I shall mention some examples of benthic sea animals which, due to the nature of their biological cycle may follow processes of allopatric speciation by intervention of marine slides or other mass movements. The selection of these examples is based on Giese and Pearse (1974-1979).

In the vast majority of Porifera the fertilization is internal (Fell, 1974). Most species are viviparous and have free-swimming larvae which swim for 3-4 hours to 1-2 days. However, some species, like *Halichondria moorei* and *Polymastia robusta*, have benthic larvae that creep along the substratum for 20-60 hours and for 18-20 days, respectively (Fell, 1974).

Fertilization is frequently internal in the Anthozoa (in the gastrovascular cavity) and ovoviparous species have larvae (planulae) which settle before metamorphosis (Campbell, 1974).

Fertilization of most Turbellaria occurs internally, and the eggs are then encased in capsules wich are embedded in a jelly mass (as for most Acoela and Polycladida), or in a shelled cocoon (Henley, 1974). In a few cases, like *Paravortex* and *Bresslauilla*, the ova develops within the parental body (Henley, 1974). Most Turbellaria have direct development, the juveniles hatching as miniature adults; only some Polycladida have free-swimming larvae (Henley, 1974).

In Gnathostomulida the mating has never been observed (Sterrer, 1974). Three types of copulation are, therefore, postulated: (1) sperm penetration into the partner's body, (2) hypodermic impregnation, and (3) injection into the vagina. Oviposition has been observed only in *Gnathostomula jenneri*: the zygote becomes spherical and sticks to the substratum. Development is direct, as in most representatives of interstitial fauna (Sterrer, 1974).

In the Nemertinea internal fertilization occurs in viviparous species such as *Prosorhochmus claparedei* and *Prostoma obscurum* and possibly in some pelagic species (Riser, 1974). Some species, like *Lineus ruber*, have nonpelagic larvae (Riser, 1974).

Reproduction in marine Nematoda is not well known and Hope (1974), who reviewed the problem, has largely based himself on the reproduction of soil and parasitic Nematoda. Fertilization in Nematoda is internal and there are oviparous and ovoviviparous species. Development is direct (Hope, 1974).

Data on the reproduction of marine Rotifera is also very restricted and much of the information given by Thane (1974) is based on freshwater. As a matter of fact the largest number of Rotifera species live in freshwater. Fertilization is internal and the eggs are dropped on the substratum or are attached to it. Some forms are viviparous. Development is direct. In general Rotifera have a short life expectancy and high reproductive potential (Thane, 1974).

Nearly one-half of the 390 species of Gastrotricha are marine, living in benthic and epiphytic meiofaunal communities where they often form one of the numerically abundant components (Hummon, 1974). Fertilization is internal and eggs are typically released to the exterior and are attached to a sand grain or other piece of substratum. Development is direct. Life expectancy is short (5.1 days in an experimental culture) (Hummon, 1974).

Mating behaviour in Kinorhyncha is not known and it is assumed that copulation takes place (Higgins, 1974). The eggs seem to be attached to the females or laid in channels excavated by the animals in the sediments. Development is direct but there are at least six juvenile stages, each derived from a molt (Higgins, 1974).

Only nine species of Priapulida are known (Van der Land, 1975). All of them are marine and benthic. Fertilization is external. The larvae are benthic and feed on detritus. The development of *Priapulus* may take as long as 2 years (Van der Land, 1975).

In many species of Polychaeta swarming accompanies spawning; other species have true copulation (Schroeder and Hermans, 1975). Polychaeta which do not spawn freely in the sea protect their young by brooding (more than 70 species), viviparity (about 19 species) and adelphophagia (about 8 species). The larvae of Polychaeta may develop directly, without differentiation of larval structures adapted to pelagic habitat and equipped to survive immediately in the adult habitat. Interstitial forms and species with large eggs seem to have direct development (Schroeder and Hermans, 1975).

In marine Oligochaeta no specific form of courtship seems relevant to the reproductive processes (Lasserre, 1975). However, the gametes are not dispersed at random into the surrounding water. The whole life cycle is benthic (Lasserre, 1975).

Sipuncula are marine benthic worms which include 320 known species (Rice, 1975). Fertilization is external and most species have pelagic larvae. However, some species, like *Golfingia minuta*, *Themiste pyroides* and *Phascolion cryptus*, have a direct development (Rice, 1975).

Fertilization is internal in Echiura Bonelliidae and the males are permanent residents in the uterus (Gould-Somero, 1975). The larvae are not swimmers, but rather creep about on the bottom (Gould-Somero, 1975).

Marine Tardigrada are limited to the order Heterotardigrada and mating is not known for this order (Pollock, 1975). However in the known cases, mating requires the proximity of males and females and, in that occasion, the animal form dense clusters. Development is direct. It is suggested a life span of 3-4 months and 1-2 months to reach reproductive maturity (Pollock, 1975).

A relativery large number of Polyphacophora brood their embryos in the pallial groove until the larvae or juveniles are released (Pearse, 1979). These species are found among about half of the families of Polyphacophora and nearly all of them are found in cold temperate or subpolar seas (Pearse, 1979).

Direct development occurs in some species of Pelecypoda, as *Liocyma fluctuosa*, *Nucula delphinodonta*, *Musculus discors*, *M.niger* some species of *Astarte*, *Loripes lacteus*, *Transenella tantilla*, *Cardita ventricosa*, *C. barbarensis*, *Lasaea rubra*, *Gemma gemma* and *Cerastoderma elegantulum* (Sastry, 1979). These species produce large eggs. Direct development is frequently accompanied by brood protection (Sastry, 1979).

Neritidae Archaeogastropoda, Mesogastropoda and Neogastropoda have, in general, internal fertilization, and deposit eggs either in gelatinous masses or in egg capsules (Webber, 1977). Some species, as *Littorina littoralis*, *L. saxatilis*, *Thais hippocastanea* and some naticids, undergo direct development. It appears that, at least in part, the occurence of direct development is related to latitude: *Planaxis sulcatus* in Iranian Gulf has direct development, while in European seas releases veliger larvae. Also, closely related species may show variable patterns: *Lacuna vincta* releases planktotrophic veligers, while *L. pallidula* has direct development; *Littorina angulifera* and *L. saxatilis* are both oviviparous, but while the first have planktotrophic veligers, the second has direct development

(Webber, 1977).

The caetognath genus *Spadella* has an entirely benthic life
cycle (Reeve and Cosper, 1975). *Spadella* larvae do not swim about,
but immediately attach themselves to the substrate where the eggs
were laid (Reeve and Cosper, 1975).

With the possible exception of *Xenopleura vivipara* fertilization
is external in the Enteropneusta (Hadfield, 1975). Forms as *Saccoglossus* which have large eggs present, typically, direct development.
Some species of *Saccoglossus*, as *S. pusillus* and *S. horsti*, have a
brief swimming period, while others, as *S. otagoensis* and probably
S. kowalevskii, take immediately after hatching a benthic existence
(Hadfield, 1975).

Ascidiacea are, like all the Tunicates, hermaphroditic organisms.
Colonial ascidians are, with one exception, viviparous (Berrill, 1975);
species of the order Enterogona have large and often very yolky eggs
whose development gives rise to larvae with a much abbreviated free-
-swimming period (from a few minutes to 2-3 hours) (Berrill, 1975).

DISCUSSION

Although most of the benthic animals present external fecundation and pelagic larval phases, which enable them to counteract the
eventual isolating action caused by the marine slides and other mass
movements, a significant number of species carry the whole of their
biological cycle at the bottom of the oceans or in the vicinity. I
have mentioned above some examples, chosen amongst the invertebrate
Metazoa, of species under these conditions. Such species may, very
plausibly, follow a process of allopatric speciation by means of
the intervention of marine slides and other mass movements.

Any speciation process, how ever fast it may be, does not complete itself in the time scale which the studies of man allow. On
the other hand, and as far as I know, the particular case of the marine slides as isolating barriers has never been specially considered.
These facts imply that this work is not more than a speculative one,
aiming at discussing, from a theoretical point of view, the possibilities that the marine slides and other mass movements may participate in allopatric speciation processes in the sea.

REFERENCES

Almaça, C., 1979, Apport de la Dynamique des populations aquatiques
    à la Génétique évolutive, in: "Dynamique des populations et
    qualité de l'eau," H. Hoestlandt, ed., Gauthier-Villars,
    Paris (in press).

 -"-   1980, Formas de especiação nos animais, Arquipélago,
    Ponta Delgada (in press).

Berrill, N. J., 1975, Chordata: Tunicata, in: "Reproduction of ma-
    rine Invertebrates," A.C. Giese and J.S. Pearse, eds., vol.
    2, Academic Press, New York.

Bush, G. L., 1975, Modes of animal speciation, Ann. Rev. Ecol. Syst.,
    6: 339.

Campbell, R. D., 1974, Cnidaria, in: "Reproduction of marine Inverte-
    brates," A.C. Giese and J. S. Pearse, eds., vol. 1, Academic
    Press, New York.

Day, J. H., 1963, The complexity of the biotic environment, in:
    "Speciation in the sea," J. P. Harding and N. Tebble, eds.,
    The Systematics Association, Publ. Nº 5, London.

Fell, P. E., 1974, Porifera, in: "Reproduction of marine Invertebra-
    tes," A. C. Giese and J. S. Pearse, eds., vol. 1, Academic
    Press, New York.

Ford, E. B, 1971, "Ecological Genetics",Chapman and Hall, London.

Gooch, J. L., and Schopf, T. J. M., 1972, Genetic variability in the
    deep sea: relation to environmental variability, Evolution,
    26: 545.

Gould-Somero, M., 1975, Echiura, in: "Reproduction of marine Inverte-
    brates," A. C. Giese and J. S. Pearse, eds., vol. 3, Academic
    Press, New York.

Hadfield, M. G., 1975, Hemichordata,in: "Reproduction of marine Inver-
    tebrates," A. C. Giese and J. S. Pearse, eds., vol. 2, Acade-
    mic Press, New York.

Henley, C., 1974, Platyhelminthes (Turbellaria), in: "Reproduction
    of marine Invertebrates," A. C. Giese and J. S. Pearse, vol.
    1, Academic Press, New York.

Higgins, R. P., 1974, Kynorhyncha, in: "Reproduction of marine Inver-
    tebrates," A. C. Giese and J. S. Pearse, eds., vol. 1, Aca-
    demic Press, New York.

Hope, W. D., 1974, Nematoda, in:"Reproduction of marine Invertebra-
    tes,"A.C. Giese and J. S. P. Pearse, eds., vol. 1, Academic
    Press, New York.

Hummon, W. D, 1974, Gastrotricha, in: "Reproduction of marine Inver-
    tebrates", A. C. Giese and J. S. Pearse, eds., vol. 1, Aca-
    demic Press, New York.

Land, J. van der, 1975, Priapulida, in: "Reproduction of marine Inver-
    tebrates," A. C. Giese and J. S. Pearse, eds., vol. 2, Aca-

demic Press, New York.

Lasserre, P., 1975, Clitellata, in: "Reproduction of marine Inverte-
    brates," A. C. Giese and J. S. Pearse, eds., vol. 3, Academic
    Press, New York.

Marshall, N. B., 1963, Diversity, distribution and speciation of
    deepsea fishes, in: "Speciation in the sea," J. P. Harding
    and N. Tebble, eds., The Systematics Association, Publ. No.
    5, London.

Mayr, E., 1942, "Systematics and the origin of species," Columbia
    Univ. Press, New York.

-"-   1963, "Animal species and evolution," Belknap Press, Cam-
    bridge, Mass.

-"-   1970, "Populations, species, and evolution," Belknap Press,
    Cambridge, Mass.

Pearse, J. S., 1979, Polyplacophora. in: "Reproduction of marine
    Invertebrates," A. C. Giese and J. S. Pearse, eds., vol. 5,
    Academic Press, New York.

Pollock, L. W., 1975, Tardigrada, in: "Reproduction of marine Inverte-
    brates," A. C. Giese and J. S. Pearse, eds., vol. 2, Academic
    Press, New York.

Reeve, M. R., and Cosper, T. C., 1975, Chaetognatha, in: "Reproduc-
    tion of marine Invertebrates", A. C. Giese and J. S. Pearse,
    eds., vol. 2, Academic Press, New York.

Rice, M. E., 1975, Sipuncula, in: "Reproduction of Marine Inverte-
    brates," A. C. Giese and J. S. Pearse, eds., vol. 2, Academic
    Press, New York.

Riser, N. W., 1974, Nemertinea, in: "Reproduction of marine Inverte-
    brates", A. C. Giese and J. S Pearse, eds., vol. 1, Academic
    Press, New York.

Sastry, An. N., 1979, Pelecypoda (excluding Ostreidae), in: "Repro-
    duction of marine Invertebrates," A. C. Giese and J. S.
    Pearse, eds., vol. 5, Academic Press, New York.

Schroeder, P.C., and Hermans, C. O., 1975, Annelida: Polychaeta, in:
    "Reproduction of marine Invertebrates", A. C. Giese and
    J. S. Pearse, eds., vol. 3, Academic Press, New York.

Sterrer. W., 1974, Gnathostomulida, in: "Reproduction of marine
    Invertebrates," A. C. Giese and J. S. Pearse, eds., vol. 1,
    Academic Press, New York.

Thane, A., 1974, Rotifera, in: "Reproduction of marine Invertebrates",
    A. C.Giese and J. S. Pearse, eds., vol. 1, Academic Press,
    New York.

Webber, H. H., 1977, Gastropoda: Prosobranchia, in: "Reproduction
    of marine Invertebrates," A. C. Giese and J. S. Pearse, eds.,
    vol. 4, Academic Press, New York.

White, M. J. D., 1978, "Modes of speciation", W. H. Freeman, San
    Francisco.

POSSIBLE BENTHIC FAUNA AND SLOPE INSTABILITY RELATIONSHIPS

Barbara Hecker

Lamont-Doherty Geological Observatory
of Columbia University
Palisades, New York 10964

ABSTRACT

Two major types of interactions between benthic fauna and
submarine slides are addressed in this paper: (1) the contribution
of biological activity to slope instability and (2) the biological consequences of submarine slope failure. Through the processes of bioturbation and bioerosion, benthic fauna can substantially alter the physical properties of their substrata.
Activities that lead to structural weakening of both sediment and
outcrop are: consistent reworking of sediment preventing consolidation; excavation of sediment and semi-consolidated clay resulting
in the net transport of material downslope; boring of outcrops
causing decrease of rock mass; and attachment to outcrops resulting in increased drag force and gravitational pull on their
surfaces. Along the continental margins located off both the
east and west coasts of the U.S., the activity of benthic fauna
appears to play a major role in contemporary submarine erosion.

In considering the effect of mass sediment movement on benthic
communities, the issue is slightly more complex. While submarine
slope failure buries existing fauna, it also opens new habitats
for recolonization. Factors that affect the process of recolonization would vary depending on the nature and extent of the disturbance and the habitat in which it occurs. On the basis of
existing knowledge concerning the rate of deep-sea biological processes, it appears that recolonization of an area disturbed by
submarine slides would be slow. This process, however, might be
hastened by slides enriching the nutrients of deep-water habitats
along continental margins. Additionally, such nutrient enrichment
could lead to high faunal abundances in areas subjected to episodic

slope failure. Such areas would also be expected to support a fauna consisting of more mobile species than areas that are tranquil since selection would favor mobility. In-depth studies of the biological recovery of areas disturbed by submarine slides could answer many questions concerning the rates and life habits of deep-sea benthic organisms.

INTRODUCTION

The relationship between benthic fauna and submarine slides encompasses two major types of interaction: (1) the role of organisms in contributing to slope instability and (2) the role of slope failure in structuring benthic communities. The activity of fauna can substantially alter the physical properties of both soft and hard substrata. Various activities such as burrowing, feeding and movement across the surface all tend to have a destabilizing effect on sediment by preventing consolidation. Conversely, tube construction by various taxa, results in consolidation of the surface layer of sediments. However, the types of activities that weaken sediment stability and the frequency of their occurrence are believed to far outnumber the activities that strengthen it (Rowe, 1974). Organisms that bore into rock outcrops and burrow into consolidated clay degrade the structural integrity of submarine canyon walls. The cumulative action of organisms on both soft and hard substrata may substantially contribute to submarine slope instability, thus making them more susceptible to mechanisms that trigger slope failure.

Catastrophic burial of benthic communities by slides is only one aspect of how fauna is affected by mass movements of sediment. As a result of annihilating the organisms along its path, a submarine slide also provides new habitats for colonization. In the deep-sea this leads to spatial heterogeneity in an otherwise rather homogeneous environment, thereby providing suitable habitats for a variety of species. Submarine slides have also been proposed as a mechanism for the transport of organically rich terrigenous sediment to the deep water habitats along continental margins (Griggs et al., 1969). This nutrient enrichment is reflected in higher faunal abundances in areas characterized by episodic sediment failure. Chronic slope instability would also have an effect on the types of organisms inhabiting such areas by selecting for different life habits than would be found in more benign habitats.

Few investigators have addressed the question of how mass sediment movements affect modern fauna. Slightly more information is available on how organisms destabilize substrata. Two of the major problems associated with answering these questions are: (1) the relative inaccessibility of deep water habitats, and (2) the long time-frames involved in reestablishment of deep-sea communities. In this paper I address the interactions mentioned above by: reviewing relevant literature, discussing ongoing investigations, presenting observations made during submersible dives, and suggesting possible areas for future study.

BIOLOGICALLY INDUCED EROSION

The fact that biological activity plays a major role in submarine erosion along continental margins has been recognized by numerous investigators (Dillon and Zimmerman, 1970; Stanley, 1971; Warme et al., 1971; Rowe et al., 1974; Cacchione et al, 1978; Ryan et al., 1978, Warme et al., 1978; Valentine et al., 1980; Malahoff et al., 1981). Figure 1 lists some of the mechanisms by which organisms destabilize substrata and possibly influence slope failure. The reworking and modification of sediment by biological activity is known as bioturbation. In contrast,the breakdown of rock outcrops by marine.borers is termed bioerosion Neumann (1966). Both processes lead to considerable structural modification of the seafloor that may adversely affect slope stability.

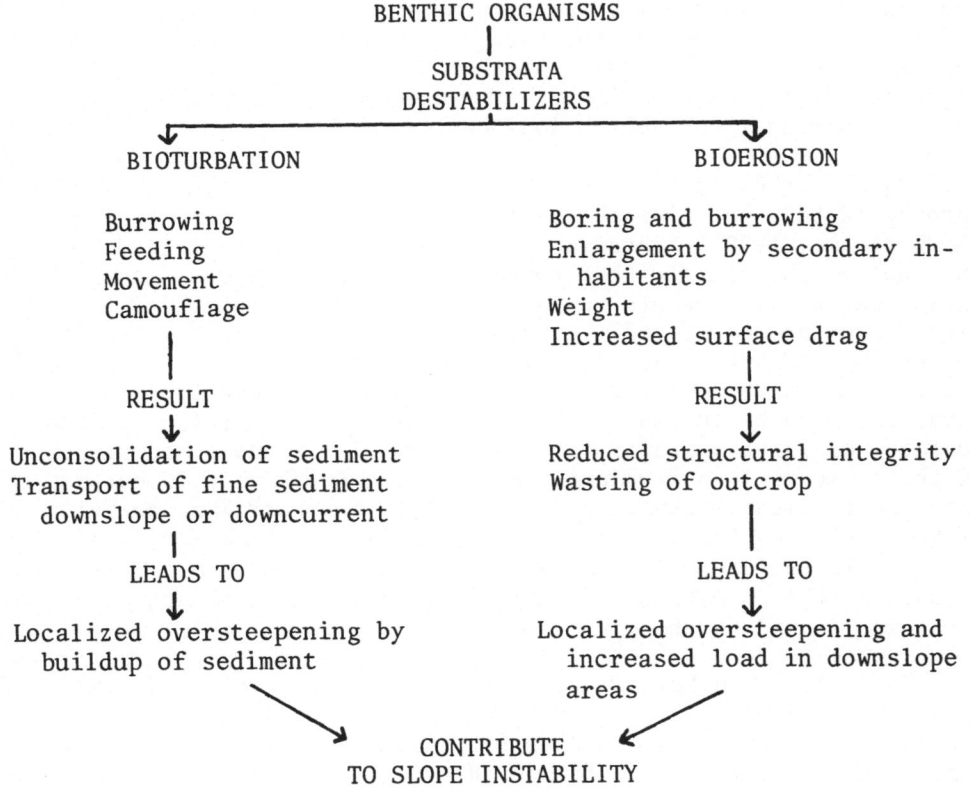

Figure 1. Activities by which benthic organisms destabilize substrata and lead to slope instability.

Bioturbation

The continual mixing and turnover of marine sediments results from a variety of biological activities. Benthic species of many taxa and sizes burrow in sediment to obtain food or escape predation. Visual evidence of these activities is readily discernible on the sediment surface of most areas that have been investigated, ranging from beaches to abyssal plains. The burrowing activity of infaunal species is usually responsible for the microtopography of sediment surfaces, while the burrowing of epifaunal species frequently results in large excavations and depressions (Heezen and Hollister, 1971; Rowe, 1974; Ryan et al., 1978; Warme et al., 1978; Malahoff et al., 1981). In addition to the physical mixing of sediment, Rhoads (1974) found that deposit-feeding organisms tended to decrease the physical stability of their substratum by forming pellets of the material they ingested.

Biological activities restricted to the surface of the sediment also play a part in this continual turnover. The simple movement of large epifaunal species frequently results in tracks and trails that may be several centimeters deep. Over time these traces are broken down by the burrowing activity of infaunal organisms (Paul et al., 1978). Another means by which epibenthic organisms disturb the sediment is by using it to camouflage themselves, causing local depressions on its surface.

Many of these activities result not only in keeping sediments unconsolidated, but also act as a downslope transport mechanism (Ryan et al., 1978). Observations made during submersible dives, and photographs taken with a towed camera sled, indicate that biological activities are responsible for redistribution of a significant amount of sediment along the continental margin of the eastern U.S. coast. Material excavated during burrow construction settles in mounds downslope of the openings. In some areas excavations made by the red crab *Geryon quinquidens* are the dominant features of large parts of the sea floor. The tunnels and arches of these excavations frequently extend to a depth of 0.75 meters into the sediment(Malahoff et al., 1981). The burrowing activity of both epifauna and infauna is responsible for the majority of small-scale features seen in many of the areas surveyed. This attests to the importance of biological activity in structuring localized seabed morphology.

In addition to burrowing, various other activities of many epibenthic organisms also result in resuspension and downslope transport of sediments. Feeding rat-tails and eels stir up sediment in their search for infauna. Rays, skates, flounders, and goosefish lie on top of the sediment or partially bury themselves to escape detection by possible prey and predators. Crabs scurry along the surface in their search for food or to escape from

predators. In their wake all of these animals leave clouds of
suspended sediment that is carried for a considerable distance
before it returns to the seabed. Over time this net transport of
sediment downslope can lead to localized oversteepening, making a
slope more susceptible to failure.

Bioerosion

The breakdown of hard substrata is another means by which
marine organisms contribute to submarine erosion. Biological
activity has been recognized as the primary cause of contemporary
erosion of submarine canyon walls (Warme et al., 1978; Valentine
et al., 1980). The surfaces of rock outcrops found in submarine
canyons, off both the east and west U.S. coasts, are frequently
riddled with holes excavated by a variety of different taxa (Warme
et al., 1971; Warme et al., 1978). Penetration into these outcrops
at times exceeds 30 cm. The passageways are also frequently en-
larged and modified by secondary inhabitants. Exposures of semi-
consolidated clay are frequently populated by a diverse community
of epibenthic burrowers (Warme et al., 1978; Valentine et al.,
1980). Crustaceans and fish excavate holes penetrating from 6 cm
to over 2 m into the substratum. Examples of small slumps resulting from these activities are presented in Malahoff et al.
(1981). The boring and burrowing activities of benthic organisms
into hard substrata play a major role in reducing the structural
integrity of outcrops. The result is substantial wasting of outcrop surfaces, as is evidenced by numerous talus deposits seen on
the walls and in the axis of submarine canyons.

However, penetration of outcrops is not the only mechanism by
which benthic organisms erode hard substrata. The attachment of
sessile taxa, such as anemones, sponges and corals, could enhance
the erosive action of currents by increasing the drag force on
outcrop surfaces. It is also conceivable that the weight of some
attached species may exert enough gravitational pull to break off
parts of outcrops that have already been weakened by boring organisms. Observations made this summer during dives with the
research submersible ALVIN indicate that this type of erosion may
be quite common. Coral skeletons are frequently found at the base
of outcrops with large pieces of rock still attached to them. A
sandstone outcrop at 400 m depth in the axis of Lydonia Canyon
showed evidence that this type of weathering may affect large areas.
Approximately 50% of the outcrop's surface was covered with large
coral colonies and sponges. A portion of the surface of this outcrop appeared to have broken off leaving an indented area that was
several meters in diameter. This indented area was colonized by
small individuals of the same coral species.

The cumulative action by boring and burrowing organisms, as well as increased weight and drag due to the higher surface area of attached forms, all serve to weaken outcrops. These processes result in substantial wasting of the surface layers of outcrops, which leads to significant accumulations of talus debris in the axes of canyons and on their walls. In some areas, particularly if the slope is steep, this buildup of talus debris may contribute to localized oversteepening and, in turn, lead to slope failure.

Biological versus Physical Erosion

While few investigators who have made visual observations of the sea floor along continental margins would argue that the activity of benthic organisms does not contribute to contemporary erosion, the question of relative importance still remains. To what extent is contemporary erosion biologically induced or mediated? Gaining a better understanding of the role of organisms in the erosion of submarine substrata requires information concerning the rates at which these processes take place. The action of organisms on their substratum frequently acts as a catalyst for erosion by physical forces, thus a separation of biological from physical factors leading to slope instability will be difficult to achieve. Additionally, in assessing the role of benthic fauna in causing some instances of slope instability, it should be noted that the net accumulation of biologically induced changes may merely set the stage for a mechanism that triggers slope failure.

EFFECT OF SUBMARINE SLIDES ON BENTHIC FAUNA

Mass movements of sediment can influence benthic fauna in a variety of ways. Experimental evidence shows that most organisms do not survive sudden burial by a substantial amount of sediment (Nichols et al., 1978). In the process of destroying existing communities, slope failure also creates new habitats for colonization. This habitat renewal results in a patchy environment more suitable to some species. Additionally, submarine slides have been proposed as one of the mechanisms by which nutrient enriched terrigenous sediment is transported deep water habitats. Recovery of a benthic community from disturbance is governed by a variety of factors such as: extent and frequency of the disturbance, composition of the new sediment cover, food supply, and biological rates which are related to depth. Figure 2 outlines some of the factors that would determine how a community recovers from catastrophic slope failure.

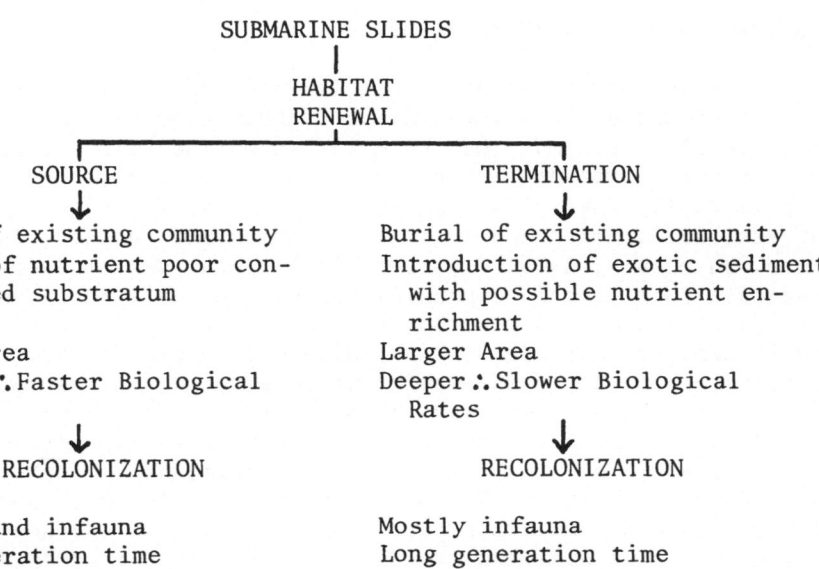

Figure 2.   Factors that may affect the biological recovery of
            areas that have been disturbed by submarine slides.

Destruction of benthic communities

     Submarine slides probably destroy most benthic organisms along
their path, as well as at their termination. As a slope fails,
loose material moves downslope and carries with it the surficial
sediments. Since most infaunal organisms inhabit the upper 5 cm
of the sediment (Jumars, 1978), the benthic community in a slide
path would also be displaced. The chances of fragile benthic
fauna surviving such abrasive action are quite small. Also,
environmental differences between source and depositional sites of
a submarine slide make it doubtful that any organisms surviving
transport could persist.

     Most experimental evidence dealing with the survival of benthic
communities subjected to catastrophic burial comes from studies on
shallow water invertebrates. Some taxa are able to migrate to the
sediment-water interface following sudden burial. Factors that
affect an organism's ability to escape after burial are: life
habits, such as feeding mode and mobility; the composition of the
deposited material; and the thickness of the deposited sediment.
Sessile filter-feeders are the most susceptible to destruction by
catastrophic sediment deposition (Kranz, 1974). Organisms buried
by exotic sediments are less likely to escape than those buried by
sediments of the type they usually inhabit (Maurer et al., 1978).
Nichols et al. (1978) found that mortality increased with increasing thickness of sediment cover and that no organisms

survived burial by 30 cm or more.

However, these studies have limited applicability to the effect of submarine slides on the benthic community at their depositional site. Many of the major submarine slides terminate in deeper water habitats where biological rates have been shown to be substantially lower than shallower water rates. (Jannasch and Wirsen, 1973; Smith and Teal, 1973; Smith and Hessler, 1974; Turekian et al., 1975; Grassle, 1977). Analyses of deep-sea cores indicate that benthic communities have frequently been annihilated by sudden sediment deposition in the geologic past (Heezen et al., 1955; Brongersma-Sanders, 1957; Ryan and Heezen, 1976). The destructive force of mass sediment movements is illustrated by sharp contacts between clay and overlying coarser deposits, indicating a sudden halt in biological activity (Griggs et al., 1969). Jumars (1977) reported the accidental burial of a benthic community at 1200 m depth in the San Diego Trough by 2-10 cm of sediment. After 24 hours, the distribution of organisms in the sediment of the buried area differed from that of the control areas and also differed with increased depth of burial. A decrease in the number of organisms reaching the surface was noted with increasing depth of deposition. While no instances of mortality were observed in the control samples, decaying individuals were found in the buried samples. Most of these decaying organisms originally inhibited the sediment-water interface. From this evidence it would appear that deep-water benthic assemblages, exposed to the sudden influx of large amounts of sediments, can not escape annihilation.

Recolonization

In the process of destroying existing benthic communities, a submarine slide also creates new habitats for exploitation. Repopulation of disturbed habitats depends on a variety of factors such as the time-scale of the disturbance, the size of the area involved, and the nature of the exposed sediment. Most of our information concerning recolonization of disturbed soft-bottom marine habitats comes from studies conducted in shallow water. A classic concept of successional stages in the recovery of a disturbed area is that "opportunistic" species invade the new habitat and are eventually replaced by "equilibrium" species. Opportunistic species are characterized by rapid growth, short generation time, large brood sizes and poor competitive ability. They are usually sessile, occupy a low trophic level and expend a high proportion of their resources on reproduction. Equilibrium species are characterized by slow growth, long generation time, small brood sizes and higher competitive ability. Opportunists can rapidly exploit a new habitat, but are eventually excluded due to their short life span and poor competitive ability. However, biological systems are rarely as simple as theory would predict and a variety of different sequences and interactions have been identified (Boesch, 1973; Grassle and Grassle, 1974; Woodin, 1974; McCall, 1977; Rhoads et al., 1977).

In situ studies of recolonization of deep-sea sediments have yielded differing results. Grassle (1977) reported very slow rates of repopulation of azoic sediment trays at 1760 m depth in the North Atlantic. After 26 months faunal densities in the experimental trays were still an order of magnitude lower than in the surrounding sediment. In terms of the species that colonized the trays, his results were quite surprising in that most of the individuals were motile forms and many belonged to higher trophic levels. In contrast, Desbruyères et al. (1980) reported rapid recolonization of azoic sediment trays by opportunistic species at 2160 m depth in the Bay of Biscay. After 6 months faunal densities in trays containing sediment that had been collected at the same locality were 5 times higher than in the surrounding area. In contrast, trays containing littoral sediment showed much lower recruitment despite the higher organic content of this sediment. The species composition of the colonizing fauna differed substantially from that of the surrounding sediment, with polychaetes numerically comprising 98% and 52% of the fauna, respectively. The recruited fauna was dominated in numbers by one opportunistic polychaete of the genus, *Pronospio* sp. (82%). The authors indicate that the lower recruitment in the littoral sediment trays might be due to compositional differences between the sediment of the two areas or an inability of *Pronospio* sp. to tolerate elevated organic material.

The conflicting results obtained from these two studies might be caused by differences in environmental parameters between the localities. These discrepancies might also be the result of differences in the procedures used to collect and handle the sediment. Desbruyères et al. (1980) pointed out that the collection method used by Grassle, an anchor dredge, severely perturbed the sediment used in the experimental trays. This disturbance may have resulted in compositional differences between the experimental sediment and that of the surrounding area. If this was the case, it would be more comparable to littoral sediment without the higher organic matter. Desbruyères et al. used undisturbed sediment collected with a box corer, but this may have been nutrient enriched by the remains of the original infauna. An increased nutrient level may allow rapid colonization and support opportunistic species in the deep-sea and is consistent with the results of Turner's (1973, 1977) study of organisms colonizing wood panels.

Preliminary results from an ongoing study at a Deep Sea Drilling Program site in the Bahamas (2765 m depth) indicate that recovery of a benthic community may require time-spans in excess of 10 years. Two box cores taken with ALVIN during 1978, 8.5 years after the disturbance, yielded 2 individuals compared to 26 in a control core

taken upslope of the buried area. Only one of these two colonizing
individuals could be identified. It was a polychaete of the genus
Glycera which is a mobile carnivore (Fauchald and Jumars, 1979). This
tentative result may support Grassle's finding that succession in
deep-sea habitats may be quite different from that usually observed
in shallower water. Possible differences between the composition
and organic content of the sediment in the disturbed and experimental areas might explain the slow rate of recolonization observed.

Despite the apparently conflicting results found in these
studies, several tentative generalizations can be made about the
recovery of a benthic community disturbed by submarine slides. If
the sediment differs in composition from that of the surrounding
area, repopulation would be very slow and the species composition
of the two areas might be similar. In cases of nutrient enrichment of the disturbed area, recolonization would be faster and
species composition might be quite different from that of the
surrounding community.

When a slope fails, nutrient-poor, partially consolidated
sediment is exposed at the source and along the path of the slide.
Recolonization of such areas would be expected to proceed rather
slowly. Infaunal organisms would be excluded from these areas until
the substratum has been broken down and modified by epibenthic
activity or physical forces. At the depositional site recolonization would be slow if the sediment composition differs significantly from that of the surrounding area. If sediments in the
source area of the slide had a high organic content, then repopulation at its termination might be faster and result in a
different species composition than the surrounding area. Additionally, faunal densities in such areas would be higher than in
the surrounding community, since increased nutrients would support
higher production (Griggs et al., 1969; Jumars and Hessler, 1976).
As the result of distance from a healthy recruitment community,
large disturbed areas would be colonized more slowly than small
ones. Because the termination will usually be lower than the
source area, recolonizing fauna in the vicinity of a slide's
termination may be characterized by slower growing, low fecundity
species as the proportion of these increase with depth (Grassle
and Sanders, 1973).

Sediment instability may also play a role in structuring some
deep-sea communities. Provocative insight into the biological
consequences of episodic slope failure has been presented by
Jumars and Hessler (1976). A box core taken at 7298 m depth in
the axis of the Aleutian Trench yielded an unusually high faunal
density and a low species diversity. Additionally, the proportion of sessile polychaetes was low in comparison to other deep-
sea areas. The authors point out that local sediment instability
as evidenced by steep walls, high sedimentation, and frequent

seismic activity could explain the structure of this trench community by selecting for motile organisms. In-depth studies of
faunal communities inhabiting regions subjected to episodic slope
failure would answer the question of how areas recover from catastrophic disturbance, as well as provide insight into the structuring mechanisms of deep-sea benthic communities and the life
habits of their constituents.

ACKNOWLEDGEMENTS

     This work was supported by the Department of the Interior
Contracts BLM AA551-CT8-49, BLM AA851-CTO-59 and NSF grants
OCE 77-27537, OCE 79-19099. R.W. Embley gave me access to photographs and transcripts taken during ALVIN dives. W.B.F. Ryan
designed and operated the camera sled used in the survey of submarine canyons. U.S.G.S. at Woods Hole invited me to participate
in a submersible program in Lydonia Canyon last summer. The Deep
Submergence Group at Woods Hole and the captain, pilots and crew
of the research vessels LULU and ALVIN have been very helpful.
D. Logan and M. Rawson offered valuable criticism on this manuscript. To all of these people I am very indebted.

Boesch, D.F. 1973. Classification and community structure of
    macrobenthos in the Hampton Roads Area, Virginia. Mar. Biol.
    21:226.
Brongersma-Sanders, M. 1957. Mass mortality in the sea. in:
    "Treatise on Marine Ecology and Paleoecology", J.W. Hedgepeth,
    ed., Geol. Soc. Amer., Mem. 67 Vol 1.
Cacchione, E.A., G.T. Rowe, and A. Malahoff. 1978. Submersible
    investigation of outer Hudson submarine canyon. in: "Sedimen-
    tation in Submarine Canyons, Fans, and Trenches", D.J. Stanley
    and G. Kelling, eds., pp. 42.
Desbruyères, D., J.Y. Bervas and A. Khripounoff. 1980. Un cas de
    colonisation rapide d'un sédiment profond. Oceanologica Acta
    3(3):285.
Dillon, W.P. and H.B. Zimmerman. 1970. Erosion by biological
    activity in two New England Submarine Canyons. J. Sed. Petrol.
    40(2):542.
Fauchald, K. and P.A. Jumars. 1979. The diet of worms: A study
    of polychaete feeding guilds. Oceanogr. Mar. Biol. Ann. Rev.
    17:193.
Grassle, J.F. 1977. Slow recolonization of deep-sea sediment.
    Nature 265(5595):618.
_____, and H.L. Sanders. 1973. Life histories and the role of
    disturbance. Deep-Sea Res. 20:643.
_____, and J.P. Grassle. 1974. Opportunistic life histories
    and genetic systems in marine benthic polychaetes. J. Mar. Res.
    32(2):253.

Griggs, G.B., A.G. Carey, Jr. and L.D. Kulm. 1969. Deep-sea
sedimentation and sediment-fauna interaction in Cascadia Channel
and on Cascadia Abyssal Plain. Deep-Sea Res. 16:157.

Heezen, B.C., M. Ewing and R.J. Menzies. 1955. The influence of
submarine turbidity currents on abyssal productivity. Oikos
6(2):170.

_____, and C.D. Hollister. 1971. "The Face of the Deep", Oxford
University Press.

Jannasch, H.W., and C.O. Wirsen. 1977. Microbial life in the
deep-sea. Sci. Amer. 236:42.

Jumars, P.A. 1977. Potential environmental impact of deep-sea
manganese nodule mining:community analysis and prediction. Final
Report prepared for Mar. Ecosystem Analysis Program Office of
the NOAA.

_____. 1978. Spatial autocorrelation with RUM (Remote Underwater
Manipulator): vertical and horizontal structure of a bathyal
benthic community. Deep-Sea Res. 25:589.

_____, and R.R. Hessler. 1976. Hadal community structure:
Implications from the Aleutian Trench. J. Mar. Res. 34(4):547.

Kranz, P.M. 1974. The anastrophic burial of bivalves and its
paleoecological significance. J. Geol. 82:237.

Malahoff, A., R.W. Embley and D.J. Fornari. 1981. Geomorphology
of Norfolk and Washington Canyons and the surrounding continental
slope and upper rise as observed from DSRV ALVIN. in: "Heezen
Memorial Volume", R. Scrutton, ed., J. Wiley and Sons, London.

Maurer, D.L., R.T. Keck, J.C. Tinsman, W.A. Leathem, C.A. Wethe,
M. Huntzinger, C. Lord and T.M. Church. 1978. Vertical migra-
tion of benthos in simulated dredged material oyerburdens.
Dredged Material Research Program Tech. Rep. D-78-35 1:1.

McCall, P.L. 1977. Community patterns and adaptive strategies of
the infaunal benthos of Long Island Sound. J. Mar. Res. 35(2):
221.

Neumann, A.C. 1966. Observations on coastal erosion in Bermuda
and measurements on the boring rate of the sponge, Cliona lampa.
Limnol. Oceanogr. 11:92.

Nichols, J.A., G.T. Rowe, C.H. Clifford and R.A. Young. 1978.
In situ experiments on the burial of marine invertebrates. J.
Sedimentary Petrol. 48(2):419.

Paul, A.Z., E.M. Thorndike, L.G. Sullivan, B.C. Heezen and R.D.
Gerard. 1978. Observations of the deep-sea floor from 202 days
of time-lapse photography. Nature 222:812.

Rhoads, D.C. 1974. Organism-sediment relations on the muddy sea
floor. Oceanogr. Mar. Biol. Ann. Rev. 12:263.

_____, R.C. Aller and M.B. Goldhaber. 1977. The influence of
colonizing benthos on physical properties and chemical diagenesis of the estuarine seafloor. in: "Ecology of Marine
Benthos", B.C. Coull, ed., Belle W. Baruch Library in Marine
Sci. #6.

Rowe, G.T. 1974. The effects of the benthic fauna on the physical properties of deep-sea sediments. in: "Deep-Sea Sediments". A.L. Inderbitzen, ed., Plenum Press.

_____, G. Keller, H. Edgerton, N. Staresinic and J. Macillvaine. 1974. Time-lapse photography of the biological reworking sediments in Hudson Submarine Canyon. J. Sediment Petrol. 44(2): 549.

Ryan, W.B.F. and B.C. Heezen. 1976. Smothering of deep-sea benthic communities, from natural disasters. Tech. Rept. prepared for Mar. Ecosystem Analysis Program office for the NOAA.

_____, M.B. Cita, E.L. Miller, D. Hanselman, W.D. Nesteroff, B. Hecker and M. Nibbelink. 1978. Bedrock geology in New England submarine canyons. Oceanologica Acta 1(2):233.

Smith, K.L., and R.R. Hessler. 1974. Respiration of benthopelagic fishes: in situ measurements at 1230 meters. Sci. 184:72.

_____, and J.M. Teal. 1973. Deep-sea community respiration: an in situ study at 1850 meters. Sci. 179:282.

Stanley, D.T. 1971. Bioturbation and sediment failure in some submarine canyons. Vie et Milieu. Suppl., 22:541.

Turekian, K.K., K. Cochran, D,P. Kharkar, R.M. Cerrato, J.R. Vaisnys, H.L. Sanders, J.F. Grassle and J.A. Allen. 1975. Slow growth rate of deep-sea clam determined by ^{228}Ra chronology. Proc. Nat. Acad. Sci. 72:2829.

Turner, R.D. 1973. Wood-boring bivalves, opportunistic species in the deep-sea. Sci. 180:1377.

_____. 1977. Wood, mollusks and deep-sea food chains. Bull. Amer. Malacological Union, Inc. 13.

Valentine, P.C., J.R. Uzmann and R.A. Cooper. 1980. Geology and biology of Oceanographer submarine canyon. Mar. Geol. 38:283.

Warme, J.E., T.B. Scanland and N.F. Marshall. 1971. Submarine canyon erosion: Contribution of marine rock burrowers. Sci. 173:1127.

_____, R.A. Slater, and R.A. Cooper. 1978. Bioerosion and submarine canyons. in: "Sedimentation in Submarine Canyons, Fans, and Trenches", D.J. Stanley and G. Kelling, eds., pp. 65.

Woodin, S.A. 1974. Polychaete abundance patterns in a marine soft-sediment environment: the importance of biological interactions. Ecol. Monogr. 44:171.

DISCUSSION, SUMMARY AND RECOMMENDATION

Much time was spent during the workshop to define, firstly the present level of knowledge, secondly the actual scientific and applied problems, thirdly the economic and environmental problems involved, and finally the most urgent problems to be solved in the near future.

It was stressed that the approach has to be inter- disciplinary, maybe with more disciplines than there were present at the workshop. It was also stressed that because the marine environment imposes heavy constraints not encountered on land, the seabottom and subbottom are virtually inaccessible, and therefore in situ test- ings are hard and rather expensive to perform. Many of the observations can only be indirectly carried out, e.g. by use of seismic methods, laboratory testing of disturbed surficial samples etc.

The workshop wants to present the following comments and recommendations:

It has become apparent in recent years that subma- rine slope movements are much more common than previously recognized, and have substantial environmental impacts. Discussion during the presentation of papers and through- out the meeting resulted in recognition of the following general points:

1) The general term submarine slide is often used to describe seafloor features which result from mass movement, but it is recognized that submarine slope failure involves a wide range of initiating factors, de- formation and flow mechanisms, seabed morphologies and materials.

2) Precision geophysical techniques are now avail- able to map seafloor morphology and subbottom character- istics in detail. Areas which have experienced submarine sliding in various forms can be identified.

3) These maps provide a fundamental data base for marine geotechnical and geological investigations.

4) In presently identified slide areas there is a lack of geotechnical and age of failure data.

5) An integrated multidisciplinary approach is essential in the investigation of the various aspects of marine slope instability, and will lead to precise description, evaluation, assessment and analysis of the processes.

6) Geotechnical models exist for evaluation slope instability problems. Despite uncertainties in the quantification of some of the relevant input parameters these models can provide useful information in geotechnical and geological investigations.

7) The integrated approach should lead to a better quantification of the important parameters, and greatly advance our understanding of different slide mechanisms and processes.

8) Concentration of interdisciplinary investigations on some recent or active slides is necessary for understanding the mechanisms, geotechnical parameters and causes of failure.

9) The factors leading to the initiation of failure cannot at this time be fully quantified.

Some Specific Studies Are Necessary:

a) Wave/Seabottom interaction effects.

b) Deformation and flow mechanisms of masses involved in submarine failures.

c) Ages of and movement rates for already identified features.

d) Rates of sedimentation and effects of sedimentary loading.

e) Effects of gas on the in situ properties of sediments.

f) In situ geotechnical properties in relation to static and dynamic (time-varying) loading.

g) Calibration of the rapidly developing remote mapping techniques via visual observation and/or seafloor sampling.

h) Implications of marine slope instability processes to interpretations of the geologic record.

i) Earthquake-induced ground motion in the seafloor environment.

j) Relevant biologic and geochemical effects.